Einstein's Struggles with Quantum Theory

Einstein's Struggles with Quantum Theory

A Reappraisal

Dipankar Home

Bose Institute, Kolkata

and

Andrew Whitaker

Queen's University Belfast

 Springer

Dipankar Home
Bose Institute
Kolkata

Andrew Whitaker
Queen's University Belfast

ISBN-13: 978-1-4419-2445-2   e-ISBN-13: 978-0-387-71520-9

Printed on acid-free paper.

9 8 7 6 5 4 3 2 1

springer.com

The authors respectfully dedicate this book to the memory of

David Bohm

and

John Stewart Bell

whose pioneering work did so much to demonstrate the
importance of Einstein's views and arguments on
quantum theory.

Contents

Foreword

It is a commonly expressed view that Einstein, despite his introducing several seminal notions that were quite fundamental to the emerging quantum theory in the early twentieth century, later turned his back on that theory, being unable to accept the essentially uncertain and apparently subjective picture of the sub-microscopic world that quantum theory seemed to demand. According to this view, Einstein was too wedded to a nineteenth-century objectivistic and deterministic picture of reality to be able to accept the then "modern" attitude to basic laws of the universe, as espoused particularly by the Copenhagen school of Bohr, Heisenberg, Born, and Pauli. That attitude seemed to demand a universe governed by purely probabilistic laws relating to fundamentally imprecise entities that were difficult to visualize and whose very objectivity was put into question. It had seemed that Einstein's stubborn insistence on deterministic realism might be likened to that of a father with fixed views of the path that his child should follow, then turning his back when that child failed to adopt the lifestyle that he had set out for it.

Accordingly, despite Einstein's being far ahead of his contemporaries in the first quarter of the twentieth century, he is viewed as being subsequently unattuned to the later developments that took place in physics, held back by an outdated philosophical standpoint unsuited to the emergent theory of quantum mechanics. The quantum world, it would be argued, demanded that Heisenberg's uncertainty principle restrict the precision of how physical entities could behave, with these entities being described by merely randomistic action, as determined by Born's probability law. Accordingly, Einstein's stance, by the mid-twentieth century, was taken to be retrograde.

Moreover, such an attitude toward his later years seemed to be supported by the apparently limited nature of Einstein's late devotion to a particular form of unified field theory, aimed at uniting gravity and electromagnetism into one geometric framework, where quantum phenomena were intended somehow to arise out of such a classically based scheme. One of the main limitations of Einstein's approach was that it seemed to totally ignore the more recent discoveries of the strong and weak nuclear forces. It must indeed be admitted that the proposals that Einstein came up with in this area have not stood the test of time, and many would argue that his superb instincts for uncovering deep underlying truths of the physical world had, by then, run their course. Yet, a strong case can be made that, at least with regard to the quantum theory, it was *Einstein* who had seen more deeply into the subject than his contemporaries, such as Bohr, by probing that theory's intrinsic weaknesses as an overriding scheme of things.

In his famous debates with Bohr, Einstein was not able to place his finger firmly on the failings of quantum theory that he instinctively felt, and it has been the consensus view that it was Bohr who "won" these debates. Indeed, the exciting new theory of quantum mechanics, which had been crafted through the combined work of Planck, Bohr, Heisenberg, Schrödinger, Born, Dirac, Pauli, and many others—not to mention the early Einstein himself—was found to encompass numerous previously inexplicable phenomena, and it resulted in many extraordinary predictions, over a broad range, these not yet having been found to be at odds with observation some 80 years after the full formulation of the quantum laws.

Yet, there remain many puzzles in the interpretation of the quantum theory, some of which border on the paradoxical, and it was Einstein's continual probing of quantum theory's foundations that have brought out most fully many of these profoundly disturbing features. Despite the extraordinary success of the quantum theory, and of Bohr's effective championing of the theory in his debates with Einstein, we can now look back from our vantage point of the early twenty-first century and see how fruitful Einstein's penetrating criticisms were. Although his was not a lone voice—for Schrödinger and Dirac also regarded the quantum theory as being in an important way "provisional"—Einstein's criticisms were made more openly, and they attained a particular weight owing to his reputation. But Bohr's arguments in support of the conventional quantum theory were not refuted by Einstein, and Einstein's attempts at knock-down arguments were all effectively parried by Bohr. Yet it was Einstein's continually innovative probing that has led to burgeoning areas of fascinating and potentially practical research, largely encompassed within the scope of what is now referred to as "quantum information theory". This area arose, most notably, from the "EPR phenomena" that emerged from the famous collaborative work of Einstein, Rosen, and Podolsky, as refined by Schrödinger, developed by Bohm, and shown to be experimentally testable phenomena, rather than mere manifestations of a philosophical conundrum, through the epoch-making work of Bell.

It is immensely refreshing, therefore, to find a book which at last pays due respect to the later views of Einstein on the quantum theory, and the developments which grew out of them. Dipankar Home and Andrew Whitaker have provided us with masterly expositions of all the issues, analyzing many different viewpoints concerning the profound questions raised by quantum theory. We find that matters of "interpretation" of quantum theory, that have for decades been regarded by physicists as "mere matters of philosophy", can lead to very significant physical effects, some even having important current commercial implications, such as quantum cryptography, and others having more remote potentialities, such as quantum computation. It is reasonable to speculate that there will be future developments of great importance arising from these ideas, perhaps involving a complete overhaul of the basic principles of quantum theory, resulting in a new perspective that might be more in keeping with Einstein's requirements.

Wherever these developments may carry us in the future, they are bound to continue to have an important input from the deliberations of Einstein and others concerning the puzzles and paradoxes that permeate the foundations of twentieth-century quantum mechanics. This book is likely to retain a very significant role for such developments for many decades to come.

Roger Penrose
Mathematical Institute,
Oxford

Preface

It is not too soon to start to dispel the clouds of myth and to see the great mountain peak that these clouds hide. As always, the myth has its charms, but the truth is far more beautiful.

Robert Oppenheimer on Albert Einstein, quoted on the back cover of *Einstein—A Hundred Years of Relativity* (A. Robinson, ed.) (Palazzo, Bath, 2005).

I have little patience for physicists who take a board of wood, look for its thinnest part, and drill a great number of holes where the drilling is easy.

Albert Einstein, as quoted by Philipp Frank in: Einstein's philosophy of science, *Reviews of Modern Physics* 21, 349-55 (1949); also in *The Expanded Quotable Einstein* (A. Calaprice, ed.) (Princeton University Press, Princeton, 2000), p. 233.

An enormous number of books on Albert Einstein have been published; yet the authors feel they may claim that this book is, to the best of our knowledge, the first to provide a comprehensive account of Einstein's encounter with quantum theory: the early unquestioned achievements, the unrecognised struggles over several decades, and the gradual realisation in more recent years that the questions Einstein had raised had not only been perceptive but had borne fruit.

In contrast, while many of the books published so far describe Einstein's life, and most give some account, full or otherwise, of relativity, very few describe in any convincing way Einstein's views on quantum theory, over and above, perhaps, discussion of his important work in the first decade of the twentieth century on the light particle, later known, of course, as the photon, and the specific heats of gases and solids, and his somewhat later contribution towards the development of quantum statistics.

From one point of view this may seem surprising, because over the course of his career Einstein devoted far more time and effort to the problems of quantum theory than to those of relativity. From another point of view, though, it is only to be expected, because for far too long the universally-told story was that, subsequent to the admittedly important work just mentioned, Einstein's approach to quantum theory was frankly unworthy of discussion. He had, it was said, failed to come to terms with the conceptual revolution that Niels Bohr had shown was required in order to solve the paradoxes and problems apparent from the early days of quantum theory. As a result, it was suggested, Einstein's work was frankly valueless, and his criticisms of Bohr's Copenhagen interpretation of quantum theory demonstrated only his own misunderstandings. (See, for example, Heisenberg's remarks in his Foreword to the *Born-Einstein letters*.[1])

Even so it is surprising to see the views of such an admirer of Einstein, and such a kind and perceptive commentator as Abraham Pais as late as the 1980s. In his biography of Einstein,[2] he gives a beautiful study of Einstein's contributions to quantum theory as far as the mid-1920s, and then dismisses all his later work as valueless. In his companion biography of Niels Bohr,[3] he suggests that, rather than bothering to debate with Einstein, Bohr could just have said that he was 'out to lunch' as far as quantum theory was concerned.

Fortunately times have changed—extremely gradually. The work of John Bell in the 1960s, and the resulting theoretical and experimental developments have helped to convince many, though certainly not all, that Einstein's ideas of the 1930s involving realism, determinism, locality and entanglement had brought many essential aspects of quantum theory into the realm of discussion. (Things were not straightforward, though; Bell was a self-confessed follower of Einstein, yet is perhaps best-known, misleadingly, as "the man who proved Einstein wrong"!) The considerable amount of interesting work now performed on foundational aspects of quantum theory, and the rather exciting development of quantum information theory over the last few years are evidence of the fact that the spirit of Copenhagen, which for many years rather stifled novelty along with criticism and discussion, is no longer so dominant.

Together with the debate on that strand of Einstein's ideas, there was a second strand of dissent between Einstein and the supporters of Copenhagen, in this case mostly Max Born and Wolfgang Pauli. This strand, which was largely though not totally independent of the first, concerned the classical limit of quantum theory, and the question of the macroscopic validity of quantum theory. For this strand of discussion also, developments in recent decades have shown that the ideas raised by Einstein have been interesting and fruitful.

Before saying more about this book, it will be helpful to mention the few previous books that have concentrated on Einstein's approach to quantum theory. That by Arthur Fine[4] is a set of previously published articles on various aspects of Einstein's approach to quantum theory; it contains much of interest and significance. We have referred to several of Fine's arguments and debated some others. The book by Jagdish Mehra[5] also contains some useful information and some worthwhile discussion. A recent book by Edmund Bolles[6] strongly supports Einstein in his debates with Bohr; this is an interesting book, rather more concerned with places and personalities than the details of the physics, and in any case it only takes the story up to the 1920s. Two other relevant popular books covering more conceptual issues and a variety of approaches are by G. Venkataraman[7] and John Gribbin.[8]

As has been said, this book is the first to present a comprehensive account and analysis of Einstein's struggles with quantum theory, linking his ideas and critical observations with modern studies on the fundamental aspects of the theory.

Part A covers the story up to Bohr's announcement in 1927 of his framework of complementarity, which was to become the philosophical centre of the Copenhagen interpretation. It includes an account of Einstein's encounter with the ideas of Ernst Mach over several decades, an encounter which set the scene for Einstein's later approach to quantum theory. It describes Einstein's own major contributions to quantum theory extending right up to his crucial advice to Werner Heisenberg that encouraged him in his formulation of the uncertainty principle. It includes an account of the conceptual problems set by the "new" Heisenberg-Schrödinger quantum theory of 1925–6, and the ways in which the "orthodox" approaches to quantum theory of Bohr and John von Neumann claimed to solve them.

Part B presents Einstein's views and arguments on quantum theory from 1927 till his death in 1955. It starts with an account of the years up to 1935, years in which Einstein studied but ultimately rejected simple hidden-variable theories, and in which there took place the early debates with Bohr about the meaning and validity of quantum theory. Then we give a full account of the Einstein–Podolsky–Rosen (EPR) argument of 1935, and of Einstein's later versions of the same basic idea. This paper was dismissed by almost the whole community of physicists at the time, but 70 years later is seen to have launched a number of exceptionally fruitful concepts. This part also includes a full account of the other strand of the debate— that on the classical limit of quantum theory and the validity of quantum theory in the macroscopic domain. The part ends with a substantial analysis of Einstein's views, arguing for Einstein as a realist, but over and above that as a pragmatist, who thought of realism as a means, not of approaching some underlying reality, but of developing useful ideas and concepts.

At the time of Einstein's death, it seemed that his opinions and ideas on quantum theory would fade away with him. That this has not been the case is explained in Part C. This part includes a full account of the work of John Bell, which slowly convinced a new generation of physicists that Einstein's work was of interest and use. Largely as a consequence of the ideas of Einstein and Bell, a number of novel interpretations of quantum theory have been produced, challenging the strangle-hold of the Copenhagen dogma, and these are discussed next. One of the most important of these novel interpretations is that of David Bohm, whose hidden variable model of quantum theory was an important influence for Bell. The work of Einstein and Bell has also played a significant part in the building of quantum information theory, and we give a brief account of this topic. The point is not just that EPR states play a large role in the subject, but that the open-minded questioning pragmatic approach of Einstein could be expected to lead to such developments, while the Copenhagen approach, which explained conceptual difficulties away rather than seeing what developments they might stimulate, might be expected *not* to do so. (Of course, while these conceptual questions had heuristically motivated the emergence of quantum information theory, many of the subsequent developments of this subject may be regarded as independent of the foundational issues. See Chapter 11.) This part also includes a sketch of recent developments on the classical limit, and tests of the macroscopic validity of quantum theory, which show that Einstein was correct in his belief that there are a considerable number of interesting aspects of this topic that were left unaddressed by the Copenhagen approach.

Part D brings the study up to date. It includes a very brief account of some topics in the foundations of quantum theory, just to show that, after 80 years and rather contrary to Bohr's views from the 1920s on, the subject is lively; fresh questions and problems arise, fresh ideas are promulgated, fresh experiments shed light on old and new ideas. We also include a brief discussion of how Einstein's own views might have been affected by the developments of Part C, and a final assessment of his contributions to our present understanding of quantum theory and its foundations.

We expect that the book should be of interest to a wide range of readers. We hope that the expert might find some novel arguments and interesting syntheses. However, anybody with an interest in Einstein and quantum theory should be able to understand and appreciate the main themes of the book. A brief account of quantum theory is provided, but in practice, for some of the more mathematical discussions, it would be useful to have some background in quantum theory, at the level, say, of an undergraduate student. However, there is a route through the book that is totally non-mathematical. The general reader more interested in the ideas and concepts than in mathematical details could very profitably study Chapters 1, 2, 8 and 14, and the less mathematical sections of other chapters.

Acknowledgements. We would like to thank colleagues and experts, too many to list here, with whom we have discussed the foundations of quantum theory, and the views of Einstein in particular, over a period in excess of 20 years. We recognise that we owe much to many.

DH would in particular like to acknowledge warmly the extremely helpful support provided by the Jawaharlal Nehru Fellowship, which was crucial in enabling him to take up the writing of this book, and to perform the necessary research work and travel. Thanks are also due to Alok Pan for help in preparing the typescript. DH remembers with gratitude the late Professor Shyamal Sengupta who introduced him to the foundational problems of quantum theory.

AW would like to thank the Bose Institute and the SN Bose Centre, both in Kolkata, for enabling useful interactions relevant to the topics discussed in this book during his visit in January 2005. He would thank Queen's University Belfast for the award of Study Leave for the academic year 2004–2005, which enabled considerable progress to be made on this book, and would also thank Joan Whitaker for helping to check references.

Both authors would like to thank Queen's University Belfast for the award of a University Fellowship to AW and an International Fellowship to DH under the University Exchange Scheme; this has funded a visit of AW to Kolkata in February 2006, during which a final version of the book was completed.

Both authors are deeply grateful to Professor Roger Penrose for kindly agreeing to write the Foreword.

Finally we would like to thank relatives and friends, but in particular Anindita Home and Joan Whitaker, for encouragement and support during the writing of this book.

References

1. W. Heisenberg, Introduction, in: M. Born (ed.), *The Born-Einstein Letters* (Macmillan, Basingstoke, 2nd edn,, 2005), pp. xxxiv–xxxvii.
2. A. Pais, *"Subtle is the Lord...": The Science and Life of Albert Einstein* (Clarendon, Oxford, 1982).
3. A. Pais, *Neils Bohr's Times: In Physics, Philosophy and Polity* (Oxford University Press, Oxford, 1991).
4. A. Fine, *The Shaky Game: Einstein, Realism and the Quantum Theory* (University of Chicago Press, Chicago, 1986).
5. J. Mehra, *Einstein, Physics and Reality* (World Scientific, Singapore, 1999).
6. E.B. Bolles, *Einstein Defiant: Genius versus Genius in the Quantum Revolution* (Joseph Henry, Washington, 2005).
7. G. Venkataraman, *What is Reality? [Quantum Revolution III]* (Univeristies Press, Hyderabad, 1993).
8. J. Gribbin, *Schrödinger's Kittens and the Search for Reality* (Orien, London, 1995).

Part A
Setting the Scene

Strenuous intellectual work and the study of God's Nature are the angels that will lead me through all the troubles of this life with consolation, strength, and uncompromising rigor.

Albert Einstein to Pauline Winteler, 1897, *The Collected Papers of Albert Einstein* (Princeton University Press, Princeton, 1998); also in *The Expanded Quotable Einstein* (A. Calaprice, ed.) (Princeton University Press, Princeton, 2000), p. 5.

I am not a positivist. Positivism states that what cannot be observed does not exist. This conception is scientifically indefensible, for it is impossible to make valid affirmations of what people 'can' or 'cannot' observe. One would have to say that 'only what we observe exists' which is obviously false.

Albert Einstein in an interview with Alfred Stern, *Contemporary Jewish Record* **8**, 245–9 (1945); also in A. Calaprice, ed., p. 253.

1
The Philosophical Background: Einstein and Mach

Introduction

Following the 25 years of the 'old quantum theory' in which results of great importance were discovered, but by methods that lacked any consistent foundation,[1] the modern rigorous form of quantum theory was produced in the mid-1920s.[2,3] Werner Heisenberg's matrix mechanics of mid-1925[4,5] was followed by Erwin Schrödinger's sequence of papers developing the mathematically analogous wave mechanics, which was published through the first half of 1926.[6,7] In September 1927, in a lecture at Como,[8] Niels Bohr responded to concerns about the conceptual structure of the new theory by announcing his ideas on *complementarity*, which were to constitute what became known as the Copenhagen interpretation of quantum theory. Einstein heard Bohr expound these ideas the following month at the fifth Solvay congress in Brussels, and, at least by that stage, had clearly concluded that Bohr's views were unacceptable[9]; Einstein would argue against them for the remainder of his life.

So in 2 years Einstein's position with respect to the new quantum theory, or at least its conceptual foundations, was settled, and, by this decision Einstein effectively terminated his career as a physicist of influence. The overwhelming majority of physicists considered that Bohr's manoeuvres were successful, that he had been able to demonstrate the conceptual future of physics, and that Einstein, and the few supporters that he did have, merely showed themselves to be too rigid in their thought-processes to be able to cope with the new intellectual demands posed by study of complementarity.

But of course Einstein's contemplation of quantum theory and its conceptual status was certainly not restricted to the period between 1925 and 1927. For the whole of his scientific career up to that point, he had been greatly concerned with the nature of scientific theories, while more specifically he had not only made major and lasting contributions to the developing quantum theory through the first quarter of the twentieth century, but, it may be argued, of all the leading physicists he had thought the most deeply about the physical implications of the theory as it developed. (Many others, Bohr in particular, had, of course, thought a very great deal about the theory, but much of that thought was devoted to the more pragmatic, though certainly not unworthy, aim of getting the theory to work, rather than analysing its conceptual implications and studying what it might tell us about the physical universe.)

In this chapter and the next we sketch the background to Einstein's decision making in 1926 and 1927 that made his final position practically inevitable. In this chapter we discuss the philosophical struggles over the previous decades, and in particular the relationship of his ideas with those of Ernst Mach. In the next chapter we discuss Einstein's contributions to quantum theory up to around 1927.

Positivism and Ernst Mach

In the famous Schilpp volume, in which a number of physicists and philosophers describe or assess Einstein's work, and Einstein himself provides an intellectual biography and also responds to the various comments, Philipp Frank[10] remarks that there are two major, but strongly conflicting positions in the philosophy of science, metaphysics and positivism, and that supporters of each position regard Einstein as its most famous advocate; interpreting Frank's rather general 'metaphysics' as Einstein's more specific 'realism' we may say that, rather amazingly, this is absolutely true.

At the time when Einstein was beginning his study of physics, Ernst Mach was a hugely influential figure in the world of physics. His considerable success in physics itself was totally overshadowed by the huge general interest in his view of what science should be,[11] and much of Einstein's own approach to the philosophy of science may be described as a response—first positive, later negative—to the position of Mach. It is true that, in his excellent biography of Einstein, Frank[12] says that David Hume was the philosopher who had the greatest influence on Einstein, Mach being in second place. While Hume's ideas may have had a large general influence on Einstein's views, it would seem that Mach's had a greater effect on the specific way in which he approached his work; indeed in Frank's book there are over twenty references to Mach, as against only four to Hume.

Mach is always described as a positivist. His beliefs are considered to follow on from those of Auguste Comte, and in turn to lead to those of the *Vienna Circle*, the logical positivists of the mid-twentieth century, which included Moritz Schlick, Rudolf Carnap and Otto Neurath. Mach himself had a strong influence on the ideas of Ludwig Wittgenstein, and through him Bertrand Russell, while logical positivism inspired the work of Karl Popper, though he was to reject its fundamental tenets, and Popper himself influenced such later philosophers of science as Thomas Kuhn and Paul Feyerabend, although again their eventual positions differed sharply from his. Thus Mach had a great influence on the history of thought throughout the twentieth century.

Mach actually rejected the term positivism, associating it with the beliefs of Comte, which were unpopular at the time. Neither incidentally did he consider himself a philosopher; he regarded his approach as methodology of science, which, he argued, was not a philosophy, and he actually had the definite aim of making philosophy, at least philosophy of science, redundant. It may be said, though, that Mach's views certainly dovetailed with those usually described as positivism, which may be defined as the belief that analysis should be restricted to studying the relations between observed phenomena, and that any discussion of mechanism or 'real entities' is meaningless.

Mach identified the physical world with what could be immediately sensed, individual sensations being referred to as 'elements'. The purpose of science, he believed, was to describe and relate these elements in the simplest way possible, and this, he believed, was by use of mathematical functions. The idea of cause had no place in his beliefs; he felt that one should not go beyond establishing and analysing the relationships between the various elements. Equally he was suspicious of attributing great significance to the idea of scientific laws; a law, he considered, was no more than a description of how elements were connected. Relations should preferentially be expressed in mathematical form because that matched Mach's desire for 'economy' in one's descriptions.

Once science has reached its final form, Mach believed that mathematical rela-tionships were all that would be required. At this point there would be no need, indeed no place, for theories. He was prepared to concede that, until one reached that point, theories could have a provisional use in providing an interim account of relations between events, again provided this account was the most economical one possible. However the question of the scope of this concession was to lead to mis-understanding, on Einstein's part in particular. Like all positivists, it may be said that Mach's fundamental aim in his discussion of science was the elimination of 'metaphysics', but he had a particularly wide definition of what metaphysics was. It was not restricted to unsubstantiated speculation, or unphysical philosophising, but included anything outside direct experience; thus he claimed to return science to the scientists.

Mach was a strong believer in the unity of science, and also the desirability of scientific and technological progress, which he conceived not as an end in itself, but as a means of enhancing the survival and welfare of the species. Mach was influential from the mid-1860s, though his period of greatest celebrity probably followed his return to Vienna in 1895, ironically enough as Professor of Philosophy. In 1898 he suffered a severe stroke and was forced to retire from his University position, but he continued working and writing up to his death in 1916.[11]

From what has been said, it is not surprising that many physicists found his position congenial, at least for considerable periods, in particular the young Max Planck from around 1879,[11] and the young Albert Einstein from around 1897.[13,14] Rejection of metaphysics may easily be interpreted as the best way to avoid the prejudices and philosophical conceits of the past; concentration on individual 'elements', and the attempt to ascertain the most 'economical' relations between them, as the open-minded approach to establishing the facts of physics. Mach's belief in the unity and primacy of science would also have been a positive influence, and Einstein in particular also refers positively to his scepticism and independence.[13]

For Einstein, if not for Planck, the general nature of Mach's beliefs would indeed have been exceedingly attractive. Seemingly one with his belief in science, and his ideas on how science should be carried out were his progressive politics, his atheism, his pacifism. While, at least in later years, Einstein was certainly not an atheist, though his God was that of Spinoza, revealing himself in the harmony of the Universe, rather than a God concerning himself with mankind,[15] and while, again at least in later years, he effectively renounced strict pacifism, he must in any case have relished Mach's honesty, and his willingness to declare and defend his beliefs, knowing them to be unconventional and generally unpopular.

If these are general reasons why Einstein would have been initially attracted by the position of Mach, there are also more specific factors, related in particular to Einstein's theories of special and general relativity, which made an initially positive relationship practically inevitable. It has been said that Mach believed theories had only a provisional role to play in physics; they might be helpful in the development of the subject, but would have no place in its final form, which would consist only of experimental results and mathematical relationships. In fact Mach actually wrote many books on theoretical physics, including important accounts of his views on mechanics, heat and optics, and particularly the history of these disciplines, but the purpose of these books was largely to criticise those whose arguments Mach felt unsatisfactory, either because they practised metaphysics or for other reasons. In any case, Mach felt that the history of science had a definite value for science. Ideas whose source or history was unknown stood isolated and unjustified and thus 'metaphysical'. By studying the history of these ideas, and developing their historical 'relations' to other ideas, they could be rendered less 'metaphysical'.

Mach's Critique of Newton

By far the most influential of Mach's historical surveys was *The Science of Mechanics*,[16] published in German as *Die Mechanik in ihrer Entwicklung Historischkritisch Dargestellt* in 1883. In this book Mach launched a sustained attack on Newton's methodology. It may be argued[11] that Mach completely misconstrued Newton's approach, confused by his claim to 'feign no hypotheses'. Mach took this as a statement of a positivistic intent, similar to that of Mach himself, and assumed that for Newton, as for Mach, a hypothesis was any concept removed at all from direct experimental perception. In fact, though, for Newton, analysis based on, but extending way beyond Mach's 'elements' was essential for the development of science; only speculations unsupported by such analysis deserved the title of 'hypotheses'.

An important example is Newton's definition of 'mass' as the quantity of matter in a body. Once mass is defined, Newton could then regard his third law as expressing one of the ways in which it was manifested. From Mach's point of view, though, Newton's definition is, at best circular, defining mass through the concept of density, and density as mass per unit volume. At worst it is meaningless.

Mach chose to define mass through Newton's third law; in an action–reaction pair, m_1/m_2 is *defined* as being equal to a_2/a_1, defined, that is, directly from measurable quantities. The supporter of Newton's general methodology will admit that Newton's definition is somewhat limp, and might certainly wish for improvement, but will argue that it is attempting to encapsulate some idea of the meaning of the word 'mass', to help us understand what it actually is. The gravitational attraction is then merely a property of the masses of the two objects, or one may perhaps say that the masses *cause* the attraction. For the supporter of Mach, though, such talk is meaningless; Mach has economically provided relations between quantities that may be directly measured, and no more is required. Certainly any talk of causation, any identification of particular elements of the situation as 'cause' and 'effect' is, Mach would say, totally illegitimate.

From Einstein's point of view, Mach's most potent criticisms concerned Newton's postulation of 'absolute space' (and also, to a lesser extent, 'absolute time'). Newton required notions such as motion, velocity and acceleration; the natural question would be 'motion, velocity, acceleration with respect to what?', and the natural answer would be 'with respect to absolute space'. Motion, velocity, acceleration with respect to absolute space would then be absolute motion, absolute velocity, absolute acceleration. For Newton absolute space was an *a priori*; it was provided by God.

For Mach, in contrast, absolute space was a monstrosity. It was metaphysical— 'a pure thing of thought, a mental construct that could not be produced by experience'.[16] Its postulation was unnecessary and wholly uneconomical. For Mach, all motion was *relative*; experimental evidence was always of *relative* positions and motions of objects. If one says that one body moves in absolute space rather that with respect to other bodies, then according to Mach, one is committing two errors; first, one cannot know how the first body would move if the others were not present, and secondly, if the others were not there, there would be no way of discussing the motion of the first. The *relational* character may also be used to dispose of the idea of absolute time; rather than saying that a phenomenon occurs at a particular time, we may relate its occurrence to that of another phenomenon correlated directly with the rotation of the earth.

Very much associated with Mach's arguments against absolute space was what has become known and famous as *Mach's principle*.[17] The argument emerged in Mach's discussion of Newton's 'bucket experiment'. In retrospect at least we may realise that the analysis of translational motion in terms of absolute space is complicated by the existence of an infinity of inertial frames, but there is no such problem for rotational motion; Newton took full advantage of that fact. He described a vessel suspended from a long cord, which is turned many times so that the cord is strongly twisted; the vessel is filled with water and held steady before being released. At first the water is unaffected; it does not turn with the vessel, and its surface remains flat; at this stage there is relative motion between the water and the vessel. However as the vessel gradually transmits some rotational motion to the water, the surface of the water becomes distorted, water moving away from the axis of rotation towards the sides of the vessel. For Newton this is clear evidence that the water responds not to a relative rotation, relative to the vessel, but to an absolute motion, relative to absolute space.

Mach responded with spirit. First he said that the lack of effect of relative motion between water and vessel may be a result of the limited thickness of the walls of the vessel; nobody knows, he said, what might happen if the walls were several miles thick. Secondly he pointed out that the motion of the water that produces the centrifugal effect is by no means absolute; it is a relative motion, relative in fact to the fixed stars. Challenging Newton, Mach inquired: What would happen if the water were at rest relative to Newton's absolute space, and the stars rotated around the bucket? Mach actually affirms that Ptolemy and Copernicus are equally correct; the ideas of Copernicus are preferred just because they are simpler and more useful.

As Norton points out in Ref. [17], there may actually be two interpretations of Mach's remarks. First there may be an actual endorsement of what has become known as Mach's principle: this implies that there is a physical mechanism by which the inertial properties of a particular object are determined by the presence and behaviour of all the other masses in the universe. The strength of the effect should be broadly independent of distance, but highly dependent on the mass of the body, so that a far off star and the vessel itself might cause comparable effects if the thickness of the walls of the vessel were increased enormously. Mach did accept, if not actually endorse, the possibility of experimental investigation of such suggestions.[17]

However the second interpretation is virtually the converse of the first. It is that Mach's actual argument stops at the point where he demonstrates the relation between the behaviour of the water and that of the stars. At this point Newton's claim to have demonstrated the absolute nature of the motion has been challenged by wholly economical analysis. According to this interpretation, the comments about increasing massively the thickness of the sides of the vessel, and of rotating the stars about the vessel, which speak of cause and effect rather than mere relation, are brought up merely as examples of uneconomical mental constructs, arbitrary fictions, speculations of the type Mach considered superfluous in scientific description. For support for this interpretation, see for example, Ref. [11], p. 275.

It is possible that there was a certain ambivalence in Mach's position. The second interpretation seems much the closer to his general approach to such matters, but he may have become intrigued by suggestions, experimental or theoretical, which took his remarks more seriously than he intended. Einstein himself took a great interest in the principle, as will be seen, but any possible inconsistency in Mach's approach to the principle may have played a part in the eventual strains between the two men.

Einstein and Mach—Part 1

It is, in any case, now time to review the relationship between Mach's approach to physics and to Newtonian mechanics in particular, and Einstein's theories of special and general relativity. We do not necessarily suggest that such a relationship, or all least all parts of this relationship, are evidence of Mach's influence on Einstein; just as relevant are aspects of Einstein's work which may have been totally independent of Mach's influence, but were perceived in a different light.

First, of course, Einstein's comment about Mach's scepticism and independence has already been mentioned. In particular this must have referred to Mach's questioning and probing of the edifice of Newtonian classical mechanics, which might otherwise have seemed totally beyond criticism at the time. This must have been a source of great encouragement to the young man. In his paper introducing special relativity,[18] Einstein's introductory remarks could scarcely have echoed Mach more strongly.[14] He commences with a conceptual analysis of the meaning of space and time, and his definition of the time at which an event occurs, based on a simultaneity between the event itself and the small hand of a watch pointing in a particular direction, is pure Mach. Time and place are only meaningful concepts, Einstein argues, if given by measurement-in-principle. In addition, for Einstein, physics here refers directly to events occurring in space and time, not to any concept beyond such direct experience. Einstein's 'events' correspond almost exactly to Mach's 'elements'. It seems unlikely that Einstein would have been able to leave behind so totally any appeals to 'common sense' or what was happening 'in reality', and to present his results in terms of rather counter-intuitive mathematics, had he not taken to heart the precepts of Mach.

Clearly the conceptual struggles involved in Einstein's creation of special relativity followed directly Mach's precepts. Just as apparently significant was that the results of his struggles also appeared, at least superficially, to relate fairly directly to Mach's vision. For Mach, as has been stressed, science consisted of establishing *relations* between elements (or events in Einstein's terms). This abandonment of the idea of the absolute, so central in Mach's critique of Newton, may be called 'Machian relativity'. Einstein's own work on 'Einsteinian relativity', in particular in renouncing the idea of absolute time, could seem to be a shining example of the Machian paradigm, even perhaps the fulfilment of Mach's critical analysis of Newton. To many, Einstein's theory of special relativity seemed the ultimate triumph of positivism in physics. As Mehra[19] points out, Einstein even follows Mach in avoiding the word 'theory', preferring to speak of an 'heuristic point of view'.

It is often remarked that Einstein's rise up the academic ladder was swift, considering how difficult even many physicists must have found it to come to terms with his remarkable papers of 1905. Part of the perceived answer [11] is that supporters of Mach and positivism smoothed his way. It is certainly true that, following his appointment as privatdocent at the University of Bern, he obtained an *extraordinary* chair at the University of Zurich in 1909 thanks to the withdrawal in his favour of the positivist and political ally of Mach, Friedrich Adler. Shortly afterwards in 1911 he became full Professor at the German University of Prague. It was quite widely believed, this belief almost certainly being shared by Einstein himself, that he had received strong backing from two members of the faculty selection board, Anton Lampa and Georg Pick, who were both Machian positivists, and, behind them, probably from Mach himself. Like all the positive aspects of the relationship between Mach and Einstein described in this section, though, this will need to be re-assessed in the following section. It should certainly be said that Einstein also received help from non-positivistic physicists who had the vision and courage to recognise the genius in these papers; these particularly included Max Planck, of whose own philosophical journey more will be said shortly. In addition the actions of Lampa, Pick and Mach himself were, at best and as we shall see, much more equivocal than has been suggested.

By this stage Mach had, in any case, given some indications of general support for special relativity. In 1909 a new edition of one of his books, *Conservation of Energy*,[20] was published. It included the words: 'Space and time are not here conceived as independent entities, but as forms of the dependence of the phenomena on each other. I subscribe then to the principle of relativity, which is also firmly upheld in my *Mechanics* and *Wärmelehre*'. Mach gave a reference to the work of Hermann Minkowski, who had provided an attractive geometrical picture of Einstein's equations, emphasising that four-dimensional spacetime was at the heart of the analysis. Mach sent a copy of this book to Einstein, who must have been delighted to receive what certainly seemed to be such definite endorsement from the man he admired so much. Though only Minkowski's name was mentioned, Einstein would have taken for granted from Mach's mention of 'the principle of relativity' and the gift of the book that Mach was supporting him personally. Nobody would, in any case, have questioned the fact that Minkowski's work, interesting and important as it was, derived totally from that of Einstein. There was obviously more than just support involved; by linking the work of Minkowski and, through him, Einstein to his own books, Mach seemed to be authenticating the view that Einstein was his trusted follower, even his successor.

Over the next few years this connection between the two scientists appeared to be strengthened. In correspondence, Einstein emphasised Mach's great influence among physicists, and fairly explicitly took the side of Mach against Planck, who had in the last few years launched savage attacks against the older man, for reasons again to be discussed in detail shortly. (This support from Einstein was quite striking, because Einstein himself owed much to Planck, whose support of his own first paper on special relativity had been crucial.) Furthermore, in two papers Mach commented[11] on Einstein, Minkowski and also Hendrik Lorentz in terms that seemed supportive. Mach's own view was that the nature of space, time and matter was still a problem. This view, he said, was supported by these younger physicists, who were making essential progress towards its resolution. The general perception of unity between Mach and Einstein was certainly shared by Philipp Frank, who was both a close philosophical ally of Mach and a strong supporter of special relativity, and who was to become Einstein's biographer.[12] In discussions with Mach in 1910, he formed the definite impression that Mach accepted both the physics and the philosophy of Einstein's relativity.

The putative link between the work of Mach and Einstein was actually considerably strengthened further in the years leading up to Mach's death in 1916, as Einstein's work on gravitation proceeded to his theory of general relativity. In general terms, it is obvious that Einstein's preliminary conceptual analysis dealing with the principle of equivalence, and the identity of inertial and gravitational mass, is Machian in general nature. More specifically, once Einstein decided that the principle of relativity should be extended to arbitrarily moving reference frames, and that gravitation is to be described as the metric tensor of a curved spacetime, Einstein's ideas focussed on the relativity of mass, and thus on Mach's principle.

As early as 1912 Einstein wrote:[21] 'This suggests the hypothesis that the whole inertia of any material point is an effect of the presence of all other masses, depending on a kind of interaction with them' giving a reference to Mach's *Mechanics*. By 1918, the same sentiment was being expressed[22] in a much more technical way, referred to by Einstein as Mach's principle: 'The G-field is without remainder determined by the mass of bodies. Since mass and energy are, according to results of the special theory of relativity, the same, and since energy is formally described by the symmetric energy tensor $(T_{\mu\nu})$, this therefore entails that the G-field be conditioned and determined by the energy tensor'. (So the actual metric of space is determined by mass–energy.) In his 1921 Stafford Little Lectures at Princeton, published as his book *'The Meaning of Relativity'*,[23] Einstein described Mach's principle as one of the founding concepts of general relativity, and incidentally used it to argue towards a closed static universe.

During the years before Mach's death, Einstein wrote to him with great enthusiasm about the progress he was making with relativity; he explicitly related his ideas to those of Mach on Newton's bucket, and his letters display absolute confidence that Mach would be as excited as Einstein himself with what was being achieved. And when Mach died, Einstein wrote a moving obituary;[24] he praised the way in which Mach's historical writings, particularly his *Mechanics*, had influenced the philosophical approach of the younger generation of physicists. He made a particular point of suggesting that, had Mach been younger when the question of the constant nature of the speed of light had arisen, he would almost certainly have discovered special relativity himself, and indeed that his analysis of Newton's bucket put him very close to the ideas of general relativity.

So this section ends with the apparent establishment of what was a practically universal belief then and for some time after, Einstein as Mach's devoted follower and successor, Mach as partial precursor and proud advocate of Einsteinian relativity, and, a part of the story still commonly upheld today, Einstein the avowed positivist, constructing his theories from direct physical measurements.

That, at the very least, it cannot be the whole story is made clear by an event that took place in 1921, five years after Mach's death and Einstein's eulogy. The first part of Mach's posthumous book, *The Principles of Physical Optics*, was at long last published. In the Preface, which had been written in 1913, Mach announced that he was cancelling his views on relativity. He had become aware, he wrote, that he was becoming regarded as the forerunner of relativity. However he denied this just as strongly as he rejected the atomic theory. He reported that he found relativity theory becoming more and more dogmatic, his reasons for this change of opinion being based on the physiology of his senses, epistemological doubts, and evidence from his own experiments.

It should be mentioned that in a recent book, Gereon Wolters[25] suggests that this Preface was forged by Mach's son, Ludwig, who had responsibility for arranging the publication of the book, and was certainly a very strong opponent of relativity. The story is perhaps not implausible, though there is no direct evidence for it, but even if it is if true, it remains the fact that around this period Einstein's views concerning Mach's approach and the attractions or otherwise of positivism changed completely; his eventual opposition to Bohr and Copenhagen perhaps became inevitable. The reasons for this shift are discussed in the following section.

Einstein and Mach—Part 2

An obvious suggestion regarding the growing dissonance between the positions of Mach and Einstein is that it was to do with atomic theory. It is well known that Mach's belief that science should be restricted to argument coming directly from experiment made him highly unsympathetic towards the idea of atoms. During the nineteenth century, atomic theory had progressed from characterisation of the atom and molecule in the chemical atomic theory of Joseph Dalton and his followers, through kinetic theory, to the statistical thermodynamics of James Clerk Maxwell and Ludwig Boltzmann. Maxwell died in 1879, and Boltzmann[26] was certainly the leading European advocate of atomic physics in the latter period of the nineteenth century and the beginning of the twentieth. (The American, Josiah Willard Gibbs also did extremely important work in the area.) Boltzmann was, like Mach, Austrian, and both men were in Vienna for most of the period from 1894. Conflict was inevitable.

Much of Boltzmann's direct struggle with the positivists was actually with the chemist, Wilhelm Ostwald, the leader of the so-called energeticists who, rejecting atomic theory, believed they could explain all of physics with the single quantity of energy. Boltzmann was able to demonstrate the limpness of energeticism in debate fairly convincingly, but in the 1890s another convinced opponent of his work was Planck, a strong supporter of the primacy of classical thermodynamics. During the nineteenth century, classical thermodynamics had developed at the hands of Rudolf Clausius, William Thomson (Lord Kelvin), Hermann von Helmholtz and many other eminent scientists, and, in many ways, Planck could be regarded as the last name in this distinguished chain.

Classical thermodynamics attempts to describe phenomena involving the exchange of energy, and to obtain general laws, with the minimum number of assumptions (in particular with no atomic assumptions). It was thus totally different in nature from Boltzmann's statistical thermodynamics, which attempted to explore the same area with the explicit use of atomic assumptions. Positivists like Mach obviously supported classical thermodynamics, but of course this subject was and is of general importance; it is certainly interesting to know what can be understood without particular assumptions, which, in the way of things, may change with time. To contribute to classical thermodynamics, it is not necessary to oppose Boltzmann and his methods.

But through the 1890s Planck certainly did oppose them. His task was to relate the second law of thermodynamics to the particle mechanics of Newton; to do this he required a rigorous mathematical approach to entropy. So did Boltzmann, but Planck strongly opposed Boltzmann's use of probabilistic arguments. For Boltzmann the second law of thermodynamics was probabilistic; entropy *could* decrease (in a given region, for a given period of time) but the chance of it happening to a significant extent was infinitesimally small. Planck insisted it should be identically zero. He was thus strongly opposed to Boltzmann, and, he subsequently felt, made an enemy of him. Then in 1900 came the famous moment when, to obtain agreement with recent experimental results on black body radiation, Planck was forced to construct a novel theoretical expression, and the attempt to justify this expression forced him to adopt totally Boltzmann's approach. For Planck it was a major volte-face. It led him to fame as the founder of quantum theory, and to an eventual Nobel prize, but it seems that he felt guilty about his many years of outright hostility to Boltzmann's ideas.

Then in 1906 Boltzmann committed suicide. It has been commonly believed that this unhappy event was directly linked to the constant attacks on his atomism from Mach and Ostwald. This is certainly at the very least a grave over-simplification of the facts.[26] Boltzmann's physical health was bad. (He suffered from heart and eyesight problems, asthma and headaches.) He also demonstrated manic-depressive tendencies. But of course one cannot deny, may even surmise, that the struggles with the positivists had played a part in disturbing his mental balance. Certainly Planck appeared to think this was the case. He had actually become rather disillusioned with Mach's ideas through the 1890s, while still at that time, of course, still resolutely opposed to Boltzmann, and his response now was to launch two startlingly strong attacks on Mach's ideas in 1908 and again in 1910.

Planck's approach was to express belief in a unified 'world picture', within which laws, matter and atomic concepts were definite and unchanging, and should therefore be regarded as real. This viewpoint was, of course, completely opposed to Mach's concentration on individual elements of experience. Planck also criticised Mach on the grounds of the lack of genuine scientific advances made on the back of his methodology. 'By their fruits ye shall know them', he stated. Boltzmann numbered among his students Svante Arrhenius, Walther Nernst, Lise Meitner, Stefan Meyer, Fritz Hassenöhrl and Paul Ehrenfest, all eminent discoverers. It is true that Mach himself had respectable and lasting achievements in physics,[11] but his devoted followers produced practically nothing of worth.

Much later Einstein was to use the same argument—the sterility of positivism, but for now he was on Mach's side, expressing sympathy over Planck's attack, albeit in fairly general terms, in letters to Mach. At first sight this seems extraordinary. From 1900 Einstein had studied Boltzmann and admired him greatly. In that year he wrote to his future wife, Mileva, 'The Boltzmann is magnificent. . . I am firmly convinced that the principles of the theory are right, . . . that in the case of gases we are really dealing with discrete mass points of definite finite size, which are moving according to certain conditions'.[27] And his Brownian motion paper,[28] one of the magnificent group of 1905, has been properly thought of as demonstrating convincingly the existence of molecules.

Yet Einstein seems to have believed that he could ride both the atomic and Machian horses together by making use of Mach's idea of economy. Frank reports[12] that, in Einstein and Mach's one-and-only meeting, Mach agreed with Einstein's suggestion that *if* use of atomic ideas was the only way of establishing relations between observable properties, the atomic hypothesis should be regarded as *economical* and hence acceptable. It was probably only much later that Einstein realised that Mach by no means accepted the antecedent in the argument. Despite some controversy, John Blackmore argues convincingly that Mach remained opposed to atoms all his life, an opinion that is backed up by the excerpt from the Preface to his posthumous *Optics* already quoted. It is still interesting, though, to see Einstein seemingly prepared to accept the 'saving the appearances' argument urged on Galileo by Cardinal Bellarmine, but rejected by him. Einstein seemed willing to accept a belief on Mach's part that the atomic hypothesis was useful, rather than demanding an acknowledgement that atoms actually existed.

Despite initial opinions, relativity was to be a much greater stumbling block for any constructive relationship between Mach and Einstein than atoms. Despite everything that was said in the previous section, fundamentally Einstein's relativity had little connection with that of Mach. Let us consider the structure of special relativity, and concentrate on the change from the Galilean transformation equations

$$x' = x - vt \tag{1.1a}$$

$$t' = t \tag{1.1b}$$

to those of the Lorentz transformation:

$$x' = \gamma(x - vt) \tag{1.2a}$$

$$t' = \gamma(t - x/v^2) \tag{1.2b}$$

where $\gamma = \left(1 - v^2/c^2\right)^{-1/2}$.

The element of 'relativity' is already present in (1.1a) in that x and x' are different. To be sure it is of a benign kind, totally unproblematic conceptually. In the Lorentz transformation, this element is amplified in that t and t' differ, and the relationship between x and x' is conceptually much less natural, and, of course, this is not easy to accept physically. Nevertheless from the formal point of view, the real change from Galileo to Lorentz is to introduce, not relativity, but, in total contrast, a new absolute, the absolute speed of light, encapsulated in (1.2) in the sense that $x'^2 - c^2 t'^2 = x^2 - c^2 t^2$.

This may be regarded as part of a general Principle of Relativity: all laws of physics are the same in all inertial frames. It may be said that Einstein's Principle of Relativity had nothing in common with the 'principle of relativity' for which Mach announced support in his book of 1909, which was actually little more than Mach's belief in the relational rather than the absolute. While Einstein was quite prepared to take courage from Mach to remove from his work the metaphysics of Newton, he actually added his own brand of metaphysics—the constant speed of light, and Einsteinian relativity.

Indeed, Gerald Holton[29] argues that Einstein did not actually appreciate the name 'theory of relativity'; '[I]f he had chosen to give it a name,' Holton suggests, 'he would have preferred to call it something like *Invariantentheorie* [theory of invariants], which is of course much more true to the method and aim of the theory,' and, in a footnote, Holton gives extensive quotations from Einstein himself, Minkowski, Arnold Sommerfeld and John Synge to back up his opinion.

Among positivists and supporters of Mach, there were different responses to special relativity.[11] Frank endorsed the theory fully, but tried to de-emphasise the constant nature of the speed of light. Joseph Petzoldt, a long-standing and very close ally of Mach, also strongly supported the theory and spent considerable effort in trying to persuade Mach of its merits, but regarded the issue of the speed of light as an unfortunate blemish that should be removed; Petzoldt suggested to Einstein how this could be done, but Einstein did not agree. Adler rejected the theory because of this 'blemish'.

It is difficult to assess Mach's position with certainty. At times, as we have already seen, he seemed to give some support to Einstein and Minkowski, but in retrospect this may have meant no more than that he was pleased in a general way with their questioning of Newtonian absolutes, rather than with what they actually put in their place. Quite against what was popularly believed, far from supporting Einstein for the Prague chair, Mach, and his fellow positivists, Lampa and Pick, actively campaigned for Gustav Jaumann, Mach's former assistant, and a close follower of his old master. Jaumann was actually offered the chair before Einstein, but declined it.

In the years in which Einstein was working towards the theory of general relativity, Einstein may have assumed that his work would be more and more pleasing to Mach; not only was there continuing elements of 'relativity', but also Mach's principle played a central role in the theory. Ironically enough, precisely the opposite seems to have been the case, and for several reasons. First, as already noted, Mach's own approach to Mach's principle was ambivalent. The more genuine Machian position would seem to be that the reference to the fixed stars is intended merely to demonstrate that motion is relative, not to propose an actual physical mechanism. Mach might not have objected to a little discussion of the latter possibility, even a little light experimentation,[17] to ascertain if, just possibly, there might be something substantive in the suggestion, but he could well have resented it becoming the unsubstantiated basis of a major theory.

Secondly, this work of Einstein was undoubtedly a 'theory'; unlike special relativity, it could scarcely be categorised as a 'heuristic viewpoint'. From Mach's point of view, far from being an attempt to sum up the relevant experimental measurements in a mathematical way, and as economically as possible, it must have appeared to be speculating well beyond the experimental results, bringing in metaphysical arguments, in particular the curvature of spacetime. Thirdly, the mathematical contents of the theory must have been particularly unpleasing to

Mach; he had long regarded the use of four or more dimensions in analysis of physical problems as meaningless. And fourthly, Mach preferred his theories to be tentative, subject to constant re-examination and revision; in his view Einsteinian relativity was becoming more and more dogmatic, and he particularly stressed this in his Preface that was eventually to make his views public. In retrospect, it may not seem surprising that this Preface was written almost immediately after he received one of Einstein's excited letters telling Mach of the progress he was making with the theory.

Einstein and Mach—the Denouement

Until 1921 Einstein was content to remain generally regarded as a disciple of Mach, and this has led Blackmore[11] to remark that 'Mach had noted the incompatibility well before Einstein'. The truth may be a little more complex. Consciously or unconsciously, right up to this date Einstein was extending what he would regard as the basic facts, or Machian 'elements', on which his theories would be built. Holton[14] quotes a letter of 1918 to Michele Besso in which ostensibly Einstein is defending Machian empiricism, but manages to include as basic facts the following: the impossibility of a *perpetuum mobile*, the law of inertia, the equivalence of heat and mechanical energy, the speed of light being constant, Maxwell's equations, and the equivalence of inertial and gravitational mass, all of which Mach would consider very far removed from his idea of basic elements of experience.

Einstein was certainly aware by this period that he was advancing beyond Mach, as evidenced by an exchange of letters in 1917 again with Besso.[30] Einstein mentioned the criticism of special relativity, already referred to above, that he had just received from Adler; this was based on Adler's strong positivism. Besso replied: 'As to Mach's little horse, we should not insult it; did it not make possible the infernal journey through the relativities? And who knows—in the case of the nasty quanta, it may also carry Don Quixote de la Einsta through it all!' Einstein replied: 'I do not inveigh against Mach's little horse, but you know what I think about it. It cannot give birth to anything living, it can only exterminate harmful vermin.'

This remark sums up beautifully the attitude that Einstein had developed towards Mach by this date. The kind of conceptual analysis encouraged by positivism could certainly uncover areas of science where quite unnecessary assumptions about the basic concepts had been taken for granted for so long that it took great pertinacity and courage to challenge them. Positivism could help to exterminate these, but could do little to aid the discovery of what must replace or generalise these concepts. For this purpose, novel ideas, laws or theoretical structures were required, going way beyond what could be deduced with Mach's economy from Mach's elements. Einstein may, at this time, have been able to convince himself that he was only extending Mach's vision of how science should be performed, rather than contradicting it. Possibly he may have realised that he could not truly be categorised as a follower of Mach, but preferred to avoid spelling this out publicly, either from the generous urge not to hurt the man he respected in many ways, or the less generous one not to lose influence and support.

In either case, the publication of Mach's posthumous preface in 1921 must have shaken him a great deal. Not only could he now speak freely about Mach, but he also initially allowed some frustration to appear,[14] describing Mach in a 1921 lecture as 'good at mechanics but deplorable at philosophy'. In 1922 he wrote that 'Mach's system studies the existing relations between data of experience; for Mach, science is the totality of these relations. That point of view is wrong, and in fact what Mach has done is make a catalogue not a system'. Later Einstein was again to speak generously of Mach's personality and his willingness to question scientific authority, though never of his philosophical position. As the years went on, Einstein also became less and less convinced of the central role, or indeed any role, that Mach's principle might play in general relativity. In 1954, in fact, he wrote to a colleague: 'As a matter of fact, one should no longer speak of Mach's principle at all'.[31]

From the point of view of the present book, one important result of Einstein's intellectual journey described above, is, of course, that the Copenhagen approach to quantum theory would be judged as to whether or not it was positivistic in nature; for Einstein it seemed so, and so he was forced to oppose it consistently. Whether he was correct in his judgement is of course, a very important question that will be discussed later in the book.

Einstein's own general philosophical position in his maturity will also be a major topic of the rest of this book. Here we shall just restrict ourselves to some comments Einstein made in the Schilpp volume. Mach's epistemological position is unsatisfactory, he says 'for he did not place in the correct light the essentially speculative nature of thought and more especially of scientific thought, in consequence of which he condemned theory on precisely those points where its constructive-speculative character unconcealably comes to light, as for example in the kinetic theory'.[13] At the end of this volume, Einstein replies to the articles in the main part of the book: to Henry Margenau, who had suggested that 'Einstein's position . . . contains features of rationalism and extreme empiricism', Einstein's interesting reply is as follows: 'This remark is entirely correct . . . A wavering between the two extremes seems to me unavoidable'.

One last point of interest is that inevitably Einstein moved very close to Planck, against whose onslaught he had earlier defended Mach; in an address in honour of Planck's 60th birthday, as early as 1918, he wrote:[32] 'The supreme task of the physicist is to arrive at those universal elementary laws from which the cosmos can be built up by pure deduction. There is no logical path to these laws; only intuition, resting on sympathetic understanding of experience, can reach them'.

References

1. ter Haar, D. (1967). The Old Quantum Theory. Oxford: Pergamon.
2. Jammer, M. (1989). The Conceptual Development of Quantum Mechanics. New York: 1st ed., McGraw-Hill. New York: 2nd edn., Tomash, Los Angeles/American Institute of Physics.

3. Mehra J. and Rechenberg, H. (1982–87). The Historical Development of Quantum Theory. New York: McGraw-Hill.
4. Heisenberg, W. (1925). Über quantentheoretische Umdeutung kinematischer und mechanischer Beziehungen, Zeitschrift für Physik 33, 879–93; English translation, Quantum theoretical reinterpretation of kinematic and mechanical relations, in Ref. 5, pp. 261–76.
5. van der Waerden, B.L. (1967). Sources of Quantum Mechanics. Amsterdam: North-Holland.
6. Schrödinger, E. (1926). Quantisierung als Eigenwertproblem (4 parts), Annalen der Physik 79, 361–76; 79, 489–527; 80, 437–90; 81, 109–39; English translation, Quantization as a problem of proper values, in Ref. 7.
7. Schrödinger, E. (1928). Collected Papers on Wave Mechanics. London: Blackie.
8. Bohr, N. (1928). The quantum postulate and the recent development of atomic theory, Nature (Supplement) 121, 580–90; reprinted in N. Bohr, Atomic Theory and the Description of Nature. Cambridge: Cambridge University Press, 1934.
9. Electrons et Photons. (1928). Reports and Discussions of the 5th Solvay Conference Paris: Gauthier-Villar.
10. Frank P.G. (1949). In: Albert Einstein: Philosopher–Scientist (P.A. Schilpp, ed.) New York: Tudor, pp. 269–86.
11. Blackmore, J.T. (1972). Ernst Mach: His Work, Life and Influence. Berkeley: University of California Press.
12. Frank, P.G. (1947). Einstein, His Life and Time. New York: Knopf.
13. Einstein, A. (1949). Autobiographical notes, In: Albert Einstein: Philosopher–Scientist. (P.A. Schilpp, ed.), New York: Tudor, pp. 2–95.
14. Holton, G.J. (1968). Mach, Einstein and the search for reality, Daedalus 97, 636–73; reprinted in Gerard Holton, Thematic Origins of Scientific Thought: Kepler to Einstein (Cambridge, Massachusetts, Harvard University Press) Ch. 8 in 1st ed. (1973), Ch. 7 in 2nd ed. (1988).
15. Jammer, M. (1999). Einstein and Religion: Physics and Theology. Princeton: Princeton University Press.
16. Mach, E. (1893). The Science of Mechanics: A Historical and Critical Account of its Development Chicago: Open Court.
17. Barbour, J. and Pfister, H. (eds.) (1995). Mach's Principle: from Newton's Bucket to Quantum Gravity. Boston: Birkhaüser.
18. Einstein, A. (1905). Elektrodynamik bewegter Körper [The electrodynamics of moving bodies], Annalen der Physik, 17, 891–921 (1905).
19. Mehra, J. (1999). Einstein, Physics and Reality. Singapore: World Scientific.
20. Mach, E. (1909). Die Geschichte und die Wurzel des Satzes von der Erhaltung der Arbeit. Leipzig: Barth. (2nd ed.); translated as History and Root of the Principle of Conservation of Energy (Chicago: Open Court 1911).
21. Einstein, A. (1912). Gibt es eine Gravitationswirkung die der elektrodynamischen Induktionswirkung analog ist? [Is there a gravitational analogue of the effect of elecrodynamic induction?], Vierteljahrsschrift für gerichtliche Medizin 44, 37–40; translation given by C. Hoefer in Einstein's formulations of Mach's principle, Ref. 17, pp. 67–86.
22. Einstein, A. (1912). Prinzipielles zur allgemeinen Relativatätstheorie [Principle points of the general theory of relativity], Annalen der Physik 55, 241–4 (1912); translation by C. Hoefer as for Ref. [21].

23. Einstein, A. (1922). The Meaning of Relativity. Princeton: Princeton University Press, and later revised editions.
24. Einstein, A. (1916). Ernst Mach, Physikalische Zeitschrift **17**, 101–4.
25. Wolters, G. (1987). Mach I, Mach II, Einstein und die Relativatätstheorie. Berlin: de Gruyter.
26. Cercignani, C. (1998). Ludwig Boltzmann: The Man Who Trusted Atoms. Oxford: Oxford University Press.
27. Stachel J. et al. (eds.) (1987). The Collected Papers of Albert Einstein. Princeton: Princeton University Press, Vol. 1, p. 230.
28. Einstein, A. (1905). Die von der molekularkinetischen Theorie der Wärme geforderte Bewegung von in ruhenden Flüssigkeiten suspendierten Teilchen [Motion of suspended particles in kinetic theory], Annalen der Physik **17**, 549–60.
29. Holton, G. (1980). Einstein's scientific programme: the formative years, In: Some Strangeness in the Proportion: A Centennial Symposium to Celebrate the Achievements of Albert Einstein (H. Woolf, ed.). Reading, Massachusetts: Addison-Wesley, pp. 49–65.
30. Speziali P. (ed.) (1972). Albert Einstein—Michele Besso Correspondance 1903–1955. Paris: Hermann.
31. Pais, A. (1982). 'Subtle is the Lord': The Science and the Life of Albert Einstein. Oxford: Clarendon.
32. Einstein, A. (1954). Principles of research, In: Ideas and Opinions New York: Crown, pp. 224–27.

2
Einstein and Quantum Theory: The Early Years

Einstein and the Development of Quantum Theory

It is well known that Einstein played a very important part in the early development of quantum theory,[1] but it is probable that this part is still very often underestimated. In this chapter we discuss his work through to the coming of the Heisenberg-Schrödinger quantum theory in 1925–1926, and also his influence, mostly on Heisenberg and Born, in the period immediately afterwards.

It was, of course, Max Planck, who initiated the quantum theory in 1900 through the study of black body radiation and its spectral dependence.[2] Until 1900 the experimental evidence had seemed to support the formula of Wilhelm Wien, which was based on a model of a radiating heated gas, buttressed by a few rather dubious assumptions. In that year, though, the first reliable measurements in the infrared were performed by Otto Lummer and Ernst Pringsheim, and also by Heinrich Rubens and Ferdinand Kurlbaum, both groups working, like Planck, in Berlin. These results showed clearly that the Wien expression failed in this region of the spectrum. Planck had already been studying black body radiation for a number of years, and was obviously in an ideal position to explain this development.

Planck's first step was to use the detailed experimental data to produce a new formula, the so-called and exceptionally important Planck radiation formula. It agreed with the Wien formula in the region where the latter itself agreed with experiment, but, of course, deviated from it in the infrared. The formula contained the new constant h, now known as Planck's constant and one of the very most important constants of physics; Planck derived a very respectable value for h. This was all somewhat *ad hoc*, and Planck now wished to develop a proof for this formula from basic principles. The model that Planck had already been using for black body radiation consisted of an array of simple harmonic oscillators in the radiating cavity, interacting with the radiation and sharing a particular energy distribution. Planck now used statistical techniques due to Boltzmann to find the most probable energy distribution among the oscillators, and then related this to the electromagnetic field density, and hence to the required frequency distribution of radiated energy. In Chapter 1 we pointed out the anguish Planck felt in being forced to admit the strength of Boltzmann's methods that he had previously pilloried.

The quantity h in Planck's radiation formula led to an element of energy hf, where f is the frequency of the radiation, appearing in his derivation. For us, a century later, it is quite clear that this represents an energy discontinuity, and the standard accounts of Planck's work due to Martin Klein[3] and Hans Kangro[4] have assumed that this was also clear to Planck himself. However, in 1978 Thomas Kuhn[2] published a famous book in which he argued that, for at least 8 years following the initial discovery, Planck did not appreciate the physical meaning of his mathematical concept of the element of energy. Kuhn argues that it was Einstein who was the first to make this clear. Einstein, Kuhn suggests, carried out a research programme on black body radiation from 1902 that was 'so nearly independent of Planck's that it would almost certainly have led to the black body law even if Planck had never lived'.

In a paper of 1906,[5] Einstein considered Planck's oscillators interacting with gas molecules, which according to classical theory must have average energy kT, and it was then straightforward to obtain an expression for the spectral density of black body radiation, different expressions, in fact, depending on particular assumptions. If no discontinuity in energy is assumed, Einstein produced what is invariably called the Rayleigh–Jeans law. This law agrees with experiment and Planck's law for low frequencies, exactly the region where Wien's law failed. However for higher frequencies the spectral density increased indefinitely, and it predicted that the total energy emitted at any temperature should be infinite; thus not only did it disagree with experiment, but its consequences appeared absurd.

Lord Rayleigh had produced this law in 1905 (though James Jeans had to provide a correction); Rayleigh's method was to calculate the density of modes of vibration in the cavity. Einstein actually obtained the same expression a little earlier, but what was more important is that he made it totally clear that it was the unique and definite answer according to classical physics, rather than just the result of plausible assumptions. What is more, Einstein also showed that, if you depart from classical ideas by imposing the discontinuity, so that $E_n = nhf$, you obtain Planck's law and agreement with experiment. Kuhn was prepared to say of this paper of Einstein that '[in] a sense, it announces the birth of quantum theory'.

Not everybody accepts Kuhn's reassessment, of course. Both Klein and Kuhn presented their own versions of the story to the Centennial Symposium[6] for Einstein, while Klein, Abner Shimony and Trevor Pinch[7] have written a set of interesting reviews of Kuhn's book. (See also[8] for a recent general study.) However it is fair to say that at the very least Kuhn shows convincingly that Planck's presentation lacked some clarity, and that Einstein's part in providing clarification in these fundamental matters was immense. To say this is no slur on Planck; clearly his was the first step and that must be a crucial one.

It is true though, as Kuhn points out, that Einstein was frank about the limitations in Planck's work. He[9] wrote: 'Delighted as every physicist must be that Planck in so fortunate a way disregarded the need [to justify his choice of energy element], it would be out of place to forget that Planck's radiation law is incompatible with the theoretical foundations which provide his point of departure.' Bernstein[10] also refers to an interesting comment by Michele Besso, long-term friend and correspondent of Einstein; in a letter of 1928,[11] he wrote that 'in helping you edit your communications on the quanta I deprived you of a part of your glory, but, on the other hand, I made a friend for you in Planck.' It would seem that in a draft of his 1905 paper, of which more shortly, Einstein must have included severe criticisms of Planck, which Besso persuaded him to tone down. This would have been beneficial to Einstein, for whom friendship with, and support from, Planck, particularly over relativity, was of great importance; their relationship was very close, at least till the coming of Hitler which led to temporary but severe strain.[12] The comment again makes it clear that one should be careful not to underestimate Einstein's contribution. Einstein certainly felt that Planck's work required clarification, if not actual correction. While Planck deserves great credit for his pioneering work, it is clear that Einstein's contributions were, to a considerable extent, independent of Planck's, and demonstrated a vastly deeper appreciation of the implications of the quantum.

The Photon and Specific Heats

By 1906 Einstein had already written, in the previous year, what in retrospect must undoubtedly be seen as his most important paper on quantum theory, the paper in which he introduced the concept of the light particle. (This was not christened as the photon till Gilbert Lewis did so more than 20 years later, but for convenience we shall usually give it this name from now on.) This paper was one of those published in his magnificent year of 1905,[13] and it was indeed his only contribution to physics that he was prepared to describe as 'revolutionary'. It was also important for containing the work on the photoelectric effect, for which he was eventually awarded the Nobel prize in 1921.

Let us stress why the photon concept was so challenging conceptually. To accept quantisation of energy inside a radiating system, as implied by Planck's argument once it is fully comprehended, was definitely completely counter to our conceptions built up on general experience of physics, but, as long as energy remains unquantised while it is propagating, there is no direct clash with experimental evidence. It is true, of course, that there must be a mismatch between radiation emitted in a discontinuous fashion and energy propagating strictly according to Maxwell's equations; as Charles Darwin stressed, this must imply that, at the point of radiation, energy is conserved only at the statistical level, not for individual systems. Nevertheless it might be felt that this difficulty could be left for further study. The concept of the light particle or photon, on the other hand, appeared in 1905 to contradict directly all the evidence gathered over the previous 100 years that appeared to show that light, indeed all branches of electromagnetic radiation, consisted of waves, evidence gathered from experiments demonstrating interference, diffraction and polarisation, and evidence which, by 1905, seemed totally beyond question.

Einstein's argument was not, as implied in many texts and popular accounts, based on the experimental observations of the photoelectric effect. Rather it came from an analysis from first principles of Wien's law. Einstein used this law to calculate the entropy of radiation; when he investigated how this entropy varied with volume, he discovered, to his surprise, that the variation corresponded to that of a gas of particles each of energy hf. (Recognising that Wien's law is only an approximation to the truth, we might ask what happens if Planck's law is studied instead. In 1909, studying the statistical fluctuations of black body radiation, Einstein showed that Planck's law gives two terms, one corresponding to particles, one to waves; Wien's law gives just the particles, and the Rayleigh–Jeans law gives just the waves.)

If one accepts the photon concept, analysis of the photoelectric effect is reasonably straightforward, and leads to the famous equation

$$E_{e(max)} = hf - \phi \qquad (2.1)$$

where hf is the energy of the incoming photon, ϕ the work function of the metal, and E_e the energy of the photoelectron. Increasing the intensity of the radiation increases the number of photons but not, of course, the energy of each, so it increases the photoelectron current but not the energy of each electron. Some features of the photoelectric equation had been discovered experimentally before Einstein's work by Philip Lenard; he had found that the current increases with the frequency, but not with the intensity of the radiation. It seems unlikely, though, that these rather fragmentary results played any part in stimulating Einstein's theory.

Experiments on the photoelectric effect were notoriously difficult to perform, because, unless an exceptionally good vacuum is available, a layer of absorbed atoms forms on the metal surface and renders the results meaningless. Nevertheless, in a series of excellent experiments in 1914 and 1916, Robert Millikan eventually confirmed Einstein's equation in detail, and also provided a value of h in excellent agreement with that of Planck. This was good physics with many fruitful consequences. It led to the invention of the industrial photocell, which was exceedingly useful technologically; it also persuaded the Nobel Academy to forget their concerns over the status of relativity and award Einstein the prize he had deserved for so long. Lastly it helped Millikan towards his own Nobel prize in 1923. From the conceptual point of view, however, the application to the photoelectric effect was unimportant compared to the argument for the photon, and it is rather unfortunate that this paper of Einstein is commonly known as his 'photoelectric effect' paper.

The very existence of the photon was not only the most important aspect of this paper, it was one of the most dramatic claims in the whole area of physics of all time. It is scarcely surprising that it attracted very great opposition; indeed for nearly 20 years after its publication, the idea received practically no support, just a lot of criticism and a fair amount of ridicule. One early exception was Johannes Stark, who for a brief period was a great supporter, not only of Einstein but also of the photon, which he used in his theories of photochemistry. However in 1913 he and Einstein clashed in a priority dispute, and by the 1920s Stark and Lenard, both Nobel prizewinners, were the leading Nazi party vilifiers of 'Jewish physics' and Einstein in particular.

For Millikan, an arch-experimentalist, it was emphatically the photoelectric law he verified; the hypothesis of the photon itself he regarded as 'wholly untenable' and 'bold, not to say... reckless'. Even Einstein's greatest supporters could not take the photon seriously. In 1913, Planck, together with Walther Nernst, Rubens and Warburg, proposed Einstein as a regular member of the Prussian Academy. They wrote a recommendation packed with practically embarrassing praise for his many achievements; of the photon they merely noted that this should not be held against him, since in any area of study it is necessary to take risks if one is to achieve real advances.[10] Bohr, incidentally, was for decades particularly dismissive of the photon concept. It was not till the 1920s and the discovery of the Compton effect that physicists would have to engage fully with the photon.

Einstein's advocacy of the photon did not receive a positive response, but quite the opposite was the case for another important paper from this period,[14] that of 1907 on the specific heat of solids. There was considerable novelty in this paper, but more in the topic of the application of quantum ideas than in the actual arguments used. At this time virtually all the work on quantum theory had been in the area of black body radiation. As well as showing himself to be supreme in this area, Einstein had also extended the scope of the quantum to the photoelectric effect, and now he extended it further to what would now be called the solid state.

There was a long-standing and major problem in this area. Just as Einstein made clear for black body radiation, there was a definite prediction from classical physics for the specific heat of solids, but this prediction was disobeyed in some cases. The prediction came from the law of equipartition of energy. For a solid of N atoms, each with six degrees of freedom, one translational and one vibrational for each dimension, and with each degree of freedom contributing an average energy of $kT/2$ at temperature T, the total energy of the solid should be $3NkT$. The specific heat at constant volume should then be $3Nk$, and this does indeed correspond to a well-known empirical rule, that of Dulong and Petit. However, the rule fails for some cases such as diamond, boron and silicon, which are hard elements, and for which the specific heat falls below the prescribed value, at least at normal temperatures.

At least in retrospect the solution of this problem by basic quantum ideas seems fairly straightforward. The difference between the laws of Rayleigh–Jeans and Planck is that the first obeys the classical law of equipartition of energy whereby the average energy of a one-dimensional simple harmonic oscillator at temperature T is kT, corresponding to two degrees of freedom. With quantisation of energy, that expression must be replaced by $hf/[\exp(hf/kT) - 1]$ which is very nearly equal to kT when f is low or T high, but takes lower values for higher values of f or lower values of T. For a particular value of f, then as T tends to zero, so does the average energy. Einstein translated this to the case of the solid, assuming that each atom vibrates independently at the same frequency; this frequency would be expected to be larger for harder solids. (A better approximation would be to have a range of frequencies but Einstein left that embellishment to others, to Peter Debye,[15] and to Max Born and Theodore von Kármán.[16,17]) Then the value of the specific heat will be very close to $3Nk$ for temperatures around or above a critical temperature Θ, different for each element, but below that it will fall off towards zero for lower temperatures. Θ increases with the characteristic frequency; for most elements it is below normal temperatures, so the Dulong–Petit value applies; however for the harder elements, Θ lies above normal temperatures, so the value of specific heat around normal temperature must be less than that predicted classically.

Nernst, who as we have seen was by 1913 one of Einstein's supporters for the Prussian Academy, was in 1907 a leading physical chemist, particularly concerned with the behaviour of substances at low temperatures. He was responsible for what became known as the Third Law of Thermodynamics, which states that it is impossible to reduce the temperature of any system to absolute zero, and he eventually was awarded the 1921 Nobel prize for chemistry for this contribution. One aspect of this law is that specific heats should tend to zero as the temperature itself tends to zero, just as Einstein suggested. Nernst was able to put considerable resources into investigating the situation experimentally, and the results were in quite good agreement with Einstein's theory. This aroused Nernst's interest in the more general aspects of quantum theory, and he persuaded the Belgian industrialist, Ernest Solvay, to support fairly lavishly the first Solvay Conference held in 1911 to discuss problems in the area.[18]

Again we see Einstein's contributions playing the major role in taking the subject forward, and indeed this had been the story at all stages, apart obviously from Planck's original proposal. From this time on, though he continued to think deeply about quantum theory, as he did for the rest of his life, his direct contributions were somewhat fewer, partly because of his intense efforts in the area of general relativity, and from around 1920 his work on unified field theories. From 1913 Bohr's influence over the subject grew dramatically. However Einstein was to make a number of crucial interventions, direct and indirect, in the 1920s. First though we look at another important piece of work he performed in 1916.

Spontaneous and Stimulated Emission; Probability and Statistics

In this section we discuss a crucial paper of Einstein[19] published in 1916. This paper was exceptionally important in numerous ways. First we may say that it provided a direct and clear connection between the two main pillars of quantum theory at this time: Planck's law, and Bohr's idea from 1913 that an atom may make a transition between energy levels with energies E_m and E_n accompanied by emission or absorption of a packet of radiation of frequency f, with $hf = |E_m - E_n|$. (Note that when describing Bohr's idea one should not think directly of photons, because Bohr was, at that time, a great opponent of that idea.)

Einstein considered a system with two energy levels, with $E_m > E_n$, in thermal equilibrium with electromagnetic radiation. While Bohr did not explicitly consider the types of transition, one would deduce from his work that two types needed to be considered. The first is from E_n to E_m caused by *absorption* of radiation from a field at frequency f with $E_m - E_n = hf$; the probability of a transition would be proportional to the energy density, ρ, of the field. The second is from E_m to E_n; this would be thought to be independent of any field and may be called *spontaneous emission*. However to obtain agreement with Planck, Einstein showed, rather surprisingly, that a second type of emission was required which would be caused by the field, and would occur with probability proportional to ρ. This type of transition would be called *stimulated emission*.

Einstein used a coefficient A to discuss the spontaneous emission, and a different coefficient B for the absorption and the spontaneous emission, and this paper is almost invariably referred to as 'the A and B coefficients paper'. The basic equations are of the form

$$
\begin{aligned}
W_{mn} &= N_m(\rho B_{mn} + A_{mn}) \quad (m \rightarrow n) \\
W_{nm} &= N_n B_{nm} \quad\quad\quad\quad\;\; (n \rightarrow m)
\end{aligned}
\tag{2.2}
$$

In this equation, W_{mn} and W_{nm} are the rates of processes from m to n, and from n to m respectively, and N_m and N_n are the number of systems in the two states.

From the theoretical point of view, Einstein's work has been the basis of all the studies of interaction between matter and radiation over the succeeding 90 years. It was also the central argument that enabled Bohr and Hendrik Kramers to develop and use the *correspondence principle* between 1918 and the birth of the new quantum theory in 1925. In this work, they attempted to deduce selection-rules for atomic transitions by ensuring that they predicted the correct frequency distribution of energy for the classical case when the appropriate quantum numbers tended to infinity.

Another important aspect of the work was that, in the course of it, Einstein required, for the first time, an expression for the momentum of the photon. Using the well-known relativistic equation, $E^2 = p^2c^2 + m^2c^4$, and with the mass of the photon equal to zero, it is easy to deduce that the momentum of the photon should be equal to E/c, which is h/λ. This result was to be used, effectively in reverse, by Louis de Broglie in his own celebrated contribution, in which the de Broglie wavelength of a particle of momentum p is written as h/p. From the practical point of view this piece of work was also of immense importance; stimulated emission is, of course, the basis of the laser, invented in the 1950s and of exceptional importance in industry and medicine.

For the present book, the most interesting feature of this paper is the use of the rate equations above. The most natural interpretation, particularly of the spontaneous emission processes which occurred independently of the applied time-dependent field, would be that these take place randomly and thus non-deterministically. If that view is taken, it is easy to accuse Einstein of lack of consistency in his later strong opposition to giving up determinism; in 1916, it may be said, Einstein not only accepted it but pioneered it! At this time, Einstein was well aware of the problem. In acknowledging it, he emphasised that it was not a new development; the basic equation of radioactivity, $\mathrm{d}N/\mathrm{d}t = -\lambda N$, produced by Rutherford in 1900, is of exactly the same nature and so has exactly the same problem. Indeed it may also be said that Einstein's pioneering of the photon also had this feature; when a radiating object emitted photons, again the natural assumption was that the emission was random in nature.

A typical comment along these lines is that of John Wheeler,[6] who writes: 'Who that has known or read Einstein does not remember him arguing against chance in nature? Yet this is the same Einstein who in 1905, before anyone, explained that light is carried from place to place as quanta of energy, accidental in time and space of their arrival; and in 1916, again before anyone, gave us in his A's and B's, his emission and absorption coefficients, the still standard mathematical description of quantum jumps as chance events. How could the later Einstein speak against this early Einstein, against the evidence and against the views of his greatest colleagues? How can our own day be anything but troubled to have to say "nay" to one teaching, "yea" to others of the great Einstein. . .?'

This argument, however, seems wrong in principle. Well before 1900 probabilistic laws had, of course, been commonly used in physics, by Boltzmann and Maxwell in particular. The use of probability in no way implied a giving up of determinism. Rather there was an assumption of classical determinism at the molecular level, if one was able to follow the behaviour of each molecule and wished to do so. It was only because one was unable to do so, and in any case one required and could make use only of suitable averages, that probabilistic methods were used, and, of course, used highly effectively. From the beginning of his career Einstein was a very great admirer of Boltzmann, and this was precisely the spirit in which Einstein would have understood his work. Indeed it is exactly the spirit of standard statistical work, in which events of different type are counted, and probabilities are constructed from the results. In this book we will distinguish between 'statistical' ideas in which probabilities are constructed from the study of individual events, and there is certainly no requirement to forsake determinism for these individual events; and 'probabilistic' ideas, where events occur randomly and thus non-deterministically, and probabilities are fundamental.

There is certainly no necessary reason to assume that radioactive decay, the emission or absorption of photons in Einstein's 1905 theory, and the processes in his 1916 theory are probabilistic rather than statistical. The implication of believing them to be statistical is that there is a lower level of deterministic variables determining the times of the decays, and of the absorption and emission processes, and also the directions of motion of the photons produced in emission processes. The actual events seem to be random and non-deterministic only because the lower level variables are not observed. The best analogy may be one very close to Einstein's heart—Brownian motion; the observed motion of the suspended Brownian particles appears random and erratic only because one does not see the underlying motion of the molecules that causes the actual Brownian motion.

To return to the case of radioactivity and the processes of emission and absorption, the unobserved variables are what would later be termed the 'hidden variables' and would excite very great passions, as we shall see later in this book. It must be admitted that, for the case of the 1905 and 1916 theories, their hypothesis would seem elaborate and graceless; one might call it *ad hoc*, unless one regards the retention of determinism to be one's top priority. Einstein's words were somewhat different; he spoke of his theories of 1905 and 1916 as 'provisional',[20] thus allowing himself to claim their successes, without having to admit any necessity for any apparent facets of the theories that he did not like. In a related way he spoke of his work on the photon as a 'heuristic point of view' rather than a theory; as well as a gesture towards Mach, this again enabled him to articulate theoretical connections and ideas without having to provide a complete theory. Apart from a very short period after the announcement of the new quantum theory in 1925, Einstein would never be a supporter of simple hidden variable theories of quantum theory, as will be discussed in Chapters 5 and 8 in particular.

The Path to the New Quantum Theory

During the second decade of the twentieth century, in fact up to 1923, with the exception of the work of the previous section, most of the developments in quantum theory were due to Niels Bohr and his followers—Arnold Sommerfeld, Kramers, Alfred Landé, and later the young Heisenberg and Wolfgang Pauli.[21] One should not omit the very important experimental work of James Franck and Gustav Hertz, who demonstrated the discrete energy levels predicted by Bohr; and that of Otto Stern and Walther Gerlach whose experiments showed quantisation of angular momentum, and whose results were later understood as establishing the existence of spin. The theoretical work started with the very great success of Bohr himself with his study on the hydrogen atom in 1913; it was after his amendment to allow for the fact that the mass of the nucleus was finite that Einstein concluded: 'This is an immense achievement.'

Much of the subsequent work relied on Bohr's correspondence principle, the highly ingenious technique mentioned in the previous section. The correspondence principle used the rather incomplete model of the atom produced by Bohr's prescription of 1913 and generalised by Sommerfeld, and attempted to complete it by using the classical limit to deduce selection-rules, polarisations and intensities of spectral lines, as well as information about the Stark effect, which examines the behaviour of spectral lines in the presence of an electric field. The correspondence principle provided many useful facts about the atom. Not everybody appreciated it: Sommerfeld, for example,[22] could not help feeling that the results he had achieved from 'good physics' were being overtaken by what he called this 'magic wand' from Copenhagen. Despite the successes of the approach, it is fair to say that it did not require, nor did it produce, fundamental understanding of the quantum theory itself. Bohr did succeed, though, in achieving a reasonably good understanding of the Periodic Table, though this understanding was based on intuition, rather than being derived from rigorous arguments.

We now move to the years between 1923 and 1925, during which period a wide range of ideas from a considerable number of people led to the establishment of the new quantum theory. One crucial event in 1923 was the discovery of the Compton effect by Arthur Compton, and its analysis by Compton himself and

Debye. Once understood,[23] the effect has a straightforward interpretation; it is a collision between an incoming X-ray photon and a nearly free electron. Normal methods of analysing a collision, including use of the expressions for the energy and momentum of the photon, yield the change in wavelength of the scattered X-ray in agreement with experiment. It was now at last clear to nearly everybody that the photon's time had come! There was little further opposition to this concept.

What opposition there was came mostly from Bohr. John Slater, a visitor from USA at Copenhagen had suggested a model of radiation in which each atom is the source of a virtual field or ghost-field interacting with all the other atoms, and oscillating with the frequencies of the possible quantum transitions; this field then guides the photons between atoms. Bohr adapted Slater's model to produce the BKS (Bohr, Kramers, Slater) idea in which the photons are removed; the consequence of Bohr's changes was that determinism was lost, and conservation of energy and momentum became statistical. However experiments of Walther Bothe and Hans Geiger, and of Compton himself and Alfred Simon soon demonstrated that the BKS concept was misconceived. At this stage, Bohr was forced to admit defeat and finally accepted the photon.

The saga of the photon was Einstein's greatest triumph in the area of quantum theory. It was his deep insight that enabled him to come up with the idea in the first place, and it was his independence of mind and determination that enabled him to stick with the idea, despite the criticism and ridicule it received, even from those who were in most respects his greatest supporters. He also showed a different type of courage in not giving in to his own concerns about the clash between wave and particle aspects of radiation. In any comparison of Einstein and Bohr, it would seem only natural to take note of the fact that, in contrast, Bohr was practically the last to come to terms with Einstein's vision of the photon.

Yet somehow within a very small number of years, established belief was that the relationship had become inverted, that it was Bohr who could see the way forward, and Einstein who felt unable to follow. And ironically it was the photon itself that became the catalyst for this process. It was generally felt that Bohr's complementarity allowed one to come to terms with wave-particle duality and move forward. Einstein did not accept this and sought other ways of making progress, ways for which at best success would be slow in arriving.

The New Quantum Theory: Bohr and Heisenberg

Let us in any case follow events forwards from 1923. One extremely important strand[24] was inspired almost wholly by Bohr. Starting from work on optical dispersion by atoms performed by Rudolph Ladenburg and Kramers, Born and Heisenberg were able to extend the ideas to interactions between electrons, but it was Heisenberg in 1925 who took the final step of constructing the square arrays of numbers that could be used to create the first rigorous form of quantum theory. Born recognised these arrays as matrices, and it was Born and Pascual Jordan, later with Heisenberg himself, who constructed the full formalism. Paul Dirac got to grips with the German work phenomenally quickly, and came up with some important and sophisticated ideas of his own. Lastly Pauli made the first major use of the new matrix mechanics; in a rather elaborate calculation he was able to solve the problem of the hydrogen atom, his results for the allowed energies of the atom duplicating, of course, those of Bohr obtained 12 years previously by the non-rigorous means of the old quantum theory.[25,26]

Two general points should be mentioned. In the analysis of hydrogen and larger atoms, two important steps had been made around 1924–5. Pauli had long recognised the need for a fourth quantum number to supplement those related to the three spatial dimensions. He now announced his famous exclusion principle: in an atom, each electron must have a different set of quantum numbers. It was Samuel Goudsmit and George Uhlenbeck who identified this fourth quantum number as being due to an intrinsic spin of the electron.

The history of the concept of spin is interesting. Though the idea worked well, and it seemed to provide a straightforward picture of the electron, a little more examination suggested that there were major problems. Study of the relation between the mass and the energy of the electron suggested that, unless it had an unreasonably large mass or an unreasonably large volume, its surface must be travelling faster than the speed of light. A full understanding of spin had to wait till Dirac's relativistic theory of the hydrogen atom, which was produced in 1928.

A more specific point on the electron spin relates to Bohr and Einstein. In their work, Goudsmit and Uhlenbeck made use of a coupling between the spin and the orbital angular momentum of the electron to explain the so-called fine structure in the energy-level diagrams of atoms. Bohr was highly supportive of their argument in general, but saw a major problem. The spin could couple only to a magnetic field, not to an electric one, but in an atom it would seem that there was only an electric field. Bohr was coming close to feeling that the idea must, after all, be rejected, when Einstein pointed out that, according to relativity, an object moving through an electric field also experiences an effective magnetic field. It is interesting to see that, even though Einstein may have been a little out of touch with the details of the work being carried out by Bohr and his co-workers, he was still able to produce telling arguments of the greatest significance.

The second general point is that, during the course of this work, the physical basis on which the old quantum theory had originally been based gradually disappeared. In the course of the work on optical dispersion, in order to relate the model used to actual experimental results, it became necessary to replace the actual oscillations by a hypothetical set of oscillators, the frequencies of which corresponded to those of the observed transitions. This may be described as a further application of the correspondence principle, and this process was to continue; Jammer[25] wrote that 'there was rarely in the history of physics a comprehensive theory which owed as much to one principle as quantum mechanics owed to Bohr's correspondence principle'.

In the end, restriction of the theory to those quantities that may be observed in experiments, frequencies of radiation in particular, became Heisenberg's main motivation in his great work. The contrast with Bohr's original orbits, highly physical in nature, was immense. At least in 1925 and for a short period thereafter, Heisenberg and Pauli in particular took great delight in the belief that the lesson of quantum theory was that no attempt should be made to work from any model of the physical process being studied; equally one should not expect that the final theory will provide any picture of the process. All that was required was that the predictions agreed with experience. Such a concept was bound to be anathema to Einstein, as we shall shortly see.

The New Quantum Theory: Einstein's Contributions

First though we shall follow an alternative route through the years 1923 to 1925 and 1926, this time describing a series of steps in which Einstein was closely involved,[1] and concluding with the alternative formulation of quantum theory, that due to Schrödinger. The first crucial idea was that of Louis de Broglie. Since it seemed clear by this date that radiation had a wave aspect as well as a particle aspect, de Broglie made the bold hypothesis that electrons, and other entities until then assumed to be straightforwardly particles, also had a wave-like nature. De Broglie inverted Einstein's formula for the momentum of a photon, $p = h/\lambda$, to produce the famous de Broglie wavelength, $\lambda = h/p$. This work comprised de Broglie's doctoral thesis, and his principal examiner, the very well-known physicist Paul Langevin, while admiring de Broglie's imagination, was not sure whether the work was significant enough to be acceptable. However, when he sent a copy to Einstein to get his opinion, the reply was not only was it acceptable for de Broglie's doctorate, it was 'a first feeble light on this worst of physics enigmas'.

Einstein's endorsement came well before the experimental confirmation of de Broglie's idea. Relevant experiments had actually been performed as early as 1921 by Clinton Davisson and Charles Kunsman, who studied the scattering of electrons by metals, and by Carl Ramsauer, using the scattering by gas atoms. However it was not until the interpretation of these experiments by Walter Elsasser that it was suspected that a wave-like nature for the electron was being demonstrated. Elsasser's paper was published in 1925, but only again after Einstein had been consulted. The conclusive demonstration had to wait till 1927 when it was provided by Davisson and Lester Germer, and also by George Thomson. Davisson and Thomson were to share the Nobel prize for physics in 1937, 8 years after de Broglie himself was awarded it.

The second development to be reported is the work of the then unknown Indian, Satyendra Nath Bose, who sent Einstein a copy of a paper in which he presented a novel derivation of Planck's law; this paper had already been rejected by one journal, but Einstein saw how important it was, translated it from English to German himself, and arranged for its publication. Essentially the derivation was a statistical mechanics of photons, in other words a replacement for Boltzmann's classical distribution for this case. Bose had made use of a number of novel assumptions. The most important of these was that the photons were treated as indistinguishable, quite contrary to the particles in the Boltzmann distribution, which are assumed distinguishable. Also the photons did not behave independently, as Boltzmann would have decreed, and neither were they conserved.

Einstein extended the work of Bose to the quantum statistical mechanics of a gas of non-relativistic particles. The main difference from Bose's treatment was that, for Einstein's case, the number of particles did have to be kept constant. What resulted was the so-called Bose–Einstein statistics, now accepted to be the correct quantum statistical mechanics for particles with no spin, or with integral spin in units of $\hbar$. The photon, incidentally, has spin 1 in the appropriate units. The analogous statistics for particles of half-integral spin, Fermi–Dirac statistics were discovered by Enrico Fermi and Dirac, but not until 1927. Incidentally, just as in 1909 Einstein had calculated the energy fluctuation of electromagnetic radiation and found a particle-like term, now he did the same for his gas, and found, not only the expected particle-like term, but also a wave-like term. It was clear that the speculations of de Broglie and Bose, both supported by Einstein, were not independent.

It was at the very end of 1925 and through the first half of 1926 that Schrödinger[27] produced his remarkable series of papers, essentially following de Broglie's idea, but arguing from de Broglie's simple wave for the free particle case, to a general wave equation, the famous Schrödinger equation from which could be obtained, for example, the allowed energies for any form of potential. (He also obtained the wave-function for each case, interpretation of which had to wait a little for the work of Born to be described shortly.) Since his predicted energies for the cases of the hydrogen atom and the simple harmonic oscillator agreed with experiment, it seemed clear that he had produced a new version of quantum theory, completely independent of that of Heisenberg, though just as successful. His work, of course, utilised the ideas of de Broglie, and Schrödinger also acknowledged the suggestion of Debye, who argued that a wave equation was required. Schrödinger's work clearly made no use of the experimental evidence for de Broglie's ideas, for this was yet to come. The other person who Schrödinger did acknowledge was Einstein for his endorsement of de Broglie. In 1926, he[28] wrote to Einstein: 'The whole things would certainly not have originated . . . (I mean, not from me) if I had not had the importance of de Broglie's ideas brought home to me by your second paper on gas degeneracy.'

It may be said that Einstein had stimulated a second route to quantum theory, directly from his photon concept and his advocacy of the ideas of de Broglie and Bose. There is an interesting comparison with the route inspired by Bohr. As has been said, this had relied largely on the correspondence principle, which essentially allowed Bohr and others to base their formalism mostly on the results at the classical limit. Van der Waerden[24] wrote that modern (Heisenberg) quantum theory may be described as 'systematic guessing, guided by the Principle of Correspondence'. This technique was supremely clever, but did not tell one anything very fundamental about physics in the quantum regime. In contrast the photon concept of Einstein, and the ideas of de Broglie and Bose did attempt actually to say something about nature. It will be readily admitted that the message coming from these developments was confusing, even contradictory, and it seemed certain that much work would have to be done to reconcile what appeared to be the seemingly paradoxical aspects of nature. One may not be surprised that it was Einstein who continued on this path of at least attempting to construct a comprehensible physics, and Bohr who again was satisfied with provision of a framework that provided only a very limited physical picture, but had the great virtue of working well.

It should also be remembered that the main motivation for Schrödinger's work was dislike of the arid formalism of the Heisenberg approach, and the desire to provide an alternative with a direct physical interpretation, particularly one dispensing with the quantum jumps of Bohr and Heisenberg that Schrödinger detested so much. In this he seemed at first to be successful, at least as he saw it himself, but, as we shall see in later chapters, the success was largely illusory. Einstein would, of course, have been broadly in support of Schrödinger on the desire for a visualisable picture of the physics.

Einstein's response to the work of Heisenberg and Schrödinger is covered in the main in Chapter 5, but in this chapter we describe briefly some definite contributions of Einstein made to the Heisenberg approach to quantum theory in the years up to Bohr's definitive statement on complementarity in 1927. The first resulted from a famous conversation between Heisenberg and Einstein in 1926. Heisenberg explained to Einstein that his technique in producing matrix mechanics was based on restricting the theory to consider only quantities that were observable; he suggested he was merely following in Einstein's footsteps from when he had produced his theory of special relativity. It will be remembered that, at this point in 1905, Einstein had been indeed, at least in his own mind, a follower of Mach. However by 1926 his approach was very different. He now replied to Heisenberg that such an idea was nonsense; rather the theory *determined* what could be observed.

This credo was to serve Einstein well in his arguments with Bohr over quantum theory. Einstein always remained unmoved by protestations of Bohr or John von Neumann that particular quantities could not exist; he regarded such statements as theory-dependent, and he retained the freedom to aim at producing his own theory. It was also to serve Heisenberg well. Rather than declaring from the very start that such quantities as the positions and momenta of electrons in an atom were unobservable, he used aspects of the quantum theory to determine the limits of their (combined) observability and obtained his famous Heisenberg relation, usually called the uncertainty principle. This was one of his most celebrated contributions to quantum theory.

This second and very much related area where Einstein contributed in this period was to Born's elucidation of the wave-function as providing probabilities for various experimental outcomes. In Born's paper[29] he wrote that his idea was based on a remark of Einstein's describing a light beam as consisting of quanta guided by a 'ghost field', which itself carried no energy or momentum, and whose sole task is to determine the probability for each photon to follow a certain path. It is not clear from what writing or saying of Einstein that Born drew this remark, though Jost[6] suggests that it may have been Einstein's lecture at the 1911 Solvay conference. It seems far from obvious that the Born of 1926 was actually in tune with the Einstein of 1911, and much less likely still that Einstein and Born would have been in agreement on these matters in 1926 or later. Indeed the Born–Einstein letters show that the statistical or probabilistic interpretation of the wave-function is the topic on which the two were most liable to talk past each other. What is significant, though, is that the depth and breadth of Einstein's thoughts were such that he was able to influence critically the path taken by the founders of quantum theory in these crucial years.

There was in fact a great deal of conceptual advance during these years, from Heisenberg's initial belief that agreement between mathematical formalism and experiment was all that was required, through the work of Heisenberg and Born described above, which acknowledged that some interpretation was required and that some knowledge of quantities such as the position and momentum of an electron was obtainable, even though such knowledge was only partial, and through to Bohr's complementarity, described in detail in Chapter 4. However once the Copenhagen position had become hardened in the mind of physicists, this process was largely forgotten.[30] Born's statistical interpretation appeared an almost obvious component of the Copenhagen synthesis. It came to be felt that Born had perhaps merely been the first to articulate what was in any case reasonably straightforward.

There were several damaging results of this tendency. Born was disheartened to feel his contributions undermined. He was already displeased to be left out of the first round of Nobel prizes for quantum theory; the 1932 and 1933 prizes were awarded to Heisenberg, Schrödinger and Dirac, although, as Born said, Heisenberg did not know what a matrix was until Born told him! Born's own Nobel prize, for 'the statistical interpretation of the wave function' did not come for more than another 20 years. In this book, it is very important to note that this collective amnesia made it easier for the leaders of Copenhagen to categorise Einstein as a once magnificently creative physicist whose good days were long over, rather than one who had been able to comment with great effect on their own theories.

As has been seen in this chapter, from 1905 right through to the birth of complementarity in 1927, Einstein had contributed centrally and triumphantly to the development of quantum theory, both, particularly in the earlier part of this period, through his own ideas, and, especially later, through his promotion of the crucial of others. Very often he had been in a very small minority in recognising the importance of a particular piece of work, sometimes in a minority of one. His identification of the significance of the photon concept, and of the speculations of de Broglie and Bose, seem to indicate a practically uncanny ability to read the secrets of nature in simple intuitive and mathematical terms. One would have thought that he had indisputably earned the right to have, at the very least, a respectful audience in the future for his own views on the meaning of the quantum theory. However, as seen in the previous chapter, he had also during these years obtained an individual conception of how science should advance and of what could constitute genuine scientific knowledge, and this conception was to lead him in entirely the opposite direction to that followed by the overwhelming majority of the other physicists. His influence in the next decades, in fact, would be miniscule.

References

1. Pais, A. (1982). 'Subtle is the Lord...': The Science and the Life of Albert Einstein. Oxford: Clarendon.
2. Kuhn, T.S. (1978). Black-Body Theory and the Quantum Discontinuity 1894–1912. Oxford: Clarendon.
3. Klein, M.J. (1962). Max Planck and the beginning of the quantum theory, Archive for the History of Exact Sciences 1, 459–76.
4. Kangro, H. (1976). History of Planck's Radiation Law. London: Taylor and Francis.
5. Einstein, A. (1906). Theorie der Lichterzeugung and Lichtabsorption [Theory of light emission and light absorption], Annalen der Physik 20, 199–206.
6. Woolf, H. (ed.) (1980). Some Strangeness in the Proportion: A Centennial Symposium to Celebrate the Achievements of Albert Einstein. Reading, Massachusetts: Addison-Wesley [Klein's contribution is on pp. 161–85, Kuhn's on pp. 186–91, Wheeler's on pp. 341–75, Jost's on pp. 252–65].
7. Klein, M.J., Shimony, A. and Pinch, T.J. (1979). Paradigm lost? A review symposium, Isis 70, 430–4.
8. Gearhart, C.A. (1909). Planck, the quantum and the historians, Physics in Perspective 4, 170–215.
9. Einstein, A. (1909). Zum gegenwärtigen Stande des Strahlungsproblems [Present position of the radiation problem], Physikalische Zeitschrift 10, 185–93.
10. Bernstein, J. (1991). Quantum Profiles. Princeton: Princeton University Press.
11. Speziali, P. (ed.) (1972). Albert Einstein—Michele Besso Correspondance 1903–1955. Paris: Hermann.
12. Heilbron, J.L. (2000). The Dilemmas of an Upright Man: Max Planck and the Fortunes of German Science. Cambridge, Massachusetts: Harvard University Press.
13. Einstein, A. (1905). Über einen die Erzeugung und Verwandlung des Lichtes betreffenden heuristischen Geischtspunk [Generation and transformation of light by a heuristic approach], Annalen der Physik 17, 132–48.
14. Einstein, A. (1906). Plancksche Theorie der Strahlung und die Theorie der spezifischen Wärme [Planck's theory of radiation and the theory of specific heat], Annalen der Physik 22, 180–90.
15. Debye, P. (1912). Zur Theorie der spezifischen Wärmen [Theory of specific heats], Annalen der Physik 39, 789–839.
16. Born, M. and von Kármán, T. (1912). Uber Schwingungen in Raumgittern [Molecular frequencies], Physikalische Zeitschrift 13, 297–309.
17. Born, M. and von Kármán, T. (1913). Zur Theorie der spezifischen Wärme [On the theory of specific heats] 14, 15–19.
18. Mehra, J. (1975). The Solvay Conferences on Physics. Boston: Reidel.
19. Einstein, A. (1916). Strahlungs-emission und -absorption nach der Quanten-theorie [Radiation and absorption of radiation according to quantum theory], Deutsche Physikalische Gesellschaft, Verhandlungen 18, 318–23.
20. Jammer, M. (1999). Einstein and Religion: Physics and Theology. Princeton: Princeton University Press.
21. Pais, A. (1991). Niels Bohr's Times, in Physics, Philosophy and Polity. Oxford: Clarendon.
22. Sommerfeld, A. (1923). Atomic Structure and Spectral Lines. London: Methuen.
23. Stuewer, R.H. (1975). The Compton Effect. New York: Neale Watson.
24. van der Waerden, B.L. (1967). Sources of Quantum Mechanics. Amsterdam: North-Holland.

25. Jammer, M. (1989). The Conceptual Development of Quantum Mechanics. (1st edn., New York: McGraw-Hill, 1966; 2nd edn., New York: Tomash, Los Angeles/ American Institute of Physics, 1989).
26. Whitaker, A. (2006). Einstein, Bohr and the Quantum Dilemma. 2nd edn., Cambridge: Cambridge University Press.
27. Moore, W. (1989). Schrödinger: Life and Thought. Cambridge: Cambridge University Press.
28. Przibram, K. (ed.) (1967). Letters on Wave Mechanics: Schrödinger, Planck, Einstein, Lorentz. New York: Philosophical Library, p. 26.
29. Born, M. (1926). Zur Quantenmechanik der Stossvorgänge [On the Quantum Mechanics of Collisions], Zeitschrift für Physik **37**, 863–7.
30. Beller, M. (1990). Born's probabilistic interpretation: a case study of 'concepts in flux', Studies in the History and Philosophy of Science **21**, 563–88.

3
Quantum Mechanics and its Fundamental Issues

Introduction

The 'new' quantum theory of Heisenberg and Schrödinger was immediately seen to be extremely successful in practice. It could be used to predict energy levels for a considerable number of systems, and Schrödinger's approach in particular was capable of describing many features of the atom, and predicting the results of a great variety of experiments. Yet it was also clear that there were many surprising and puzzling aspects to the theory, aspects that were certain to cause controversy.

In particular, the theory appeared to violate two Newtonian virtues, those of determinism and realism. The first tells us that, if the state of a system is known at time t, we should be able to predict its behaviour with certainty at all later times. The aspect of the second that is stressed here is that a system should have definite values of such observables as position and momentum at all times, even if these are not measured. We may mention that the theory also appears to violate a third Newtonian virtue, that of locality, which says that an event at one point in space cannot have an immediate effect at a distant point. However, while the problems with determinism and realism were immediately obvious and there was much discussion of them in the early days of quantum theory, that with locality received practically no attention until it was mentioned by Einstein and his collaborators in the famous EPR paper of 1935; we therefore delay discussing it until we meet EPR itself in Chapter 6.

Incidentally in this chapter we shall often remark along the lines that the theory 'appeared to' have certain difficulties. What is meant by this apparently rather vague form of words is that the theory has these difficulties if the formalism is interpreted in a fairly straightforward and natural way, without adding extra elements beyond its explicit content. In particular no so-called hidden variables are added, which might give extra information about an individual system, over and above that contained in the wave-function. That there should be no hidden variables was an essential part of the so-called orthodox interpretation (or interpretations) of quantum theory, as will be discussed in Chapter 4. Einstein's terminology, to be discussed in Chapter 6 in particular, would be to say that the theory has these difficulties if it is assumed that, without the addition of hidden variables, it is *complete*, an assumption Einstein did not accept. Much of the rest of the book will be concerned with attempts, particularly many using hidden variables, to go beyond this basic approach, and hence to remove the difficulties, or, as Einstein would say, to complete the theory.

The loss of realism was one aspect of the fact that, unlike clasical theory, quantum theory provided no clear picture of the system. Even from the days of the old quantum theory, wave-particle duality, as exemplified in Einstein's light-particle and de Broglie's wave nature of the electron, had shown that simple classical pictures had, at best, a restricted value. It could, of course, be argued that this was not a genuine problem. One might suggest that, while the loss of a classical picture, and even the loss of determinism and realism, might well be disturbing for many and even anathema for some, it was not necessarily a conceptual failing of quantum theory; some, in particular Heisenberg and Pauli, could claim it as a liberating advance.

However there were aspects of this loss of realism that did present genuine conceptual difficulties. Before measurement, a straightforward application of the formalism suggests that the measured quantity will not usually have a specific value, but when the measurement is performed a definite result is obtained; yet the standard quantum evolution provides no mechanism whereby this may happen. This difficulty may be termed the measurement problem of quantum theory; the wave-function appears to change from a so-called pure state to a mixed state, but quantum theory does not allow this to happen. Also, while quantum theory, as has been said, suggests that microscopic systems do not demonstrate realism, we usually take it for granted that objects of normal size, so-called macroscopic objects, do behave in a realistic fashion. It must clearly be a challenge to explain how classical realism occurs, or at least apppears to occur, at the macroscopic level when it does not exist at the microscopic level.

In this chapter, we describe briefly the formalism of quantum theory, drawing particular attention to these aspects that appeared to present difficulties either in conception or in interpretation, and thus lay at the heart of Einstein's struggles with quantum theory.

Some Preliminaries

Like all physical theories, quantum mechanics has two distinct but inter-related components: a mathematical formalism and a physical interpretation. A mathematical formalism is an algorithmic structure consisting of equations and calculational recipes. In order to be able to compare the computed results with empirical observations, an interpretation is required to enable one to connect the formal language of a theory to actual experimental results. Thus an interpretation adds physical content to a theory in terms of some key concepts; for example, in Newtonian mechanics the concept of 'force' as a *causal agent* giving rise to 'acceleration' is crucial in ascribing physical meaning to the basic equation of motion. In this book we are assuming the reader to be familiar with the basic ideas of the formalism of non-relativistic quantum mechanics. However, for the sake of completeness we begin by briefly recapitulating some of the central features of this standard formalism.

The discussion in this entire book will be in terms of what is known as the *Schrödinger picture* in which a complex function of position coordinates evolves in time according to an equation known as the *Schrödinger equation*. This function $\psi(\vec{r}, t)$ is called the *wave-function* which is taken to characterise the state of a system. (The question of whether $\psi(\vec{r}, t)$ describes the state of an individual system, or that of an ensemble of identically prepared systems is discussed later in this book; here we shall use the phrase 'state of a system' without addressing this issue). If $\psi(\vec{r}, t)$ is known at any one instant, its values at all other instants are uniquely computable, provided the Hamiltonian characterising the system is completely known. Thus we may say that the evolution of the wave-function itself is deterministic. It should be mentioned that the Hamiltonian itself cannot be deduced from theory; it is specified on the basis of our knowledge about the system in question. Then if H, the Hamiltonian, is assumed constant, the Schrödinger evolution is given by

$$\psi(t) = \exp(-iHt/\hbar) \, \psi(t = 0) \tag{3.1}$$

or the time-dependent Schrödinger equation has the form

$$H\psi = i\hbar \left(\frac{\partial \psi}{\partial t} \right) \tag{3.2}$$

(In (3.2), as distinct from (3.1), H of course does not need to be constant.)

Using the coordinate representation, which will be the one mainly used in this book, the equation for a one-dimensional system becomes:

$$-\left(\frac{\hbar^2}{2m} \right) \frac{\partial^2 \psi}{\partial x^2} + V\psi = i\hbar \left(\frac{\partial \psi}{\partial t} \right) \tag{3.3}$$

where V represents the potential energy. Thus the Hamiltonian has two distinct roles in the quantum formalism; as well as representing the energy of the system, it determines the evolution in time of a wave-function representing the state of a system.

One can immediately see from the above equations that the Schrödinger equation is linear in ψ. This entails an important property known as the *superposition principle*; if ψ_1 and ψ_2 are both valid solutions of the Schrödinger equation, then *any* linear combination of ψ_1 and ψ_2 is also a valid solution. The superposition principle is at the heart of the quantum theory; for example, it is central in producing the basic diffraction phenomena which established de Broglie's ideas, and led directly to those of Schrödinger. Indeed it may well be described as the jewel of the theory, as the central techniques of quantum theory rely on it, and physicists would certainly be extremely reluctant to lose it. However it is also a cause of the major difficulty of quantum theory, the measurement problem; in this way it becomes less of a jewel and more of a curse! As we shall see in later chapters, some now believe that it may have to be sacrificed as a fundamental component of quantum theory, and merely retained as an extremely good approximation in a very wide range of circumstances, but not in the measurement situation.

It is interesting that, while in classical mechanics it is the basic dynamical variables, those that are directly measurable, that occur in the equation of motion, in quantum theory it is an abstract quantity $\psi(\vec{r}, t)$, which has no direct physical meaning, that obeys the fundamental equation and evolves deterministically. Hence the need for an appropriate interpretation becomes more acute in the case of quantum mechanics—the Schrödinger equation can become physically meaningful only if one knows how to relate ψ to the measurable quantities. The set of basic rules for this purpose is as follows, giving, for convenience, those for the one dimensional case:

(a) The *probability* of finding the particle between x and $x + dx$ in a measurement at a particular instant t is given by

$$P(x, t)dx = |\psi(x, t)|^2\, dx \tag{3.4}$$

where $P(x, t)$ is known as the probability density; Eq. (3.4) is called the *Born Rule*. Note that $P(x, t)$ remains positive for any form of ψ. From the Schrödinger equation (3.3), and using the boundary condition for a bound state, that $\psi \to 0$ as $x \to \pm\infty$, it is easy to confirm from the form of $P(x, t)$ given by (3.4), that the total probability is conserved, as it must be:

$$\frac{d}{dt} \int_{-\infty}^{\infty} P(x, t)dx = 0 \tag{3.5}$$

Therefore one can normalize ψ and thus arrange that

$$\int_{-\infty}^{\infty} |\psi(x, t)|^2 dx = 1 \tag{3.6}$$

Once Eq. (3.4) is postulated, the programme of quantum mechanics turns out to be entirely different from that of classical mechanics. Instead of uniquely predicting a particular outcome of a given measurement, or describing what a particle will actually do, the Schrödinger equation can only predict the possible results of a process that a particle may undergo. Then immediately the question arises of how to obtain the various probabilities of the results of measurements of the measurable dynamical variables. For this, one has the following rule:

(b) A measurable dynamical variable, say A, is represented by a Hermitian operator, $\hat{A}$. (Whether this condition is *necessary* for a physical quantity to be measurable is debatable; here we assume it to be a *sufficient* condition.) The eigenvalues of the operator are taken to be the possible results of measuring the dynamical variable, and the orthonormal eigenfunctions form a complete set, in terms of which *any* wave-function can be expanded. In other words, $\langle \phi_i | \phi_j \rangle = \int_{-\infty}^{\infty} \phi_i^* \phi_j dx$ is equal to 1 if $i = j$, for normalisation; it is equal to 0 if $i \neq j$, for orthogonality; any function ψ can be expressed as $\sum_n c_n \phi_n$, where $c_n = \int_{-\infty}^{\infty} \phi_n^* \psi \, dx$

If $|\phi_i\rangle$ is an eigenfunction of $\hat{A}$ corresponding to an eigenvalue α_i, or in other words, $\hat{A}|\phi_i\rangle = \alpha_i |\phi_i\rangle$, then the probability of obtaining the result α_i on measuring A on a system prepared in the state $|\psi\rangle$ is given by

$$P(\alpha_i) = |\langle \phi_i | \psi \rangle|^2 = \left| \int_{-\infty}^{\infty} \phi_i^* \psi \, dx \right|^2 \tag{3.7}$$

Here the notion of probability is used in the following sense: if one repeats the preparation of the same state $|\psi\rangle$ a number of times, thereby generating an ensemble of systems in an identical state $|\psi\rangle$, the probability of a given outcome is its relative frequency. Expanding $|\psi\rangle$ in terms of $|\phi_i\rangle$, in other words writing $|\psi\rangle = \sum_i a_i |\phi_i\rangle$, Eq. (3.7) reduces to

$$P(\alpha_i) = |a_i|^2 \tag{3.8}$$

Eq. (3.7) or (3.8) is sometimes referred to as the generalised *Born–Dirac rule*. From (3.8) it follows that the mean or expectation value of the results of measurements of A on state $|\psi\rangle$ is given by

$$\langle A \rangle = \sum_i P(\alpha_i)\alpha_i = \langle \psi | \hat{A} | \psi \rangle = \int \psi^* \hat{A} \psi \, dx \tag{3.9}$$

the integral being over all values of x.

Thus a definite measurement on a system with a particular wave-function may lead to any one of a number of different results. Let us suppose that we make the orthodox assumption that ψ gives us all existing information about the system, in other words that there are no extra hidden variables. (The possibility that there *are* hidden variables is discussed at length in later chapters.) It is then clear that, although the theoretical quantity ψ itself evolves deterministically in the absence of measurement, when a measurement is performed the system behaves in a non-deterministic way, since a number of different measurement results may be obtained. This is one of the main differences between classical and quantum physics; from a classical point of view, it may be described as one of the main problems of quantum theory.

Also, since one and the same wave-function may lead to different values in a measurement of a particular physical observable, and if again we assume that there are no hidden variables which might lead to these different measurement results, we must suspect that, prior to measurement, this physical observable just does not have a unique value. This may be said to be one aspect of the loss of realism in quantum theory, and we discuss it further later in this chapter in a comparison of pure and mixed states.

For completing the statement of some of the very basic elements of the standard quantum formalism, and reiteration of its fundamental problems, we now turn to discussion of what is known as the uncertainty principle, which is considered to be a cornerstone of the formalism and interpretation of quantum mechanics.

The Uncertainty Principle

The proof and mathematical statement of the uncertainty principle is fairly straightforward. We work with two non-commuting Hermitian operators $\hat{A}$ and $\hat{B}$, representing observables A and B respectively. We assume that

$$[\hat{A}, \hat{B}] = i\hbar\hat{C} \tag{3.10}$$

Then we can say something fundamental about the product of the variances ΔA, ΔB of A, B where

$$(\Delta A)^2 = \langle A^2 \rangle - \langle A \rangle^2, \quad (\Delta B)^2 = \langle B^2 \rangle - \langle B \rangle^2 \tag{3.11}$$

The actual result, demonstrated in textbooks, is that

$$(\Delta A)^2(\Delta B)^2 \geq \frac{\hbar^2}{4}\langle C \rangle^2 \tag{3.12}$$

which is the generalised Heisenberg uncertainty principle. If A, B are canonically conjugate, for example, the position and momentum operators p, q, then

$$[\hat{p}, \hat{q}] = i\hbar \tag{3.13}$$

and the uncertainty principle takes its familiar form

$$(\Delta p)^2(\Delta q)^2 \geq \frac{\hbar^2}{4} \tag{3.14}$$

It is interesting to note that the preceding derivation uses no input from quantum dynamics. Even if one uses a wave-function having the wrong symmetry and violating the Schrödinger equation, the uncertainty relation (3.14) will not necessarily be violated. The uncertainty principle is thus *insensitive* to any modification of the Schrödinger equation.

We now turn to the much more subtle matter of the interpretation of the uncertainty principle. First, let us clarify its operational meaning. The interpretational significance of the uncertainty principle may be stated in one of the two following ways. The first corresponds to Heisenberg's *gedanken* experiments which he used in his original proof or justification[1,2] for his principle. By use of this analysis he claimed to show that the product of uncertainties in the simultaneous measurement of two non-commuting canonically conjugate dynamical variables, such as momentum and position, of a single particle cannot have a value lower than $\hbar/2$.

This approach to the Heisenberg principle has been quite widely held right up to the present day. In particular it was taken for granted by both Bohr and Einstein in their discussions in the 1920s, the so-called Bohr–Einstein debate, which is discussed in detail in Chapter 5. Indeed it was looked on as a critical component of quantum theory by both men. In the debate, Einstein attempted to violate this approach to the principle, by demonstrating the possibility of simultaneous measurement to a greater degree of accuracy than the principle allows. Bohr regarded it as absolutely essential to repudiate each of Einstein's suggestions.

However it is now recognised that the Heisenberg principle does not, as such, relate to measurements on individual systems. The second approach to interpreting the uncertainty principle, derived from the commutation rule, states that it is impossible to prepare an ensemble of identical particles, all of which are in the same state, such that the product of the root mean square deviations of the values obtained by measuring any two non-commuting dynamical variables has a value less than the lower bound given by the relation (3.14). This approach recognises that uncertainty in the value of a dynamical variable, as embodied in the above statement, refers to the statistical spread over the measured values for the various identical members of the ensemble of systems. To test the uncertainty relation (3.14) we therefore require a repeatable state preparation procedure leading to an ensemble of identically prepared particles, all corresponding to the state being studied. Then on each such system one needs to measure one or other of the two dynamical variables A and B. The statistical distributions of such measured results of A and B would be characterised by variances satisfying (3.14).

The operational significance of the two approaches is totally different. In the second, there is no question of simultaneous measurement of the dynamical variables related to a single particle, while this concept forms the essence of the first approach, where the uncertainty in a single measurement is interpreted as the estimate of imprecision in the measured value of a dynamical variable for a single particle. The arguments involved in Heisenberg's original gedanken experiments are in fact *semi-classical*, while the relation itself is a rigorous consequence of the formalism of quantum mechanics. Thus Heisenberg's thought experiments in the original form should certainly not be regarded as providing a proof of the uncertainty principle. At most, they might be regarded as suggestive arguments which could serve to encourage more rigorous analysis.

'Time-Energy Uncertainty Principle'

The uncertainty principle that we have been discussing is quite general; in a particular case it may be called the position-momentum uncertainty principle. Its mathematical meaning is clear, and although, as we have shown, there has been confusion over its interpretation, resulting from the way the relation was initially deduced by Heisenberg, it is today clear how it should be used.

A somewhat analogous principle, the so-called time-energy uncertainty principle, was also deduced by Heisenberg, and has also played a large part in discussions on the nature of quantum theory. It has a form broadly analogous to the momentum-position uncertainty principle:

$$\tau_A \cdot \Delta E \geq \hbar/2 \tag{3.15}$$

However we have specifically written τ_A rather than Δt to emphasise the point that, unlike p, x and E, time cannot be represented in a straightforward way by an operator in non-relativistic quantum theory. Rather its status is exactly as in classical physics. Thus one cannot have an uncertainty or indeterminacy in time, or a spread in values of time. The quantity τ_A must rather represent a characteristic time for the change of a quantum mechanical state.

There has been a vast amount of discussion of precisely how τ_A should be interpreted. In our main discussions in this book, this is a topic we may avoid. The time-energy uncertainty principle did play an important part in the discussions of Einstein and Bohr, but the discussion of both uncertainty principles was in terms of the original prescription of Heisenberg. For Einstein and Bohr in their discussions, the two uncertainty principles stood on the same footing, and the two men talked specifically of uncertainties in simultaneous measurement. As we shall see in Chapter 5, time came into the argument in particular in the so-called photon box thought-experiment. A hole in the wall of the box is covered by a shutter which is opened for a time Δt. A photon enters during this period, so its time of entry is said to have an uncertainty of Δt. As we have said before this quantity is not strictly a quantum mechanical uncertainty or indeterminacy, but Einstein and Bohr both treated the quantity as though it were one. However more recently, there has been considerable discussion of the concept of time in quantum theory, and in particular the time-energy uncertainty principle, and this is one of our topics when we look at modern developments in Chapter 13.

Pure and Mixed States

At this stage, we will clarify what is meant by the state of a system within the standard framework of quantum mechanics. In particular we shall stress the difference between so-called pure states and mixed states, a difference important in all conceptual discussion of quantum theory, and one that is central in the quantum measurement problem, to be discussed in detail in the following chapter. Margenau[3] has suggested that a quantum mechanical state may be taken to be synonymous with a probability distribution. This may be said to be true only in the sense that the standard concept of a quantum mechanical state is a description which embodies the probabilities for all possible measurements on the system. The state of a system is a symbolic representation of the ensemble to which the system belongs. A quantum state may be specified by an ensemble whose members are all identically prepared so that they correspond to a single wave-function. This is known as a *pure state*. Two different pure states cannot be described by the same wave-function, and conversely a given pure state cannot relate to two different wave-functions. However we must also consider an ensemble whose member systems are represented by different wave-functions; this is called a mixed state. A mixed state is thus made up of *sub-ensembles* in *different* pure states.

The wave-function structures of ensembles corresponding to both pure and mixed states can be specified on the basis of the knowledge of the preparation method. In a preparation procedure, a given ensemble of systems undergoes a specified interaction with a suitable device such that the ensemble emerges in a definite known state. Examples of such preparation methods are electron guns (producing electron beams), momentum selectors (selecting particles within a definite momentum range), and slit arrangements (selecting collimated beams of particles).

Statistical Properties of Pure States

The basic prescription of quantum mechanics concerning prediction of reproducible statistical properties mentioned earlier will now be expanded. As already mentioned, the measured values of a dynamical variable, say, α, in a given state, say, ψ, are its eigenvalues, α_i. These measured values may be regarded as a set of *random variables*. The relative probabilities of these random variables pertaining to the state ψ are given by $|\langle\psi|\phi(\alpha_i)\rangle|^2$ where $\phi(\alpha_i)$ is the eigenfunction corresponding to the eigenvalue α_i. The mean value $\bar{\alpha}$ and the variance σ^2 for this set of random variables are given by

$$\bar{\alpha} = \sum_i \alpha_i |\langle\psi|\phi(\alpha_i)\rangle|^2 = \langle\psi|\alpha|\psi\rangle \tag{3.16}$$

$$\sigma^2 = \sum_i \alpha_i^2 |\langle\psi|\phi(\alpha_i)\rangle|^2 - \bar{\alpha}^2 \tag{3.17}$$

We now discuss the operational significance of the above prescriptions. Let us consider a sample of N systems, chosen *arbitrarily* from the ensemble represented by ψ, and let the dynamical variable α be measured on each of them. The mean value of these results is, say, $\bar{\alpha}_N$. Then the Central Limit theorem[4,5] of statistics for random variables implies that if such measurements are repeated over many samples of N systems, and N is taken to be sufficiently large, then the various $\bar{\alpha}_N$ values will form a normal distribution characterised by the mean value $\bar{\alpha}$ and standard deviation $\sigma' = \sigma/N^{1/2}$. In such a distribution, the probability that $\bar{\alpha}_N$ will lie between $\bar{\alpha} \pm 3\sigma'$ is 0.997. Thus one is able to determine $\bar{\alpha}$ with any desired accuracy from measurements on a single sample of N systems by suitably choosing a sufficiently large value of N so that the variance σ' is negligibly small.

One can also determine the probability distribution function $|\langle\psi|\phi(\alpha_i)\rangle|^2$ similarly, by reducing it to a mean value measurement. This can be done, for example, by considering the observables $f(\alpha)$ whose eigenvalues are $f(\alpha_j)$ such that

$$\begin{aligned} f(\alpha_j) &= 1 \quad (\alpha_j = \alpha_i) \\ &= 0 \quad (\alpha_j \neq \alpha_i) \end{aligned} \tag{3.18}$$

The mean value of $f(\alpha)$ is given by

$$\bar{f}(\alpha) = \langle \psi | f(\alpha) | \psi \rangle = | \, \psi | \phi(\alpha_i) \rangle |^2 \qquad (3.19)$$

Thus the mean value of $f(\alpha)$ can specify the probability distribution corresponding to any particular eigenvalue α_i. Here, we may mention the oft-quoted von Neumann[6] definition of the pure state by the property that if the pure state ensemble is divided into sub-ensembles, the expectation values of all dynamical variables are the same for all these sub-ensembles, and identical with those pertaining to the original ensemble.

Statistical Properties of Mixed States

Let us consider an ensemble represented by the mixture of wave-functions $\psi_1, \psi_2, \ldots \psi_i$ with weight factors $a_1, a_2, \ldots, a_i$ where $\sum_i a_i = 1$. The expectation value of a dynamical variable α for this ensemble is given by

$$\bar{\alpha} = \sum_i a_i \langle \psi_i | \alpha | \psi_i \rangle = \sum_{i,j} a_i | \langle \psi_i | \phi(\alpha_j) |^2 \alpha_j \qquad (3.20)$$

where $\phi(\alpha_j)$ is the eigenfunction of $\hat{A}$ corresponding to the eigenvalue α_j. Here the random variables α_j have the relative probabilities P_j given by

$$P_j = \sum_i a_i | \langle \psi_i | \phi(\alpha_j) \rangle |^2 \qquad (3.21)$$

The variance of these random variables is of the form

$$\sigma^2 = \sum_j P_j \alpha_j^2 - \bar{\alpha}^2 \qquad (3.22)$$

Observable Distinction between a Pure and a Mixed State

Let us consider a pure state, say, of a spin (1/2) particle described by a linear superposition of the form

$$\psi = a\chi_+^{(z)} + b\chi_-^{(z)} \qquad (3.23)$$

where $\chi_+^{(z)}$ and $\chi_-^{(z)}$ are states with z-projection of spin equal to $\hbar/2$ and $-\hbar/2$, respectively. It is tempting to infer that each particle prepared in the state (3.23) *actually* had a definite value of the z-projection of spin, $\hbar/2$ or $-\hbar/2$, even *before* any measurement, and that a measurement process merely reveals what was already the value. Such an interpretation would be legitimate for a mixed state but definitely not for a pure state. Nevertheless, one could have maintained this interpretation if the pure state turned out to be equivalent to a mixed state. That this is not so will now be illustrated with the help of a simple example[7] showing how a pure state can be operationally distinguished from a mixed state.

Let us take $a = b = 1/\sqrt{2}$ in (3.23). It is then straightforward to see that

$$\psi = \left(1/\sqrt{2}\right)\left(\chi_+^{(z)} + \chi_-^{(z)}\right) = \chi_+^{(x)} \tag{3.24}$$

since

$$\chi_+^{(z)} = (1/2)^{1/2}\left(\chi_+^{(x)} + \chi_-^{(x)}\right) \tag{3.25}$$

$$\chi_-^{(z)} = (1/2)^{1/2}\left(\chi_+^{(x)} - \chi_-^{(x)}\right) \tag{3.26}$$

where $\chi_+^{(x)}$ and $\chi_-^{(x)}$ denote states for which a measurement of the x-component of spin is certain to produce the result $\hbar/2$ and $-\hbar/2$, respectively. Thus if the x-component of spin is measured by passsing a beam of spin-1/2 particles in the state specified by (3.24) through a Stern–Gerlach apparatus with its inhomogeneous magnetic field oriented along the x-axis, all the particles would emerge in the 'up' beam (corresponding to $\chi_+^{(x)}$). However suppose the state is a mixed state where each system is either in state $\chi_+^{(z)}$ or in state $\chi_-^{(z)}$ given by (3.25) and (3.26). For this mixed state the measurement of the x-component of spin on the entire ensemble should yield a probability 50% for the result $\hbar/2$, and 50% for $-\hbar/2$. Hence it is clear that the linear superposition (3.28) *cannot* be interpreted as saying that any one particle was either in the state $\chi_+^{(z)}$ or in the state $\chi_-^{(z)}$ before a measurement was made.

It should also be stressed that, under the type of development of the wave-function in time given by Eqs. (3.1) or (3.2), which is known as *unitary* evolution, a pure state will always remain pure; an initially pure state can never become mixed.

It is true that in certain cases it may be exceedingly difficult in practice to discriminate between a pure and a mixed state. However the fundamental interpretational distinction *in principle* between a pure and a mixed state is an ineluctable feature of the standard framework of quantum mechanics. This point will play a crucial role in our discussions concerning the quantum measurement problem. But before proceeding to explain the quantum measurement problem, we will briefly discuss the related question of realism in quantum mechanics by first recapitulating the key elements of the notion of realism in classical mechanics. In the rest of this book we will find that Einstein's arguments with orthodox approaches to quantum theory were of two types. The first type was to do with approaches to realism, determinism, locality, measurement, hidden variables and so on and are discussed in detail in Chapters 5 and 6. The second was to do with the relation between quantum mechanics and classical mechanics, and is discussed in Chapter 7. How both types of discussion progressed in the years after Einstein's death is discussed in Part C of the book.

Classical Realism or Macrorealism

The concept of classical realism hinges on notions derived from our everyday experience with the familiar macroscopic world. The basic tenets may be summarised as follows:

C1. All the physical attributes of an individual object have definite values associated with them at any instant of time, *irrespective of whether or not they are being measured or observed.*

C2. Measurements on a classical system may be *non-invasive*; this means that they may *not* affect the premeasurement values of the physical attributes.

It therefore follows that, according to classical realism or macrorealism, an object is at all times in a definite state characterised by a set of sharp values for *all* its physical attributes, and all these attributes may be measured non-invasively. One can cite examples from the macroworld where the straightforward applicability of the above notion may appear to be questionable. A tossed coin, a thrown dice, or a spinning roulette wheel all have unknown and effectively unknowable attributes during the period of motion. But the reality of the attributes of these objects during motion is never questioned by a classical realist, who would refer to the *counterfactual assertion* that with sufficient knowledge and effort, these properties *could* be known as accurately as we want. Here it is important to stress that this counterfactual statement is at the core of classical realism, and one is permitted to have recourse to it because it does not lead to any incompatibility with the laws and facts of classical physics. This is in sharp contrast to the situation in quantum physics.

Quantum Realism

A simple instructive way of discussing the paradox of realism in quantum mechanics is with reference to the Stern–Gerlach experiment for spin-$1/2$ particles. Suppose we have a beam of identical particles of spin-$1/2$ prepared in a state which is described quantum mechanically by Eq. (3.23). Now, using the Stern–Gerlach arrangement to measure the z-component of spin, the probability of getting the value $+\hbar/2$ is $|a|^2$ and that of obtaining $-\hbar/2$ is $|b|^2$. Here, if one adheres to the spirit of *classical realism*, one is inclined to contend that each of the particles in the ensemble described by Eq. (3.23) actually 'had' a definite intrinsic value of its z-projection of spin, and that the measurement process was non-invasive and merely revealed what was already the case. This would be equivalent to a hidden variable theory since there is additional information about each spin over and above what is in the wave-function. However, this outlook is not valid from a quantum mechanical point of view. This can be shown very easily.

For instance, let us choose $a = b = 1/\sqrt{2}$ in Eq. (3.23). Then we have the state given by Eq. (3.24). As mentioned earlier, if one measures the x-component of the spin of a beam of particles in this state, it is predicted that all the particles would have their spin projection $\hbar/2$ on the x-axis. On the other hand, if one assumes that each particle of the ensemble was *either* in the state $\chi_-^{(z)}$ or in the state $\chi_+^{(z)}$ *prior* to measurement, then we recall that from Eqs. (3.25) and (3.26) it follows that one should expect a probability 50% for the result of $\hbar/2$ and 50% for the result of $-\hbar/2$, for a measurement along the x-axis. This contradicts what is inferred from Eq. (3.24).

The upshot of the above discussion is the moral that it is quantum mechanically *not* permissible, even in principle, to assume that each particle belonging to the ensemble described by Eq. (3.24) *actually possessed* a definite value of spin projection *before* any measurement was made. This feature of quantum mechanics may be construed as implying that the process of measurement creates a reality associated with the dynamical attributes of a microphysical object that did not exist prior to the measurement.

In analogy to the characterisation of classical realism, the concept of quantum realism may be envisaged in terms of the following tenets:

Q1. Reality cannot be associated with the *unobserved dynamical attributes of* microphysical entities. Reality lies *only* in the existence of the entities themselves, and observer-independent definite values of their *innate static attributes* such as mass, charge, spin, magnetic moment.

Q2. It is in general *not* possible to determine the state of a quantum system or the values of its dynamical attributes without affecting the system's subsequent time evolution. A measurement on a quantum system is thus in general *invasive*.

The above picture implies a drastic qualitative distinction between the *microscopic* and *macroscopic* levels of reality. The standard framework of quantum mechanics inevitably requires this bizarre discontinuity in description, and totally evades the central problem as to *how* and *at what stage* the quantum mechanical description at the microscopic level becomes converted into the classical one known to be applicable at the macroscopic level. One often speaks of a 'cut', or sometimes of a 'Heisenberg cut' between microscopic and macroscopic. However the very *legitimacy* of the 'cut' is at issue, if one believes (as most working physicists seem to) that quantum mechanics is a universal theory, and that one is *prima facie* entitled to extrapolate the quantum mechanical description to the scale of the *macro-world*. That such an extrapolation leads to an acute conceptual incompatibility with macrorealism is the content of the quantum measurement problem which is now discussed.

The Quantum Measurement Problem

This problem reveals a subtle inner inconsistency within the standard framework of quantum mechanics. Weinberg has called it 'the most important puzzle in the interpretation of quantum mechanics'.[8] The underlying basic question is how to ensure within the standard framework of quantum mechanics the very occurrence of a definite individual outcome in a single measurement. In quantum mechanics, the results of measurements can only be predicted probabilistically. Hence the statistical frequencies of outcomes are the quantities whose predictions can be verified by measurements, essentially on an ensemble of identically prepared systems. Thus it is required to guarantee within the theory the occurrence of individual outcomes, enumeration of which will enable us to test the computed probabilities. It is, thus, logically impermissible to speak of the probability of a particular outcome, unless the *definiteness* of that outcome and its distinguishability from other outcomes in the ensemble can be ensured. Now, let us examine what happens in quantum mechanics.

A generic feature of all examples of measurements analysed quantum mechanically is as follows. It is noted that inclusion of the measurement apparatus is a central feature. Consider a system initially in a state ψ which is a superposition of two states, say ψ_1 and ψ_2, which are the eigenstates of the dynamical variable which is to be measured. Let the initial state of the system be:

$$\Psi = a\psi_1 + b\psi_2 \tag{3.27}$$

The interaction of this system with a measuring device then results in a final state of the form

$$\Psi = a\psi_1\Phi_1 + b\psi_2\Phi_2 \tag{3.28}$$

where Φ_1 and Φ_2 are mutually orthogonal and macroscopically distinguishable states of the device. (For a review of various specific models of quantum theoretic treatment of measurements, see for example, Refs. 9, 10.) An ineluctable feature of the linearity of quantum mechanics as applied to any measurement process is that the final state of a measured system coupled to a measuring apparatus is *entangled* (or nonfactorisable; it is not given by a function, ψ, multiplied by another function, ϕ). The question of entanglement was central in Einstein's work on quantum theory, in particular in the EPR paper, and is discussed in detail in Chapter 6. How this entanglement gives rise to the quantum measurement problem, is briefly explained below. (For a detailed exposition see, for example, Refs. 9–14.)

First, note that, because of the entangled nature of the wave-function of Eq. (3.28), no separate state can be ascribed to any apparatus. But Eq. (3.28) represents a pure state. A pure state in quantum mechanics corresponds to an ensemble of identical members. This means that, according to Eq. (3.28), each member (in this case, a system coupled to an apparatus) of the postmeasurement ensemble is described by the *same* wave-function Ψ. Thus such an ensemble is *homogeneous* and its members are *indistinguishable*. However this is very different from what we expect for the result of a quantum mechanical measurement. An outcome of a measurement is registered essentially by associating it with some property of the apparatus (like a pointer reading). We would thus expect that each system would be *either* in state $\psi_1\Phi_1$ *or* in state $\psi_2\Phi_2$. Such a postmeasurement ensemble is *heterogeneous*.

A heterogeneous ensemble is required to be represented by a *mixed state* in quantum mechanics, but within the formalism of standard quantum mechanics it is impossible for a pure state to evolve into a mixed state.[9–11] Then the question arises of how to accommodate within quantum mechanics the occurrence of definite outcomes. Note that the definiteness of an individual outcome necessarily implies that it is distinguishable from other outcomes. Hence the distinguishability between different outcomes needs to be ensured within the theory; this requirement is definitely not satisfied with a final state of the form given by Eq. (3.28). In other words, there is *no* element in the quantum mechanical description of the postmeasurement state given by Eq. (3.28) which corresponds to a definite outcome of an individual measurement. This, in essence, leads to the quantum measurement problem or paradox, the acuteness of which was highlighted by Schroödinger in his famous 'cat paradox',[15] described in Chapter 6.

To sum up, the measurement paradox implies that it is inherently enigmatic in quantum mechanics to compute the probabilities of various outcomes when the very occurrence of an individual outcome is not ensured. One is often tempted to put this problem aside as nothing different from that in classical statistical mechanics. For example, one may refer to an apparent analogy between this and the purely classical problem of the tossing of a coin where, because of the unknown and uncontrollable ingredients inherent in the specification of the relevant initial conditions, one can predict only that the probability for heads/tails is 50%. Once, say, heads, is observed, the probability for, say, heads becomes 100%. There is, however, nothing problematic about this feature because the formal structure of classical mechanics ensures that, at least *in principle*, *if* the initial conditions *were* exactly known, an individual outcome for a single tossing is definitely *either* heads *or* tails, irrespective of whether it is actually observed. On the other hand, in quantum theory a measurement gives rise to a superposition of various possibilities where the possibility of distinguishing an individual outcome from other possible outcomes is not allowed, even in principle.

For completeness, here one may add that the key point of the measurement paradox remains unaffected even if the initial state of the macroscopic apparatus is taken to be a mixture of different states rather than the pure state that we have considered for the sake of simplicity.[16] The acuteness of this paradox makes one inclined to suspect, as Bell[12] put it, that the standard formulation of quantum theory is 'intrinsically incomplete or ambiguous at a fundamental level.' Various versions of the standard approach have tried to alleviate or minimise this paradox, but all those remain unsatisfactory. We shall comment on them in Chapter 4. Here we next proceed to discuss what is known as the classical limit problem of quantum mechanics.

The Classical Limit Problem of Quantum Mechanics

The problem of the classical limit of quantum mechanics is a particular example of a broader issue concerning the relationship between two physical theories, one of which is empirically valid in a subdomain of the other. It is a basic requirement that a theory which supercedes the previous one must be compatible with the former in a suitable limit. As a prelude to discussions of the classical limit of quantum mechanics, it will be helpful to recall some of the relevant aspects of this general issue.

Let us consider two theories A and B where A has superceded B; in other words, B has a restricted domain of validity. Here the key question is not so much whether A actually reduces to B in some appropriate limit, in the sense that A and B become completely equivalent, formally as well as conceptually, but, more importantly, whether A allows for the validity of B in the domain where B is known to be true. In other words, the question of *consistency*, not of derivability, is crucial to this issue. If under suitable conditions, the precise specification of which may not always be straightforward, the implications of A agree (without any logical discontinuity) with the *observationally* valid features of B, A and B may be viewed as mutually consistent within the range of our present empirical accessibility.

However, even in this limit there could, in principle, be certain residual unobserved features of A which are not compatible with B. Such results usually escape being noticed because of the extremely sensitive measurement required for their detection. It is thus important to bear in mind that *complete equivalence* between the theoretical frameworks of A and B, including *all* the conceptual features and conceivable consequences of the formalism, is, in general, not ensured in the common domain of their applicability, though A and B may be *consistent* in the sense stated earlier. In other words, there could be a realm in which one may view a certain range of phenomena in terms of two distinct theories which are conceptually independent, yet connected by a numerical consistency condition. This point will now be illustrated with reference to a few specific examples from pre-quantum physics.

The Limit Problem: Wave and Ray Optics

The restricted domain of validity of ray or geometrical optics is typically characterised by the condition that dimensions, say d, of the objects (including all the relevant optical systems) encountered by the light, be large compared to the wavelength, λ, of light. Then the diffraction effects are negligible, and rectilinear propagation of light is *effectively ensured*. Under this condition, ray and wave optics may be regarded to be consistent in the domain where ray optics is observationally valid, though the conceptual frameworks of these two theories are obviously entirely different. From a different consideration, the 'Eikonal equation' of ray optics is shown to be consistent with Maxwell's equations, under the assumption that the field amplitudes change very slowly over a distance comparable to λ.[17] However, the concept of sharply defined shadows used in ray optics requires *abrupt* changes in field amplitudes at the shadow boundary. The wave nature of light has therefore fine residual effects in the domain of ray optics which generally escape detection at the level of usual observations.

The Limit Problem: Special Relativity and Newtonian Mechanics

The well-known condition under which the special relativistic results (Lorentz transformations and dynamical equations) become consistent with those of

Newtonian mechanics is when the speed, v, of the relevant system becomes negligible compared to the speed of light in vacuum, c, or when $v/c \ll 1$. Subject to this condition, all special relativistic effects are believed to become negligibly small in the observationally verified domain of Newtonian mechanics. In this case, conceptual consistency is also automatically ensured because both special relativity and Newtonian mechanics describe particle mechanics in terms of the action of forces on particles having well-defined trajectories.

Still there are hidden subtleties. For example, consider a pair of *spacelike separated* events occurring at, say, (x_1, t_1) and (x_2, t_2) such that $(x_2 - x_1)/c(t_2 - t_1) = \beta \gg 1$. The relativistic transformation equation of this time interval gives

$$(t_2' - t_1') = (t_2 - t_1)(1 - (v/c)\beta)(1 - v^2/c^2)^{-1/2} \qquad (3.29)$$

In the limit $(v/c) \ll 1$, Eq. (3.29) reduces to

$$(t_2' - t_1') = (t_2 - t_1)(1 - (v/c)\beta) \qquad (3.30)$$

remembering $\beta \gg 1$. The time interval between such spacelike separated events thus remains *frame-dependent* in the usual non-relativistic limit $(v/c) \ll 1$. This means that the time-ordering of two such independent events may get reversed for some observers even when Newtonian mechanics with Galilean invariance is expected to hold good. (The non-Galilean low-velocity limit to Lorentz transformations for spacelike intervals has been discussed by Bacry and Levy-Leblond.[18]) For every (v/c), however small, one can thus find suitable values of $(x_2 - x_1)$ and $(t_2 - t_1)$ for which the above feature persists, though it would be empirically hard to detect in the non-relativistic regime. This illustrates that in spite of the fact that all usually observable features of Newtonian mechanics are consistent with special relativity, subject to an appropriate limiting condition, *complete equivalence* between these two theories is not ensured in their common domain of applicability.

The Classical Limit Problem in Standard Quantum Mechanics

In the light of the preceding discussions it should be clear that the whole question of the classical limit of quantum mechanics comes down to examining whether, under suitable approximations, the quantum mechanical results are *consistent* with the well-known classical results which are observed to be valid in the macroscopic regime. This should not involve any abrupt discontinuity in the way quantum mechanics is applied or interpreted. Whether the theoretical structure of quantum mechanics actually *reduces* to classical mechanics in any limit is not centrally important. Note that by the term 'classical limit' of quantum mechanics, we shall mean the *classical-like* behaviour of *macrophysical* systems—we will not include in our discussions the classical-like behaviour of microphysical systems under suitable conditions.

We begin by outlining the key classical features of *macrophysical* systems which need to be *allowed* by quantum mechanics in an appropriate limit. The specification of such a limit would characterise the notion of macroscopicity within the context of quantum mechanics.

A. Dynamical evolutions of macrosystems obey classical laws of motion.

B. A macrosystem is found to be *localised* at a particular position at *any* instant. In other words, position localised states emerge as *preferred* states for any typical micro object.

C. Measurements on a classical macroscopic system may be *noninvasive* in the sense that the results of *subsequent* measurements on it remain *unaffected*. In other words, a macrosystem may be *insensitive* to measurements.

It is at once clear that feature B cannot be consistently accommodated within the framework of standard quantum mechanics unless additional ingredients are introduced. The symmetry of Hilbert space with respect to different dynamical observables and basis states is an ineluctable feature of quantum mechanics. The required emergence of a *preferred* set of states localised with respect to a *particular* dynamical variable is therefore a serious problem in the classical limit of quantum mechanics. The scheme specifically suggested to address this problem will be discussed in Chapter 4.

At this stage, we note that a stronger statement of feature B can be made in the macroscopic context. It seems legitimate to associate an *ontological reality* with any localised macro-object—in the sense that it *has* position *irrespective* of whether it is actually observed or not. This feature *cannot* emerge from standard quantum mechanics by whatever limiting procedure one may adopt. As sharply put by Holland:[19] 'How can one pass from a theory in which ψ merely represents statistical knowledge of the state of a system to one in which matter has substance and form independently of our knowledge of it?' This aspect of the problem of the classical limit, the question of how to ensure a smooth conceptual connection between quantum mechanics and *macrorealist descriptions* of classical mechanics, is left entirely unaddressed by the standard framework of quantum mechanics. The extent to which standard quantum mechanics can accommodate features A and C will be discussed in Chapter 4.

Wave-Particle Duality

The issue of wave-particle duality hinges on the sense in which one uses the conceptual ideas of wave and particle. If one remains confined within the formalism of quantum theory without demanding a visualisable understanding, the problem of wave-particle duality ceases to have any relevance. In particular, if we consider optical experiments, the rules of quantum optics are well defined and sufficient to predict correctly all observable results. The electric and magnetic field operators are the basic dynamic variables in this formalism. The notion of the photon enters the theory only as a secondary entity, defined as an excitation associated with the normal modes in terms of which any electromagnetic field can be expanded.

From this point of view, the *particle* aspect of radiation can have meaning only when a detection process is considered—the quantised decrease in field energy resulting from a detection process may be described in terms of the removal of photons from the field. Here it is interesting to note that the phenomenon of the photoelectric effect, which is frequently cited as evidence of the *necessity* of regarding a propagating radiation as a bundle of photons, can indeed be explained in terms of a classical (unquantised) field. This is by invoking the *quantised exchange of energy* between matter (treated quantum mechanically so that it has discrete energy levels) and radiation (described classically). It has been shown in detail that the discrete nature of the photodetection current can be perfectly explained by a classical description of the field, provided that the linear detector is regarded as a quantum system.[20]

Dirac's overworked metaphor 'each photon interferes only with itself'[21] has added to the confusion—strictly speaking, photons do not interfere, either with themselves or with each other, but rather the interference pattern is embodied in the linear superposition of electric field amplitudes which are *not* equivalent to wave-functions; for example, wave-functions need to be normalised, but the electric field amplitudes cannot be. (What manifests itself in light-beam interference experiments is the relative phase of the electric field amplitudes, in fact of their associated quantum mechanical operators, and *not* the relative phase of the state vectors, as is the case with interference of 'particles' with nonzero rest mass.) Within the standard formalism of quantum mechanics, it is *superfluous* to invoke the notion of 'photons' in order to describe optical interference effects.[22-24]

The problem of wave-particle dualism emerges only if one insists that, apart from formal predictions of the observed results, some *intuitive understanding* is also required in terms of classically visualisable pictures of particles and waves. In this context, Bohr's idea of wave-particle *complementarity* plays a central role in formulating this problem. We discuss this in the following chapter.

References

1. Heisenberg, W. (1927). Über den anschaulichen Inhalt der quantentheoretischen Kinematik und Mechanik [The actual content of quantum mechanical kinematics and mechanics], Zeitschrift für Physik **43**, 172–98.
2. Heisenberg, W. (1930). The Physical Principles of the Quantum Theory. Chicago: University of Chicago Press, Ch. 2.
3. Margenau, H. (1936). Quantum mechanical description, Physical Review **49**, 240–2.
4. Feller, W. (1966). An Introduction to Probability Theory and its Applications. New York: Wiley, Vol. 2, p. 487.
5. Gillespie, D.T. (1983). A theorem for physicists in the theory of random variables, American Journal of Physics **51**, 520–32.
6. von Neumann, J. (1955). Mathematische Grundlagen der Quantenmechanik. (Berlin: Springer, 1932); English translation: Mathematical Foundations of Quantum Mechanics, (Princeton: Princeton University Press, p. 307.
7. Leggett, A.J. (1984). Schrödinger cat and her laboratory cousins, Contemporary Physics **25**, 583–98.
8. Weinberg, S. (1993). Dreams of a Final Theory. London: Vintage, p. 64.

9. Peres, A. (1993). Quantum Theory – Concepts and Methods. Dordrecht: Kluwer, pp. 373–429.
10. Home, D. (1997). Conceptual Foundations of Quantum Physics: An Overview from Modern Perspectives. New York: Plenum, pp. 67–78.
11. d'Espagnat, B. (1994). Veiled Reality: An Analysis of Present-Day Quantum Mechanical Concepts. Reading, Massachusetts: Addison-Wesley, pp. 138–85.
12. Bell, J.S. (1987). Speakable and Unspeakable in Quantum Mechanics. Cambridge: Cambridge University Press, pp. 119–27.
13. Leggett, A.J. (1987). Reflections on the quantum measurement problem, In: Quantum Implications: Essays in Honour of David Bohm. (Hiley, B.J. and Peat, D., eds.) London: Routledge & Kegan Paul, pp. 85–104.
14. Leggett, A.J. (1986). Quantum mechanics at the macroscopic level, In: The Lesson of Quantum Theory (Niels Bohr Centenary Symposium). (de Boer, J., Dal, E. and Ulfbeck, O., eds.) Amsterdam: Elsevier, pp. 35–7.
15. Schrödinger, E. (1935). Discussion of probability relations between separated systems, Proceedings of the Cambridge Philosophical Society 31, 555–63.
16. Wigner, E.P. (1963). The problem of measurement, American Journal of Physics 31, 6–15.
17. Born, M. and Wolf, O. (1980). Principles of Optics. Oxford: Pergamon Press, Ch. 3.
18. Bacry, H. and Levy-Leblond, J.M. (1968). Possible kinematics, Journal of Mathematical Physics 9, 1605–14.
19. Holland, P. (1993). The Quantum Theory of Motion. Cambridge: Cambridge University Press, p. 222.
20. Kidd, R., Ardin, J. and Anton, A. (1989). Evolution of the modern photon, American Journal of Physics 57, 27–35.
21. Dirac, P.A.M. (1958). The Principles of Quantum Mechanics. Oxford: Oxford University Press, pp. 7–10.
22. Agarwal, G.S. and Simon, R. (1990). Berry phase, interference of light beams and the Hannay angle, Physical Review A 42, 6924–7.
23. Sudarshan, E.C.G. and Rothman, T. (1991). The two-slit interferometer re-examined, American Journal of Physics 59, 592–5.
24. Lamb, W.E. (1995). Anti-photon, Applied Physics B 60, 77–84.

4
The Standard Interpretation
of Quantum Mechanics

Introduction

In Chapter 3 we explored the various conceptual problems and indeed apparent paradoxes present in the new quantum theory of Heisenberg and Schrödinger. For a comparatively short period, discussion on how to resolve these difficulties was quite open; indeed during this period some of those who were later to support the Copenhagen interpretation without question, such as Werner Heisenberg and Max Born, showed quite clear divergences from Bohr's position. However this period ended quite abruptly after Bohr's famous Como lecture[1] of 1927, the content of which was presented in similar form shortly afterwards at the fifth Solvay conference. At this lecture, Bohr expounded his approach on the way in which these problems should be resolved with the use of the conceptual framework he called *complementarity*, which was the basis of what was to become known as the Copenhagen interpretation of quantum theory.

At this stage, those who might be termed Bohr's followers, including Heisenberg and Born, appeared to recognise and acknowledge that Bohr had answered their objections to his ideas, and they gave him their full support for the rest of their careers, though it might be mentioned that Heisenberg's actual statements sometimes took a rather individual line. Often, though, we may neglect this latter point and couple together the contributions of both men, and we shall then talk in general terms of 'the ideas of Bohr and Heisenberg'. The effect of the falling into line behind Bohr of the great majority of physicists was that those who could not agree with Bohr, in particular Einstein and Schrödinger, were left quite isolated, a state that was to become dramatically pronounced as Bohr's position swiftly became further and further entrenched in the physics community. For several decades, to question Bohr's 'orthodox' or 'standard' interpretation of quantum theory was to mean effective professional suicide.

There was actually another strand of ideas to what was thought of as the 'standard interpretation'. Bohr's approach was entirely verbal, and it was admittedly conceptually subtle. (Whether this made it philosophically deep, as his followers would certainly have believed, is another matter! The opinion of the authors on this question will become clear in the remainder of the book.) A more mathematical approach was provided by the famous mathematician John von Neumann.[2] While the two approaches were really quite distinct in many ways, probably most physicists felt that von Neumann's argument was effectively putting the admittedly rather abstruse ideas of Bohr in mathematical form, and felt quite happy to pledge verbal support to Bohr, while at the same time following the more straighhtforward prescription of von Neumann.

In this chapter we present the approaches of both Bohr and von Neumann. We also describe some related work of von Neumann, his famous (or infamous) theorem which claimed to show that quantum theory could not be augmented by a system of hidden variables. This theorem will come under intense scrutiny later in the book, in particular in Chapter 9. The standard interpretation (or interpretations) of Bohr and von Neumann constituted, of course, the view of quantum theory that Einstein was to oppose for the rest of his life, and this same view, as well as Einstein's own criticisms, have continued to be discussed and criticised over the 50 years since his death.

It should be stated that, probably not surprisingly, there are many further versions of the so-called standard interpretation, sub-divisions, one might say, of the main types. We will not go into the philosophical nuances of the differences between these various versions. Concise accounts of the subtleties and some detailed analysis may be found in the articles of, for example, Hanson,[3] Stapp,[4] and Home and Whitaker.[5] All its versions are characterised by the common statement: a wave-function is assumed to be a *complete description* of the quantum mechanical state of either an individual system or an ensemble of identical systems. This categorisation includes the versions advocated by Bohr and also Heisenberg, in which a wave-function corresponds to the state of an individual system, and the so-called *ensemble interpretation*, often discussed by Einstein, and later expounded in detail by Ballentine,[6] in which a wave-function is regarded as describing completely only the statistical properties of an ensemble and so not necessarily representing the state of an individual member of the ensemble. (Einstein's own views on ensemble intepretations are discussed in detail in Chapter 8.)

The key phrase in the statement underpinning the various versions of the standard interpretation is 'complete description', which has the following meaning. Knowledge of the wave-function for a given system allows one to compute all the experimentally verifiable predictions about the probabilities of measurements of various quantities on the system. Any standard interpretation then states that no further specification of the quantum mechanical state by additional parameters (hidden variables), which might enable one to predict individual outcomes rather than probabilities over an ensemble of systems, is possible. While what we have given here is a declaration, it is precisely this statement, of course, that the von Neumann theorem mentioned above claimed to prove. If this position is accepted, this immediately implies accepting an *inherently probabilistic* description in the microphysical domain.

The Bohr–Heisenberg Version

We start our discussion of variants of the standard interpretation by studying the position of Bohr[1,7,8] and Heisenberg.[9,10] Bohr's views in particular have been discussed by very many authors.[11–14] The argument that Bohr and Heisenberg put forward to justify the general point of view expressed in the previous section is that a physical theory is concerned only with predicting *reproducible* results which are empirically testable. The result of any individual measurement of a quantity relating to a particular wave-function is, in general, not reproducible, unless the wave-function happens to be an eigenstate of the measured dynamical variable. The repetition of the same measurement of a particular variable over identically prepared systems with the same wave-function yields, in general, different values, whose probability distributions and ensemble averages are the *only* empirically reproducible quantities. Bohr and Heisenberg thus conclude that the probabilistic character of quantum theory is a necessary and irreducible feature of the theory. In the words of Bohr:[7] 'The fact that in one and the same well-defined experimental arrangement we generally obtain recordings of different individual processes makes indispensable the recourse to a statistical account of quantum phenomena'.

Another cardinal feature of the standard interpretation stressed by Bohr is that a wave-function is looked upon as a mere *abstract symbolic device* from which the testable statistical predictions are derived by the prescribed definite mathematical operations. As Bohr[8] put it: 'The entire quantum formalism is to be considered as a tool for deriving predictions ... as regards information obtainable under experimental conditions described in classical terms and specified by means of parameters entering into the algebraic or differential equations. ... These symbols themselves, as is indicated already by the use of imaginary numbers, are not susceptible to pictorial interpretation.' This implies that a wave-function itself is not to be thought of as providing an objective and realist description of the microphysical world—'reality' is assumed to consist only of the results of measurements.

Throughout this book we use the term 'objective' in the *intersubjective* sense of being the same for all perceiving subjects; 'realism' is taken to mean, very broadly speaking, the philosophical premise that the physical world and its attributes have a 'real' existence that transcends direct experience, and that statements about them are true or false independent of our ability to discern which they are. In this book we use, in general, the term 'realism' to imply the existence of dynamical properties of a system, or the occurrences of a physical event, *independent* of any observation. In other words, in the standard interpretation, the formalism of quantum mechanics or the 'quantum algorithm' does not reflect a well defined underlying reality, but rather it constitutes only *knowledge* about the statistics of the observed results. It may be mentioned that objectivity in the sense just of intersubjectivity does not necessarily entail realism. Realism will be discussed further in many of the later chapters of the book, and Einstein's approach to realism is described in Chapter 8 in particular.

A similar viewpoint to these statements of Bohr is to be found in the writings of Heisenberg. For instance, Heisenberg[9,10] emphasised that in the standard interpretation, the notion of objectivity is restricted to measured statistical properties. He wrote (Ref. [9], pp. 27–28), '[T]he state of a closed system represented by a wavefunction is indeed objective, but not real, and . . . the classical idea of "objectively real things" must here, to this extent, be abandoned . . . The ontology of materialism rested upon the illusion that the kind of existence, the direct "actuality" of the world around us, can be extrapolated into the atomic range. This extrapolation, however, is impossible'.

Note that 'ontology' is a notion closely related to what we have called 'realism'—it means, in general, the theory or description of 'being' in contrast to what is known as 'epistemology', which means the theory or description of how knowledge is obtained. In this book we will be specifically using the term 'ontology' in connection with descriptions of *objectively real* properties of particles or localised entities.

It is clear from the writings of Bohr and Heisenberg that they refused to attribute any form of physical reality to the dynamical properties (such as position, velocity, energy) of a quantum system *unless* they are actually measured. However, they did not deny the objective reality of innate properties of atomic entities, such as mass, charge or spin. Such a position may be thought to indicate what may be called an *instrumentalist* (or operationalist) insistence on quantum theory being *merely* a tool for predicting the reproducible experimental results obtained in measurements of the dynamical properties of a system. This type of viewpoint has also a positivist slant. (As discussed in Chapter 1, the term positivism entails denial of any kind of reality or existence being associated with entities *not* accessible to direct observation.) The question of whether the apparent instrumentalism of the Bohrian interpretation is identical to or different from the usual philosophical tradition of positivism has been discussed by various authors; see, for example, Folse,[11] Murdoch,[12] Beller[13] and Mackinnon.[14] Einstein's views will be described in Chapter 8.

In fact, one may say that Bohr and Heisenberg interpreted the uncertainty principle as imposing a prohibition on the *simultaneous existence* of definite values for the noncommuting variables. Bohr repeatedly stressed the point about 'an inherent element of ambiguity involved in assigning conventional physical attributes to atomic objects'. (See, for example, Ref. [8], pp. 313–314.) This aspect of their ideas has been critically analysed with relevant quotations by Beller and Fine.[15] One should remember that this is essentially a *metaphysical input*, going beyond what is implied by the formal justification of the uncertainty principle, which is concerned essentially with measured values and embodies no statement about the pre-measurement existence of definite values. This was discussed in the previous chapter.

This view led to Bohr's favourite recurrent theme about the impossibility of having simultaneously a 'causal' description (in the sense of satisfying the conservation laws involving sharply defined values of energy and momentum), and a 'space-time' description (in terms of definite pre-measurement locations in space-time) of quantum mechanical processes.[16] (Note that here Bohr has used his own definition of causality.) This theme demonstrates a crucial aspect of his approach, which should be regarded as rather arbitrary and certainly not derived directly from the formalism. This aspect is *mutual exclusivity* (ME) between these two descriptions. Either of the two descriptions, which are compatible classically, excludes the other in quantum theory. (This is an example of *complementarity*, discussed at length later in this section.)

For Heisenberg, a wave-function does not correspond to an actual course of events occurring in space-time, but rather it represents the tendency for various events to occur with different probabilities. (Heisenberg used the notion of *potentia* for one actual event or another.) His contention was that the quantum formalism 'does not allow a description of what happens between two observations' and that 'therefore the transition from the "possible" to the "actual" takes place during the act of observation' (Ref. [10], p. 54). For further discussion of the philosophically interesting subtleties of the differences between the views of Bohr and Heisenberg, we refer the reader to the account of Stapp,[4] which contains an illuminating correspondence with Heisenberg.

Next, we mention only briefly the views of Wolfgang Pauli, another prominent figure associated with Bohrian school. His views differed form those of Bohr and Heisenberg in that he considered a wave-function not merely to be a mathematical symbol, but also to represent an objective reality, as well as an observer's 'information' about that reality. Note that the notion of 'observer' used by Pauli was a *generalised* one, not necessarily some conscious mind. For example, he included the instruments of observation; the term represented, in his words, a sort of 'generalized impersonal subjectivity'. However, Pauli did not elaborate on the details of the precise kind of 'microrealist' interpretation he had envisaged. He once wrote to Fierz that: 'The elaboration of a new idea of reality seems to be precisely the most important and extremely difficult task of our time'.[17] For careful evaluations of Pauli's position, see Laurikainen[18] and Nair.[19]

Max Born's name is also often mentioned as one of the eminent followers of the standard interpretation, though Born's philosophical commitments were not so well defined. It was Born whose papers[20–22] on the quantum mechanical treatment of scattering introduced the interpretation of a wave-function as a probability amplitude; this work was eventually to gain him the Nobel prize, after a wait of very nearly 30 years. A well-researched analysis of the development of Born's ideas has been given by Beller,[23] who makes the point that Born was essentially 'a mathematical man' who introduced 'formal symbolic solutions with the minimum of interpretational commitments'.

As a brief historical interlude, it may be interesting to recall that Born had initially regarded a wave-function as controlling only 'the probabilities of energetic transitions of an atom, and the energy and the directions of motion of colliding electrons' (Ref. [23], p. 571). It was Pauli[24] who went further by interpreting $|\psi|^2 dx$ as the probability of finding a particle in a given region. (For relevant historical details see Ref. [23], pp. 571, 575.) Subsequently Dirac[25] took the final step of generalising this interpretation in terms of the formal language of quantum mechanics by regarding $|\langle\psi|\phi\rangle|^2$ as giving the probability of obtaining a particular eigenvalue of an observable corresponding to its eigenstate $|\phi\rangle$ when that observable is measured on a system prepared in the state $|\psi\rangle$. (This is why in the present book this prescription is referred to as the Born–Dirac rule.)

To summarise, we note that Paul Feyerabend[26] succinctly expressed the essence of the Bohr–Heisenberg position with the remark: 'A quantum system does not possess any properties over and above those that are derivable from its wave-function description.' Abner Shimony[27] aptly concluded that this approach essentially sought 'to do epistemology without an ontology'.

The Ensemble Interpretation

Another example of what we have called the standard interpretation is the so-called *ensemble interpretation* already alluded to. Here it is assumed that the wave-function does not represent the state of any individual system, merely that of the ensemble. This immediately raises the question as to whether this form of interpretation allows one to envisage dynamical properties, definite but unknown, of an individual system over and above those that are described by the wave-function. If this is so, the ensemble interpretation can characterise itself as maintaining a somewhat agnostic attitude as to whether a wave-function constitutes a 'complete' description of the quantum mechanical state.[28]

However, if one attributes definite but unknown values to the dynamical variables of an individual system, it is clear, although not always necesssarily recognised by those promoting them, that such theories are hidden variable theories. It needs to be specified whether these values are simply those revealed by the relevant measurements, or whether they are pre-measurement values that could, in general, be different from the post-measurement values. Such a specification would in turn involve spelling out the specifics of the hidden variable theory. (For an in-depth discussion of this issue see Home and Whitaker;[5] see also d'Espagnat[29].) Since the proponents of the ensemble interpretation have not provided such details, the ensemble interpretation as usually presented remains 'internally incomplete'.

The alternative position for an advocate of an ensemble interpretation is to stick emphatically to the basic point that a wave-function offers a 'complete' description of the quantum mechanical state pertaining to an ensemble. This statement, though, is actually equivalent to the usual Copenhagen theory. This is because, if the wave-function relates to the ensemble, but there are no hidden variables to distinguish between different members of the ensemble, one might just say that the wave-function relates to each member of the ensemble. This is why we have categorised the ensemble interpretation as one of the versions of the standard interpretation. Einstein spoke very often of ensemble interpretations, and we will report on his approach in Chapter 8.

Single System vis-à-vis Ensemble Interpretation

A common argument defending the ensemble interpretation is that, since it is not possible to carry out a measurement on a single system that will determine its wave-function, a wave-function cannot be taken to denote the objective state of an individual system. (For a particularly emphatic statement on this see Blokhintsev,[30] quoted with approval and elaborated in Ref. [6], p. 379.) This section will concentrate on examing this point more carefully. We first note that there are of course objective physical effects which are realized even for a *single* member of the ensemble characterised by its wave-function; for example, in a double-slit interference experiment, any *single* particle has the *objective* property of never being found at the points where the total wave-function vanishes and destructive interference is complete. This property of an individual particle is determined by the associated wave-function. Here we may also mention the following interesting argument put forward by Penrose.[31]

If the state of a system be denoted by $|\psi\rangle$, let us consider a Hermitian operator $\hat{Q} = |\psi\rangle\langle\psi|$ associated with observable Q. Let us follow the assumption by Dirac (Ref. [25], p. 37), that *in principle* one may perform the measurement of an observable associated with *any* Hermitian operator whose orthonormal eigenstates form a 'complete set', though it is admitted that many such measurements may not be feasible in practice. One may then infer that, apart from phase factors, the state $|\psi\rangle$ is the *only* one for which a measurement of Q, *if* made (a 'counterfactual' statement) on any single system in that state, would *certainly* give the outcome *unity*. (The question of whether $|\psi\rangle\langle\psi|$ is actually measurable concerns an issue distinct from the question of objective reality as a matter of principle.) It is thus possible to argue that the property of an individual system *being* in the state $|\psi\rangle$ has an 'objective reality' in the sense of ensuring the measured value of Q to be always unity.

Nevertheless, it is also inescapable that an unknown quantum state of a single system cannot be directly measured. There is a simple general argument[32–34] showing this. Since a collection of identical systems in the state $|\psi\rangle$ is required for determining an unknown $|\psi\rangle$, if one were given only a single system, one would need what is referred to as a 'cloning device' that is capable of producing an ensemble of copies of a given system all in the state $|\psi\rangle$. For such a single device to be useful in determining *any* unknown $|\psi\rangle$, it must work for all possible quantum states. However, a cloning device that is *not* state-specific is forbidden from consideration based on the requirement of *unitarity*. The argument is as follows.

Let a cloning device be composed of several independent constituents in the mutually orthogonal states $|w_1\rangle, |w_2\rangle, \ldots$ so that the state of the entire device is $|w_1\rangle \otimes |w_2\rangle \otimes |w_3\rangle \ldots |w_n\rangle$. If a device faithfully clones an input state $|\psi\rangle$, the transformation is represented by

$$|w_1\rangle \otimes |w_2\rangle \otimes \ldots \otimes |w_n\rangle \otimes |\psi\rangle \rightarrow |\psi\rangle \otimes |\psi\rangle \otimes \ldots \otimes |\psi\rangle \qquad (4.1)$$

which means that $(n + 1)$ copies of the system in the state $|\psi\rangle$ are produced. Here, for simplicity, it is assumed that *all* the component states of the device are converted to the states identical to the input state $|\psi\rangle$. If the device is not state-specific, the transformation Eq. (4.1) should be *independent* of $|\psi\rangle$; for any other input state, say, $|\phi\rangle$, we should have

$$|w_1\rangle \otimes |w_2\rangle \otimes \ldots \otimes |w_n\rangle \otimes |\phi\rangle \rightarrow |\phi\rangle \otimes |\phi\rangle \otimes \ldots \otimes |\phi\rangle \tag{4.2}$$

Now, if the transformations (4.1) and (4.2) are both unitary, the scalar product of the left-hand sides of (4.1) and (4.2) should equal the scalar product of the right-hand sides of (4.1) and (4.2). Then taking each of the wave-functions $|w_1\rangle, |w_2\rangle, \ldots |w_n\rangle$ to be normalised, it follows that one must *necessarily* have $\langle\phi|\psi\rangle = 0$, so the two states must be orthogonal. This means that cloning any two non-orthogonal states using a single device is not permitted by the rules of quantum mechanics subject to the condition of unitarity. This no-cloning theorem is extremely important throughout the topic of quantum information theory, as will be explained in Chapter 11.

Within the constraint of this impossibility of measuring the wave-function of a single quantum system, an ingenious scheme[35–37] has recently been discovered to show an objective and physical manifestation of the wave-function of a *single* system, through the empirical verification of the expectation value of an arbitrary observable by a measurement process applied to the single system in question, provided it is prepared in a *known state*. There are two main components of this procedure: (i) There is a coupling of the measured system with a measuring device whose state changes with time. (ii) The state of the measured single system is prevented from changing appreciably during a measurement process characterised by the coupling mentioned in (i) with the help of an additional interaction which is called the 'protective' interaction. Note that in order to prescribe a suitable 'protective' interaction one must already know the state of the measured system.

What is essentially demonstrated in this scheme is that, provided the state of the measured system is appropriately 'protected', a gedanken example can be envisaged where a measurement interaction gives rise to a change in momentum of the measuring device, the rate of change of which is related to the expectation value of the measured observable. Let the initial state of the measured system and the measuring device be $|\psi(o)\rangle$ and $|\alpha(o)\rangle$ respectively so that the combined state may be written as $|\alpha(o)\rangle|\psi(o)\rangle$. After a measurement interaction is switched on along with a suitable 'protective' interaction, the combined state is $|\alpha(t)\rangle|\psi(t)\rangle$ where $|\psi(t)\rangle$ is approximately the same as $|\psi(o)\rangle$. Now, to compute the change in momentum of the measuring device, we note that a measurement coupling may be represented by $g(t)QA$ where Q is the position coordinate of the measuring device, A is the measured observable and $g(t)$ represents switching on and off the measurement interaction. Thus $g(t)$ is non-zero only during some finite time interval, say $[0, T]$ where $g(t) = g_0 f(t)$, where g_0 is a constant, and

$$\int_0^T f(t)\mathrm{d}t = 1$$

Then the rate of change of momentum, P, of the measuring device is given by

$$d\langle\alpha(t)\psi(t)|P|\alpha(t)\psi(t)\rangle/dt = (i/\hbar)\langle[H,P]\rangle = -g(t)\langle\psi(t)|A|\psi(t)\rangle \quad (4.3)$$

where $\psi(t) \cong \psi(o)$ because of the 'protective' interaction, so that $\langle\psi(o)|A|\psi(o)\rangle$ can be determined by registering the change in momentum of the measuring device, and by finding out the mean change over successive time periods of observation. (That is to say that $g(t)$ is switched on and off successively along with the 'protective' interaction.) Note that this registration process is necessary to 'complete' the measurement process in question.

To sum up, $\langle\psi|A_i|\psi\rangle$ can be determined in this way for a sufficiently large number of observables A_i from which $|\psi\rangle$ can be *verified* (up to an overall phase) to be the known prepared state of the given single system. Thus the empirical manifestations of both the expectation value of any arbitrary observable and the wave-function itself are measurable (at least, in principle) for an *individual* quantum system, thereby reinforcing quite strongly the arguments in favour of associating objective physical reality with the wave-function description of a single system.

The Bohr–Heisenberg 'Solution' to the Quantum Measurement Problem

We now return to Bohr. He recognised[8] that if a measuring device was described quantum mechanically, its interaction with a measured system could not lead to a definite result in a single run of any given measurement. He sought to avoid this problem by simply decreeing that the interaction between an object and an apparatus is an 'unanalysable whole' and the apparatus 'must be' described in classical terms.

An obvious criticism of this is that Bohr made the very concept of a 'measuring instrument' depend on an ill-defined limiting procedure of a 'cut' between quantum and classical domains. In expressing what Bohr called 'a simple logical demand', he assumed *by fiat* that the composite system in a measurement process is composed of two distinct parts: the measured system and the measuring apparatus. However, there is no precise criterion definable within standard quantum mechanics that delineates the borderline between the two. Is it that we merely find it convenient to consider some parts of the global system as parts of the instrument, or is it determined *a priori* in a more physical way? Is it consistently possible within the framework of quantum mechanics to assign a classical description to the measuring apparatus? Such questions remained unanswered in Bohr's writings as well as in the advocacy of his position by others. (For a review, see, for example, Ref. [12].)

We now turn to Heisenberg's position[38] on the quantum measurement problem. Heisenberg admitted: 'It is not possible to decide, other than arbitrarily, what objects are to be considered as part of the observed system and what as part of the observer's apparatus.' Once the macroscopic level is reached, Heisenberg conceded, a *cut* is necessary to avoid the problem of infinite regress in ensuring a definite outcome. However, he argued that it is of no practical importance *where* we put the split. As Bell put it, the Heisenberg dictum was to 'put sufficiently much into the quantum system so that the inclusion of more would not significantly alter practical predictions'. Though this recipe is useful, it is ambiguous in principle. There is no fundamental reason why the physics involved in measurements should differ from the way other physical interactions are described. Hence the very *legitimacy* of such a conceptual discontinuity, not so much its precise location, is what that matters. In other words, the founders of quantum mechanics consciously preferred not to tackle this problem at a deeper level—a pragmatic standpoint apparently justified in the years when quantum mechanics was in the formative stage.

Position-Momentum Complementarity

The last section takes us to the heart of Bohr's conceptual position, complementarity, in particular position-momentum complementarity. There is detailed discussion of this subtle idea in very many publications;[11-15] here we concentrate on the central points. In the previous section we saw that measurement should be regarded as an unanalysable whole. This means that, according to Bohr, one should not talk of the position or the momentum of a particle. That would be to separate out a property of the particle from the apparatus set up to measure it. Rather one may talk of the result obtained in a measurement of position, or in a measurement of momentum. A different way of stating the same idea is to say that we may *only* discuss the value of a particular property of a system in the context of a measurement actually performed to measure that property. (A slightly weaker condition sometimes mentioned instead is that the context is of an apparatus set up to measure the property, even if the experiment is not actually performed.)

Let us now suppose we wish to discuss simultanous values of the position and the momentum of the particle. In order to discuss the position of a particle, we would need to have in place an apparatus whose task is to measure the position; similarly, to discuss the momentum, we would need to have in place an apparatus whose task is to measure the momentum. But a little thought may seem to suggest that it is quite impossible to measure both these properties simultaneously. We immediately see that a central component of standard interpretations of quantum theory, the fact that one may not discuss the position and the momentum of a particle simultaneously, emerges from the ideas of complementarity in a very straightforward way. Supporters of Bohr would claim that this helps to establish the depth of his ideas. Opponents, on the other hand, could equally complain that nothing is actually explained; one has merely argued to a desired result using an arbitrary restriction on the way one is allowed to discuss the physical situation.

This kind of argument was at the basis of much of the debate between Bohr and Einstein discussed in Chapters 5 and 6. It might be objected that a rejoinder to our rather loose statement earlier beginning: '[A] little thought . . .' could be 'Whose thought?' or 'How much thought?' In these discussions with Bohr, Einstein used every effort in his thinking to show that position and momentum could be measured simultaneously, and was forced to admit failure, as we shall see in Chapter 5. (Sometimes energy and time were discussed instead; it will be remembered from Chapter 3 that at the time this combination was assumed to behave in an exactly analogous fashion to that of momentum and energy.) However there is another important point relevant to the Bohr–Einstein debate. It may indeed be the case that momentum and position cannot be measured simultaneously, and, as Bohr at least would say, may not be discussed simultaneously. Does this actually mean that they do not exist simultaneously? A positivist would certainly say that it does. A supporter of complementarity would agree; after all, if the two properties cannot be discussed simultanously, one presumably cannot discuss whether they exist simultaneously! Einstein did not agree, and this led to his argument in the EPR paper, for which see Chapter 6.

We should now move to an explanation of the term 'complementarity'. Let us first make the point that position and momentum are conjugate variables, and the discussion involving these particular variables of a system is just an example, though obviously a very important one, of complementarity between any two conjugate variables. (Two observables, a and b, are said to be conjugate if the associated operators obey the commutation relation, $[\hat{a}, \hat{b}] = -i\hbar$, exactly that, of course, obeyed by $\hat{p}$ and $\hat{q}$.)

In normal language, to say that facts or approaches or information may be 'complementary' means that, for some reason, a single experiment or calculation or study cannot provide a complete description of the phenomenon under consideration, but one may undertake a series of such experiments or calculations or studies, each one of which provides a different part of the required description. These parts of the description may be assembled and they then provide a complete description. An example might be that a number of observers might see a novel type of object from different vantage points; each could provide a partial description, and these partial descriptions could then be put together to give a complete description.

Bohr's complementarity is similar in one way, but strikingly distinct in another. Again a number of partial descriptions of a system are possible; for a particle these could be measurement results of momentum and position. *If* it were possible to form both partial descriptions, that is to say to measure both quantities simultaneously, and *if* it were possible to put together these two descriptions, they would form a complete description of the phenomenon. This may be known as *joint completion. But*—and this is the crucial point—it is only possible to obtain one, either one in fact, of the partial descriptions, at a particular time, so clearly it is not possible to form a complete description. As Bohr[1] put it, quantum theory 'forces us to adopt a new mode of description designated as *complementary* in the sense that any given application of classical concepts precludes the simultaneous use of other classical concepts which in a different connection are equally necessary for the elucidation of phenomena'. We may say that such sets of classical concepts are *mutually exclusive* (ME).

The type of complementarity just discussed, position-momentum or time-energy, is essentially identical to that mentioned earlier in this chapter—complementarity between a space-time description and a 'causal' description. A space-time description implies retaining a knowledge of the values of position and time, while a causal description, in the sense implied by Bohr, requires knowledge of the values of momentum and energy. However a very different type of complementarity is also discussed very widely, that between wave and particle pictures of a system, and this is described in the following section.

Wave-Particle Complementarity

Complementarity between conjugate variables clearly fits into the prescribed pattern—values of position and momentum would give a complete description of the system, they cannot be measured simultaneously, and moreover it is clear that there could be no single attribute that would, on its own, provide a complete description. It is not quite so clear that wave-particle complementarity is such a good example. While the pictures are clearly mutually exclusive, it is not clear that, if they *could* be added, they would give the complete picture required for joint completion, and indeed it is not absolutely clear that one could not imagine a single picture that would be totally sufficient. Nevertheless, from the earliest days of Bohr, it has been traditional to talk of wave and particle as complementary pictures.

Bell[39] described the philosophy of complementarity as an attempt to go beyond pragmatism. Pragmatism might dictate the restriction to either wave or particle picture. Complementarity went further by attempting to explain or to justify this procedure. Bell remarked: 'Rather than being disturbed by the ambiguity in principle, ... Bohr seemed to take satisfaction in it. He seemed to revel in contradictions for example between "wave" and "particle", that seem to appear in any attempt to go beyond the pragmatic level. Not to resolve these contradictions and ambiguities, but rather to reconcile us to them, he put forward a philosophy which he called "complementarity".'

Once Bohr recognised that wave-particle dualism was inescapable, he concentrated not on overthrowing the ideas of wave and particle but rather on removing the paradoxical consequences by *limiting* their use. According to Bohr, the validity of the quantum mechanical formalism limited but did not rule out the necessity of using classical pictures. Bohr's refusal to abandon classical pictures as the necessary models of conceptual comprehension of physical theory (as a means of interpreting the 'symbolic' or purely formal description provided by the mathematical formalism) is emphatically reflected in his many remarks, such as: 'The language of Newton and Maxwell will remain the language of physicists for all time,'[40] and: 'It must above all be recognized that, however far quantum effects transcend the scope of classical physical analysis, the account of the experimental arrangement and the record of the observations must always be expressed in common language supplemented with the terminology of classical physics.'[8]

Here it is interesting to note that Heisenberg believed that 'the concept of complementarity introduced by Bohr into the interpretation of quantum theory has encouraged the physicists to use an ambigious rather than an unambiguous language... When this vague and unsystematic use of the language leads into difficulties, the physicist has to withdraw into the mathematical scheme and its unambiguous correlation with the experimental facts' (Ref. [10], p. 179). However, he also recognised that some use of langauge is inescapable '...[J]ust by discussions with Bohr I learned... that is, one cannot go entirely away from the old words because one has to talk about something ... So I saw that in order to describe phenomena one needs a language' (quoted in Ref. [11], pp. 96–97).

Of course, for many physicists wave-particle dualism might indicate that *classical* wave and particle concepts must be totally abandoned, and that radically new pictures need to be invoked for comprehending conceptually the peculiarities of microphysical phenomena. In contrast, Bohr's programme was to show that: '... [W]e are not dealing with contradictory but with complementary pictures of the phenomena, which only together offer a natural generalization of the classical mode of description,' so that: 'The two views of the nature of light are rather to be considered as different attempts at an interpretation of experimental evidence in which the limitation of the classical concepts is expressed in complemon tary ways.'[41]

The literal meaning of complementarity is 'mutually or jointly completing'. It is in this sense that two angles are said to be complementary if they make up a right angle. As was explained in the previous section, Bell[42] pointed out that Bohr's use of the term complementarity implied going beyond its usual meaning. Consider, for example, Bell suggested, the different descriptions of an elephant from the front, from the back, from the side, from the top and from the bottom. These various descriptions are complementary in the usual sense. A key point is that they are consistent with one another and not mutually exclusive; hence together they provide a single whole image of an elephant. In contrast, Bohr's wave-particle complementarity is based on elements which are *inherently incompatible* with one another.

It is important to note that within Bohr's wave-particle complementarity, classical pictures are used in the epistemic sense, devoid of any ontological significance. The seemingly paradoxical procedure of using particle representations in some parts of the description of physical reality, and wave representations in others, is justified by implying that such descriptions are 'idealized representations of our knowledge'. They do not correspond to the actual physical reality. A classical picture, wave or particle, acts essentially as a prop for visualizing the behaviour of micro-objects in a specific context. In other words, within the Bohrian framework, the images of wave and particle are merely 'shadows' of 'real' objects.

(This is something like the chained prisoners in the cave imagined by Plato in The Republic, Book VII, where the prisoners facing the wall of their prison are doomed to see only the shadows of objects outside the cave.) Though not 'real', such images are considered useful for describing the 'relationship between empirical evidences obtained under different experimental conditions'—herein lies the distinction between wave-particle complementarity and the truly realist framework (Refs. [11], p. 117; [43]).

It is interesting that Bohr did not try to justify his 'mutually exclusive' or ME hypothesis as a consequence of any rigorous general argument based on the mathematical formalism of quantum mechanics. His strategy was to defend this hypothesis by illustrative analysis of specific examples which were confined to interference effects. In such analysis, an interference pattern that can be explained by a classical wave model is viewed as a signature of wave-like propagation. If, on the other hand, the experimental arrangement provides results that can be accounted for by imagining *which* of the possible paths a single photon follows *all the way* from a source to a detector, this is taken to signify particle-like propagation. Recent studies have highlighted the fact that it is *essential* for any interference effect to be observed in position space that the quantum mechanical formalism guarantees the validity of ME.

Lastly we would stress that for Bohr, the real importance of wave-particle complementarity lay in providing an ingredient for a *general epistemological principle* of fundamental significance. Bohr regarded complementarity, not as a device of limited applicability required to solve problems in atomic physics, but as a philosophical framework of the widest possible scope, capable of providing insight into conceptual problems of the greatest depth and across the whole field of knowledge.[11] For an analysis of philosophical nuances involved in the relationship between complementarity and epistemology, see the book by Folse (Ref. [11], p. 195).

Bohr and the 'Disturbance Interpretation'

So far we have discussed Bohr's formal ideas. As we have seen, they are quite philosophical and subtle. While those comparatively few physicists who enjoyed conceptual argument would be fully able to comprehend Bohr's position, the majority, whether experimentalists or more computationally minded theoreticians, would probably only have been able, and indeed only expected, to give lip-service to orthodoxy.

So it was convenient that, occasionally at least, Bohr used language that was much closer to that of the ordinary physicist. He spoke of 'disturbance'. He implied that it was impossible for momentum and position to be measured simultaneously because the measurements would disturb each other. There then followed the positivistic argument that since they could not be measured simultaneously, it was pointless, and in fact wrong, to assume that they actually had simultaneous values. This must certainly have seemed to many a good simple way of understanding Bohr's basic ideas.

The early rounds of the Bohr–Einstein debate, discussed in Chapter 5, were essentially structured round this point. Einstein tried to show that p and x, or E and t could be measured simultaneously, but Bohr was able to convince him that this could not be done. However Einstein's next step, in the EPR paper discussed in Chapter 6, was a clever device to try to show that, though p and x could not be measured simultaneously, nevertheless they both existed with precise values at all times. Thus it challenged what we called in the previous paragraph the positivistic notion that, if a quantity cannot be measured, then it should not be discussed in scientific theories, because scientific theories should merely be a way of summarising experimental results.

For detailed analysis of EPR, we must wait for Chapter 6. However one clear point made in the paper was that the disturbance idea had to be abandoned. As in the previous discussions between Einstein and Bohr, EPR had two measurements in their scheme, but their two measurements were on separated sub-systems and so could not interfere with each other, at least as long as one wished to maintain the locality requirement, which both Bohr and Einstein did. Bohr admitted this point. His supporters maintained that this was not an important admission since, they said, the disturbance idea was never a central part of Bohr's scheme, more a way of providing some sort of physical picture of his formal ideas.[11] However, for physicists, it was the loss of what seemed a simple and convincing way of appreciating Bohr, which left only the difficult ground of complementarity. In the end this probably led to some loss of support for Bohr from the physical community, though it was to be many years after 1935 that this effect would begin to be noticeable.

von Neumann and the Projection Postulate

We have seen that the argument of Bohr is heavily conceptual. As we said earlier, von Neumann was a mathematician, and his approach could scarcely be more different. While Bohr hoped to exorcise the quantum demons by the sophistication of his conceptual analysis, for von Neumann the difficulties were plainly in the mathematics, and they then had to be explained away.

The heart of von Neumann's analysis of measurement is the so-called 'collapse postulate' or 'projection postulate'. Actually while the postulate is virtually always attributed to von Neumann since it appeared in his famous book,[2] it was common to several workers, including Dirac,[44] so the credit or blame should be spread around somewhat.

In Chapter 3 we introduced the measurement problem or paradox. To be explicit, let us say that at time t_0 we wish to measure an observable O on a particular system, and the Hermitian operator associated with O is $\hat{O}$; the eigenfunctions of $\hat{O}$ are ψ_n with eigenvalues O_n.

Let us suppose for a start that, at time t_0, the wave-function of the system happens to be equal to one of the eigenfunctions of $\hat{O}$, say ψ_1. Then the effect of the measurement is to form a correlated state of the microscopic system and the appropriate macroscopic state of the measuring device, which we may write as

Φ_1. This state is of the form $\psi_1\Phi_1$. It is then straightforward for the experimenter to examine the state of the measuring device. The fact that it is in state Φ_1, corresponding to O having the value O_1, would agree with the statement that the microscopic system was originally in state ψ_1, though we shall actually see that, if we did not know the initial state of the microscopic system, all we may deduce is that it is in state ψ_1 *at the conclusion of the measurement.*

We may represent this process schematically as

$$\psi_1\Phi_i \rightarrow \psi_1\Phi_1 \tag{4.4}$$

where Φ_i is the initial arbitrary state of the measuring device.

Similarly, if the measured system is initially in state ψ_2, the measurement process may be represented by:

$$\psi_2\Phi_i \rightarrow \psi_2\Phi_2 \tag{4.5}$$

Here ψ_2 is orthogonal to ψ_1, and Φ_2 to Φ_1.

We see that, as long as the initial state of the measured system is an eigenfunction of $\hat{O}$, the mathematical description of the measurement procedure is straightforward, and totally without problem. The macroscopic measurement device finishes the process in a state Φ_n corresponding to the initial state ψ_n, and the reading on this device may be taken in a natural way. In turn this tells us that the value of O is O_n.

Unfortunately the simplicity and lack of problem of this account is lost immediately and drastically when the initial state of the measured system is not a single eigenfunction of $\hat{O}$, but a linear superposition of two or more. The simplest case is where it is an equal superposition of ψ_1 and ψ_2, in other words $(1/\sqrt{2})(\psi_1 + \psi_2)$. In this case the linear nature of quantum theory tells us that we may merely add together (4.4) and (4.5), with the appropriate factor of $(1/\sqrt{2})$, and obtain:

$$(1/\sqrt{2})(\psi_1 + \psi_2)\Phi_i \rightarrow (1/\sqrt{2})(\psi_1\Phi_1 + \psi_2\Phi_2) \tag{4.6}$$

The general case may be represented by:

$$\left(\sum_n c_n\psi_n\right)\Phi_i \rightarrow \sum_n c_n\psi_n\Phi_n \tag{4.7}$$

In the cases of (4.6) and (4.7), the mathematics tells us that the macroscopic measuring device is left, not in a single macroscopic state, but in a linear superposition of two or more. This appears to make no physical sense at all. This is the measurement problem or measurement paradox of quantum theory, discussed in Chapter 3.

The prescription of von Neumann is bold: at the time of measurement, the right-hand side of (4.6) or (4.7) 'collapses' down to a single term (or is 'projected' or 'reduced' to a single term). We can represent the case of (4.7) by:

$$\sum_n c_n\psi_n\Phi_n \rightarrow \psi_m\Phi_m \tag{4.8}$$

The probability of the state of the system collapsing to $\psi_m \Phi_m$ is given by $|c_m|^2$. In this case we will say that the final state of the measuring device is Φ_m. This tells us that the state of the measured system is ψ_m at the end of the measurement; the value of O is O_m. All this single case tells us about the state at the beginning of the measurement is that c_m was not equal to zero.

Pragmatically it may be said that the collapse procedure is extremely effective. It predicts, of course, a unique measurement result, which is what we expect and require. Its predictions for measurement of O over an ensemble of such systems agree with experiment; in other words, the relative frequency of a particular measurement result corresponds to the probability just given. It may be said that the vast majority of physicists use the projection postulate in appropriate cases without thought or worry, and the results will be correct.

Another nice feature, and one that was important for von Neumann and Dirac, was that an immediately repeated measurement yields the same result as the initial measurement. For if the final state of the measuring system is Φ_m, this tells us that the final state of the measured system is ψ_m, so the value of O is O_m. An immediately repeated measurement will again produce the final state $\psi_m \Phi_m$, and the value of O is again O_m. That an immediately repeated measurement of position should give the same result as the initial one is reassuring; otherwise there would be a drastic clash with relativity. The result may not seem such a direct requirement for measurements of other observables, but von Neumann and Dirac felt strongly that it was the way the world should work.

So pragmatically the collapse postulate was a success. Conceptually it was quite the reverse. For, as von Neumann admitted, it meant that a system behaved completely differently at a measurement from its mode of behaviour at other times. Von Neumann called these types of behaviour the first kind and the second kind, respectively. Among the contrasts, behaviour of the second kind was reversible, while that of the first kind was irreversible. While behaviour of the second kind obeys all the laws of quantum theory, including maintaining any initial superposition, that of the first kind violates many of them. For example, it allows a superposition to evolve into a mixture. For an ensemble of systems, before the collapse each state is in the same superposition and this is a pure state; after the collapse, however, each system is in a definite state, and we have a mixture of different states. (Pure states, and mixed states or mixtures are discussed in the previous chapter.) In contrast, the usual unitary evolution of quantum theory never allows a pure state to evolve into a mixture.

We thus see that von Neumman's collapse or reduction or projection postulate appears totally at odds with the whole of the rest of quantum theory. There is another point which appears obvious but does not seem to have been pointed out clearly until this was done by John Bell[45] many years later. This is that 'measurement' is not a primary term; it does not denote a process intrinsically different from a process of the second kind. Indeed a measurement procedure may be spelled out in terms of individual sub-processes of the second kind. It therefore seems inconveivable that a measurement process should have properties so very different from processes of the second kind.

While, as we have said, physicists gave at least lip-service to Bohr, they generally followed von Neumann for the mathematical analysis of measurement. Collapse was, and in fact, still is, the pragmatic procedure to get the correct results. This, for example, EPR (Einstein–Podolsky–Rosen) used it, without question, without even mentioning the fact, in their seminal paper discussed at length in Chapter 6. (It might be mentioned that Einstein was almost certainly not directly responsible for this aspect of the paper.) However, inasmuch as the projection postulate gives only probabilities for particular measurement results, so that it violates determinism and does not predict the behaviour of individual systems, it was actually very much a central part of standard quantum theory which Einstein made it his task to stand against.

von Neumann's Impossibility 'Proof'

von Neumann's other important contribution to the foundations of quantum theory, though in the end it definitely proved a hindrance rather than a help, also appeared in his famous book.[2]

In this chapter we have stated that an essential input to any variety of standard interpretation is the idea that one cannot add additional structure to the wave-function. One cannot add extra hidden variables or parameters which might give extra information about individual systems over and above that which may be obtained from the wave-function. This may be called a 'no-go' theorem.

Bohr and Heisenberg had provided no proof of this idea. However it was at the heart of their concept of complementarity, and once the latter became established as orthodox and not open to challenge by respectable physicists, the same, it was generally presumed, went for the no-go statement. As it happened, the early experience of those who tried to produce hidden variable theories in the late 1920s, de Broglie and Einstein among them, was not positive. It seemed by no means easy to generate such theories; we shall describe this early work in the following chapter.

There the matter rested. The orthodox were quite sure they were right but had no proof. The questioners were subdued, and perhaps suspected that, in this instance, the orthodox were correct. Then came von Neumann's theorem. Von Neumann admitted that it was logically possible, though obviously extremely unlikely, that a new theory could be produced, which agreed with the results of quantum theory within experimental uncertainties, and yet retained realism and perhaps determinism. But what his theorem claimed could *not* be done was to keep the form of quantum theory as we have it today, and to add extra structure in the form of hidden variables.

The proof and the reaction to it are discussed by Max Jammer.[46] We shall not describe the proof in detail. It was not complicated mathematically, nor exceptionally long. The difficulty in assessing it lay in the fact that it was based on four rather general propositions, and it was not easy to evaluate the significance of these propositions. Of these propositions we shall mention only the fourth.

This is stated in terms of the expectation values $\langle R \rangle$ and $\langle S \rangle$ of two observables R and S. If a third observable Q is given by $aR + bS$, where a and b are constants, the postulate says that the expectation value of Q, $\langle Q \rangle$ must be given by $a\langle R \rangle + b\langle S \rangle$. The truth of this is far from clear since, in general, R and S will need quite different experimental arrangements for their measurement, so the addition of expectation values is difficult to justify. Nevertheless the postulate is true for functions of the usual variables, almost, one might say, by coincidence. However von Neumann extended the scope of the postulate so that it applied, also, to functions of the hidden variables that he was trying to rule out.

The denouement will occur in later chapters. In 1952 David Bohm was to produce a successful hidden variable theory, as we shall discuss in Chapters 8 and 10, though his own account of how he violated von Neumann's theorem was unconvincing. Finally in the early 1960s, John Bell was to identify the fault in von Neumann's theorem, or, to put things more accurately, the hidden assumption involved in the fourth postulate. Bell's work is discussed in Chapter 9.

However in this chapter, we are discussing the situation in the 1930s. While it is true[46] that among the uncommitted there was some scepticism about the theorem, among supporters of Copenhagen there was little short of euphoria. They hailed von Neumann for raising the idea on which they based so much from the realm of argument and speculation to that of mathematical proof. For example, in his pleasant popular book,[47] Max Born commented that supplementation of quantum theory with hidden variables *seemed* an impossible task. He added, though, that one did not need to remain content with this 'seeming' impossibility, for von Neumann had shown rigorously by mathematical proof that it was actually impossible.

The 'theorem' was to block genuine progress in the consideration of hidden variables essentially for 30 years, since it prevented adequate study of Bohm's work of the 1950s. Einstein himself seems to have taken little notice of the theorem, perhaps because he was justifiably suspicious of its very general but very abstract postulates. The reception of the EPR paper, though, suffered as a result of the theorem, because the paper appeared to require hidden variables; it must be said, though, that this was only one of several things the EPR paper had going against it! The theorem, in fact, overshadows Part B of the book, which covers the remainder of Einstein's life. It was not until Bell's papers, after Einstein's death, that the work of Einstein himself on the foundations of quantum theory could begin to be appreciated. This growing appreciation will be studied in Part C.

References

1. Bohr, N. (1934). The quantum postulate and the recent development of atomic theory. Nature (Supplement) **121**, 580–90 (1928); reprinted in Bohr, N. (1934). Atomic Theory and the Description of Nature. Cambridge: Cambridge University Press.
2. von Neumann, J. (1955). Mathematical Foundations of Quantum Theory. Princeton: Princeton University Press. [Translation of: Mathematische Grundlagen der Quantenmechanik. (Berlin: Springer, 1932).]

 3. Hanson, N.R. (1959). Copenhagen interpretation of quantum theory, American Journal of Physics **27**, 1–15.
 4. Stapp, H.P. (1972). The Copenhagen interpretation, American Journal of Physics **40**, 1098–116.
 5. Home, D. and Whitaker, M.A.B. (1992). Ensemble interpretation of quantum mechanics: a modern perspective, Physics Reports **210**, 223–317.
 6. Ballentine, L.E. (1970). The statistical interpretation of quantum mechanics, Reviews of Modern Physics **42**, 358–81.
 7. Bohr, N. (1963). Essays 1958/1962 on Atomic Physics and Human Knowledge. New York: Wiley, p. 25.
 8. Bohr, N. (1948). On the nature of causality and complementarity, Dialectica **2**, 312–19.
 9. Heisenberg, W. (1955). In: Niels Bohr and the Development of Physics. (Pauli, W., ed.) Oxford: Pergamon Press.
10. Heisenberg, W. (1958). Physics and Philosophy. New York: Harper and Row.
11. Folse, H.J. (1985). The Philosophy of Niels Bohr. Amsterdam: North-Holland.
12. Murdoch, D. (1987). Niels Bohr's Philosophy of Physics. Cambridge: Cambridge University Press.
13. Beller, M. (1992). The birth of Bohr's complementarity: the context and the dialogues, Studies in the History and Philosophy of Science **23**, 147–80.
14. Mackinnon, E. (1994). Bohr and the realism debates, In: Niels Bohr and Contemporary Philosophy (Faye, J. and Folse, H.J., eds.) Dordrecht: Kluwer, pp. 279–302.
15. Beller, M. and Fine, A. (1994). Bohr's response to EPR, In: Niels Bohr and Contemporary Philosophy (Faye, J. and Folse, H.J., eds.) Dordrecht: Kluwer, pp. 1–31.
16. Bohr, N. (1934). Atomic Theory and the Description of Nature. Cambridge: Cambridge University Press, pp. 54–56, p. 108.
17. Pauli, W. (1988). Letter dated 12 August 1954, quoted in Laurikainen, K.V. Beyond the Atom: The Philosophical Thought of Wolfgang Pauli. Heidelberg: Springer-Verlag, p. 20.
18. Laurikainen, K.V. (1988). Beyond the Atom: The Philosophical Thought of Wolfgang Pauli. Heidelberg: Springer-Verlag.
19. Nair, R. (1990). Journal of Scientific and Industrial Research (India) **49**, 532–42.
20. Born, M. (1926). Zur Quantenmechanik der Stossvorgänge [On the quantum mechanics of collisions], Zeitschrift für Physik **37**, 863–70.
21. Born, M. (1926). Quantenmechanik der Stossvorgänge [Quantum mechanics of collisions], Zeitschrift für Physik **38**, 803–27.
22. Born, M. (1927). Das Adiabatenprinzip in der Quantenmechanik [The adiabatic principle in quantum mechanics], Zeitschrift für Physik **40**, 167–92.
23. Beller, M. (1990). Born's probabilistic interpretation: a case study of 'concepts in flux', Studies in the History and Philosophy of Science **21**, 563–88.
24. Pauli, W. (1927). Zur Quantenmechanik des magnetischen Elektrons [On the quantum mechanics of magnetic electrons], Zeitschrift für Physik **43**, 601–23.
25. Dirac, P.A.M. (1930). The Principles of Quantum Mechanics. Oxford: Oxford University Press, Chs. 2 and 3.
26. Feyerabend, P.K. (1964). Problems of microphysics, In: Frontiers of Science and Philosophy. (Colodny, R.G., ed.) London: George Allen and Unwin, pp. 189–283.
27. Shimony, A. (1993). Physical and philosophical issues in the Bohr-Einstein debate, In: Proceedings of the Symposia on the Foundations of Modern Physics 1992. (Laurikainen, K.V. and Montonen, C., eds.) Singapore: World Scientific, pp. 79–96.

28. Ballentine, L.E. (1972). Einstein's interpretation of quantum mechanics. American Journal of Physics **40**, 1763–71.

29. d'Espagnat, B. (1995). Veiled Reality: An Analysis of Present-Day Quantum Mechanical Concepts. Reading, Massachusetts: Addison-Wesley, pp. 297–302.

30. Blokhintsev, D.I. (1968). The Philosophy of Quantum Mechanics. Dordrecht: Reidel, p. 50.

31. Penrose, R. (1987). Quantum physics and conscious thought, In: Quantum Implications: Essays in Honour of David Bohm. (Hiley, B.J. and Peat, D., eds.) London: Routledge, pp. 105–20.

32. Wootters, W.K. and Zurek, W.H. (1982). A single quantum cannot be cloned, Nature **299**, 802–3. 33.

33. Yuen, H.P. (1986). Amplification of quantum states and noiseless photon amplifiers, Physics Letters A **113**, 405–7.

34. D'Ariano, G.M. and Yuen, H.P. (1996). Impossibility of measuring the wave function of a single quantum system, Physical. Review Letters **76**, 2832–5.

35. Aharonov, Y. and Vaidman, L. (1993). Measurement of the Schrödinger wave of a single system, Physics Letters A **178**, 38–42.

36. Aharonov, Y. and Vaidman, L. (1993). The Schrödinger wave is observable after all!, In: Quantum Control and Measurement. (Ezawa, H. and Murayama, Y., eds.) Amsterdam: Elsevier, pp. 99–106.

37. Aharonov, Y., Anandan, J., and Vaidman, L. (1993). Meaning of the wave function, Physical Review A **47**, 4616–26.

38. Heisenberg, W. (1930). Physical Principles of the Quantum Theory. Chicago: University of Chicago Press (reprinted New York: Dover), p. 64.

39. Bell, J.S. (1987). Speakable and Unspeakable in Quantum Mechanics. Cambridge: Cambridge University Press, p. 189.

40. Bohr, N. (1931). Maxwell and modern theoretical physics, Nature (Supplement) **128**, 691–2.

41. Bohr, N. (1934). Atomic Theory and the Description of Nature. Cambridge: Cambridge University Press, p. 56.

42. Bell, J.S. (1987). Speakable and Unspeakable in Quantum Mechanics. Cambridge: Cambridge University Press, p. 190.

43. Selleri, F. (1990). Quantum Paradoxes and Physical Reality. Dordrecht: Kluwer, pp. 100–101.

44. Dirac, P.A.M. (1930 and many later editions). The Principles of Quantum Mechanics. Oxford: Clarendon, p. 36.

45. Bell, J.S. (1990). Against 'measurement', Physics World **3**(8), 33–40.

46. Jammer, M. (1970). The Philosophy of Quantum Mechanics. New York: Wiley, Ch. 7.

47. Born, M. (1949). Natural Philosophy of Cause and Chance. Oxford: Oxford University Press, pp. 108–9.

Interlude

From the earliest days of quantum theory right through to the work of Heisenberg and Schrödinger in 1925 and 1926, Einstein had provided many of the crucial pieces of the final picture, in particular the photon and the bringing together of the work of Planck and of Bohr in his 1916 paper introducing stimulated emission. He had also encouraged the important developments of de Broglie and Bose, and stimulated the work of Schrödinger. His 1916 paper had also been central in the formulation of Bohr's correspondence principle, the main vehicle of the development of the matrix approach to quantum theory.

His work was in part the utmost in radicalism, for example, the advocacy of the photon and the de Broglie speculation. Yet there were also elements of careful study of the radicalism of others—his analysis bringing out the basic meaning of the work of Planck and Bose. There was also a conservatism; for all Einstein's practically single-handed advocacy of the photon over many years, he remained intensely concerned about the relation between wave and particle pictures.

Between 1925 and 1927 Einstein's influence helped to move the interpretation of matrix theory from its original starkness, provision of correct results being the sole criterion, to an acknowledgement that something should and could be said about the state of the system before measurement, at least giving some information on what the result of any measurements might be. Thus Einstein's approach played a part in the statement of the Heisenberg principle and the Born interpretation of the wave-function; indeed Bohr's approach to complementarity was, at least in part, a recognition of Einstein's concerns about the nature of observation.

Orthodoxy suggests that Einstein's failure to accept complementarity was a result of intellectual inability to appreciate the requirement for fundamental conceptual change. Our reply would be: Is it remotely conceivable that he could have become senile so abruptly as to change from being leader of the pack to unable even to keep up? We would argue that Einstein's position was a clear continuation of his previous approach of combining bold speculation with a requirement for developing a coherent view of nature. Sadly his views were to be judged not just unhelpful but not even worthy of being considered.

Part B
Einstein Confronting Quantum Theory from 1925

A theory is the more impressive the greater the simplicity of its premises, the more different kinds of things it relates, and the more extended its area of applicability.

Albert Einstein, Autobiographical notes, in P.A. Schilpp (ed.), *Albert Einstein: Philosopher-Scientist* (Tudor, New York, 1949), pp. 1–95.

The important thing is not to stop questioning. Curiosity has its own reason for existing. One cannot help but be in awe when one contemplates the mysteries of eternity, of life, of the marvelous structure of reality. It is enough if one tries to comprehend only a little of this mystery every day.

Albert Einstein, quoted in the memoirs of William Miller, *Life*, May 2nd 1955; *The Expanded Quotable Einstein* (A. Calaprice, ed.) (Princeton University Press, Princeton, 2000), p. 5.

Part B
Einstein Confronting Quantum
Theory from 1925

5
Einstein's Approaches to Quantum Theory 1925–1935

Initial Impressions

Heisenberg's seminal paper initiating quantum theory was written in July 1925. It is sometimes suggested that Einstein was initially unclear in his views on the theory; Pais[1] suggests that he 'vacillated.' It seems more likely that he combined at much the same time great excitement about the mathematical content and potential of the theory, with growing concern that philosophical conclusions were being drawn from the theory that he found unacceptable. In March 1926 he wrote[2] to Max Born's wife that: 'The Heisenberg–Born concepts leave us all breathless, and have made a deep impression on all theoretically oriented people. Instead of dull resignation, there is now a singular tension in us sluggish people.' Indeed it must have been as exciting for Einstein as for any other physicist to perceive a clear mathematical route forward after the conceptual turmoil of the previous quarter-century. In 1912, Einstein[1] had written: 'The more success the quantum theory has, the sillier it looks', and his feelings had probably not changed much during the 12 years before 1912 and the 13 afterwards up to 1925. Now at last it seemed that one could hope to study atoms and light from a well-defined theory rather than by guesswork and subterfuge, however brilliant.

As with his 'heuristic' study of photons, Einstein was well able to study the consequences of a theory, putting on one side his reservations about its fundamental meaning. Bose paid a visit to Berlin in November 1925 and reported on Einstein's great interest in the theory, and suggestions that he made for Bose to study its consequences.[3] Indeed it is clear that he regarded Heisenberg's work as a great feat and continued to do so. From 1928 he proposed both Heisenberg and Schrödinger for a Nobel prize; indeed in 1931 he added[1] that: 'The achievements of both men are independent of each other and so significant that it would not be appropriate to divide a Nobel prize between them.'

However, while Einstein may have been willing to consider the Heisenberg theory as 'provisional', it was clear that this was not the generally held view. At least initially Heisenberg's position[4] was that nothing further was required than the mathematical scheme that already provided all possible results for any experiment. He believed that no picture of physical reality should be required in order to produce the theory, and none should be demanded from the theory. Indeed it was this position that he believed was in some ways the most profound result of his work. (There has been considerable discussion[5,6] of whether the approach of Heisenberg and other younger physicists such as Pauli was related to socio-political attitudes in the Weimar republic.) Einstein would obviously have been highly suspicious of this view.

Thus he must have been delighted at the emergence of Schrödinger's own theory[7] in late 1925 and 1926. This theory was based on de Broglie's concepts, which had been actively promoted by Einstein, and Schrödinger himself fully acknowledged Einstein's work on the Bose–Einstein statistics as also providing the spur for his work. More importantly, though, like Einstein, Schrödinger disliked the totally abstract nature of Heisenberg's work, and his own theory was aimed at the provision of a picture of what was actually happening at a quantum level. This picture, though, was very different from the classical one, being based fundamentally on wave-like notions. Schrödinger believed that the wave-function was essentially a distribution of charge. He also claimed that his theory eliminated the 'quantum jumps' of Bohr that Schrödinger disliked so much.

There was an interesting series of letters between Schrödinger and Einstein, in one of which the latter spoke approvingly of various features of the Schrödinger equation, in particular the fact that it contained system additivity—if E_1 and E_2 are allowed energies for two non-interacting systems individually, then $E = E_1 + E_2$ is an allowed energy for the combined system. Interestingly in this letter[8] written in April 1926, Einstein actually suggested that the system additivity requirement did *not* hold in Heisenberg quantum mechanics. In making this statement he was clearly wrong, since the formulations of Heisenberg and Schrödinger are now known to be identical mathematically; it is interesting, though, that at this stage, despite the acknowledged successes of the Heisenberg approach, Einstein still hoped that it might require modification. He wrote to Schrödinger: 'I am convinced that you have made a decisive advance with your formulation of the quantum condition, just as I am equally convinced that the Heisenberg–Born route is off the track.'

For some time it indeed seemed that the Schrödinger approach might supplant that of Heisenberg. In Born's important work on collisions that introduced the statistical interpretation of the wave-function,[9] already referred to in Chapters 2 and 3, and to be discussed thoroughly shortly, Born, one of the founders of matrix mechanics, actually used Schrödinger wave-functions, much to the disgust of Pauli. Indeed there was actually little dispute that Schrödinger methods were useful in performing computations. Schrödinger's own solution of the hydrogen atom using his own methods was vastly simpler than that of Pauli using matrix methods. But the question of conceptual superiority is a very different one. Bohr and Schrödinger had exceedingly intense discussions, in which the latter had eventually though extremely reluctantly to admit that his method did not avoid the quantum jumps. Then several physicists, including Schrödinger himself, demonstrated the mathematical identity of the two very distinct formalisms.

As such this was not totally destructive of the desire of Einstein or Schrödinger to believe in the realism of the Schrödinger wave-function. It might seem natural to argue that the wave-function provided a full treatment of any problem including a physical picture, while the matrix methods merely extracted and used the mathematics, producing measurement results but leaving out the physical picture. Einstein perhaps retained at least a vestige of that view; in the 1931 letter to the Nobel prize committee,[1] he gave his opinion that the prize should be awarded to Schrödinger before Heisenberg, because 'I have the impression that the concepts created by him will carry further than those of Heisenberg.'

However it became increasingly difficult to take the wave-function seriously as a distribution of anything directly physical. It became clear that the wave-function of an N-body system did not consist of N functions in a three-dimensional space as might have been hoped, but as a single function in a $3N$-dimensional configuration space. Also Schrödinger had hoped that a group of waves or wave-packet would represent a particle, but it was pointed out that, in nearly every case, the wave-packet would disperse rather than remaining localised. In addition it was a fact that the wave-function oscillated at a frequency equal to E/h, where E is the appropriate eigenvalue of energy. But energy is defined only with respect to an arbitrary zero, which means that the frequency is also arbitrary, which again makes the oscillation seem mathematical rather than physical. And lastly the wave-function was in general complex, which made it difficult to identify it with a distribution of anything real.

A crucial step was taken by Born in the paper already mentioned. He identified the wave-function, or actually the real function $|\psi|^2$, as the probability density, telling us the probability that the particle is to be found in a given region. In his analysis of collision experiments, the values of $|\psi|^2$ give the probabilities that the process takes various routes. In a general measurement case, let the eigenfunctions and eigenvalues of the operator $\hat{A}$ representing the observable A being measured be $\alpha_i(x)$ and A_i. Then if the wave-function is equal to $\psi(x)$ at time t_0, we should expand $\Psi(x)$ as $\sum_i a_i \alpha_i(x)$, and the probability of a measurement of A at time t_0 giving the result A_i is $|a_i|^2$.

The Born paper made it still more difficult, if not impossible, to believe in the wave-function as a charge distribution. Even more significantly it led to Einstein becoming quite sure that quantum theory, at least as it was interpreted at the time, was not acceptable to him conceptually. At the time it was accepted as the orthodox belief that one must rule out any possibility of hidden variables. In other words one insisted that the wave-function represented each system, and that there were no additional variables that might differ between different systems. Then it was clear that one and the same initial wave-function could lead to different measurement results, or in other words determinism, perhaps the most central doctrine of physics since the days of Newton, had to be relinquished. In the terms introduced in Chapter 2, the situation is probabilistic rather than statistical. This was undoubtedly anathema to Einstein, though whether it ended up as being his main concern with orthodox interpretations of the quantum theory is not so obvious, and will be discussed in Chapter 8 and then in the final chapter of the book.

It is clear that the abandonment of determinism might be avoided if one were to relax the prohibition on hidden variables. If, for example, one were to say that, even before the measurement of A, each system already possessed a value of A, and that a fraction $|a_i|^2$ had the value A_i, it seems that measurement merely records values that are already present, and that determinism has trivially been restored. With these hidden variables included, the situation is no longer probabilistic but statistical, as we defined the terms in Chapter 2. One must, of course, remember that we might perform a different measurement from A, say B, and the associated operator $\hat{B}$ may not commute with $\hat{A}$, so we will have to deal with a set of different eigenfunctions and eigenvalues, $\beta_i(x)$ and b_i. Clearly we must expect that establishing a full hidden variable theory will probably be complicated, but it would certainly seem a natural idea to try.

That it seemed highly unnatural to Bohr and Heisenberg is probably because their ideas were based on radioactive decay, or the kind of transitions used by Bohr in 1913 and developed by Einstein in his 1916 paper; in these situations, the hypothesis of hidden variables dictating the time of each decay or transition admittedly seemed awkward. Once hidden variables were rejected in those places, it perhaps seemed a backward step to reintroduce them in quantum theory. This was to ignore the fact that while, for spontaneous transitions, the task of hidden variables would be to restore determinism, for the quantum case the main achievement of hidden variables would be to restore realism, providing values for quantities not permitted them by exclusive use of the quantum formalism. At least in the simplest hidden variable ideas, regaining realism would automatically restore determinism as well.

Another reason for their rejection of hidden variables was that, particularly for Heisenberg, but to a considerable extent for Bohr also, new and surprising features in the interpretation of the new quantum theory constituted not so much difficulties to be avoided as conceptual lessons to be learned. It seems, though, that Einstein did, for a brief period, take great interest in hidden variable models, as will be seen in the following section.

Let us conclude this section, though, by noting his final reaction to quantum mechanics as presented by Heisenberg and Schrödinger (without any elaboration such as hidden variables). In December 1926 he wrote to Born[2] that: 'Quantum mechanics is certainly imposing. But an inner voice tells me that it is not yet the real thing. The theory says a lot, but does not really bring us any closer to the secret of the "old one". I, at any rate, am convinced that *He* is not playing at dice.' We briefly note Born's comment on this letter written more than 40 years later: 'Einstein's verdict in quantum mechanics came as a hard blow to me: he rejected it not for any reason, but rather by referring to an 'inner voice'. This rejection ... was based on a basic difference in philosophical attitude, which separated Einstein from the younger generation to which I felt that I belonged, although I was only a few years younger than Einstein.' Though in this letter Einstein may have used shorthand of the 'inner voice', Born must have been aware that Einstein had genuine arguments against the orthodox approach to quantum theory and against Bohr's complementarity, but sadly regarded his job as being merely to argue against Einstein's position, rather than to attempt to appreciate it.

Einstein's Unpublished Paper on Hidden Variables

By the time of this statement, Einstein was about to make his sole attempt at the construction of a hidden variables theory.[10,11] He referred to this in a note sent to Born[2] in April or May of 1927; in this note he said that he had handed in a short paper to the Prussian Academy in which: 'I show that one can attribute quite *definite movements* to Schrödinger's wave mechanics, without any statistical interpretation.' In other words Einstein hoped he could restore determinism by a model that integrated wave and particle notions. (In Born's commentary of 40 years later, he acknowledges that Einstein's arguments were thus more than a result of his 'inner voice'.) The paper was titled 'Does Schrödinger's wave mechanics determine the motion of a system or only in the sense of statistics?' In this paper, Einstein obtained an expression for kinetic energy using any solution of the Schrödinger equation, and used this to define $\dot{q}$, a velocity component of an individual particle. Cushing[11] describes in detail how he was able to obtain a unique value for $\dot{q}$ in terms of the wave-function ψ, using a non-Euclidean metric in configuration space.

Unfortunately Walther Bothe pointed out a problem. When one considered a system containing two sub-systems, which may be separated in space, the kind of situation that we saw Einstein had praised Schrödinger's theory for handling successfully, the wave-function for the whole system, ψ, could be separated out as a simple product of wave-functions for the individual sub-systems, $\psi_1 \psi_2$; however, the hidden variables obtained for the two sub-systems are mutually dependent. Einstein deemed this unacceptable, and, as a result, withdrew the paper from publication. This effect may seem to be have similarities with entanglement, which will be met in our discussion of EPR in the next chapter, and it will play a central role in the work of Bell on non-locality, and also in quantum information theory. Recently Darrin Belousek[12] and Peter Holland[13] have discussed Einstein's paper. Holland's detailed analysis shows that while the effect Einstein objected to could be removed by appropriately modifying Einstein's theory, there are other grounds for rejecting this specific theory because of its restricted domain of applicability. Holland has also argued that there can be other type of trajectory theories which are compatible with quantum predictions while displaying the mutual dependence of the motions of particles even when the many-body wave-function factorises.

Einstein's model was just one of many using hidden variables and constructed during this period.[11] Others were produced by Erwin Madelung, Earle Kennard and Reinhold Fürth. As is well-known, de Broglie presented a model involving both wave and particle at the 1927 Solvay Conference, the wave being said to 'guide' the particle. In presentation, the models fell into two rather distinct groups. Those of Einstein, de Broglie and Kennard were best described as hidden variable models of quantum theory, while those of Madelung and Fürth were attempts to provide classical (though stochastic or statistical) analogies to quantum theory, based on hydrodynamics or Brownian motion. None were successful, all of them either remaining somewhat incoherent, or failing in what we will call locality— suffering from the kind of problem we glimpsed in Einstein's model, where two spatially separated sub-systems do not behave independently. As will be discussed briefly in Chapter 10, de Broglie's model was attacked viciously by Pauli at the Solvay Conference, and he abandoned it at that time, and thereafter became a supporter of Copenhagen for a considerable period.

It seems certain that, for Einstein, the failure of all these models, not just his own, convinced him that such simple hidden variable models could not represent quantum theory in a satisfactory way. A factor that may not have been as important as one might at first guess was the publication by John von Neumann[14] in 1932 of his famous 'proof' that this would be the case; he claimed that there could be no hidden variable model of quantum theory. While this theorem may have been particularly effective in convincing those who already supported Copenhagen that they were in the right, it was never referred to by Einstein; it may be that he was (rightly) unconvinced that its fairly general arguments really had the power to deal effectively with any general type of hidden variable theory.

Anyway, after the failure of his own hidden variable model, Einstein not only abandoned any programme of producing such models himself, but appeared to have no interest in those produced by others, notably the model of David Bohm, which is mentioned two paragraphs down, and then discussed in some detail in Chapter 10. Formally there is an exception to this statement. As we shall see in Chapter 8, Einstein often discussed the advantages of ensemble interpretations of quantum theory. While the actual nature of his ensembles is not clear, there is at least some evidence that they may have been Gibbs ensembles, and theories involving Gibbs ensembles are, in fact, hidden variable theories. However, it seems that Einstein did not make this identification, and indeed the tradition of those supporting Gibbs ensembles has been to regard such theories as distinct from those using hidden variables. All this is discussed in detail in the later chapter.

Einstein's aspirations turned from simple hidden variable theories to a vastly more grandiose scheme. Following the success of his general theory of relativity in 1915, he had become interested in producing a unified field theory that would unite gravitation with electromagnetism. This was to become his main labour for the rest of his life[1,15]; following some years of study and thought, he published his first paper on this topic in 1922. By 1927 it would seem that his approach to quantum theory[15] had become a denial of any hope of making progress by tinkering with the formalism presented by Heisenberg and Schrödinger, by such simple means as hidden variables. Rather the way forward was to work for and achieve his unified field theory. Then he hoped that it would become clear that quantum theory was an exceptionally good approximation to this theory, at least in all the areas explored in experiment so far, but the conceptual riddles and paradoxes found in today's quantum theory would not be present in his desired unified field theory. This theory, of course, never materialised.

To conclude this section we will sketch the history of the hidden variables and classical analogue approaches beyond the 1920s. In so doing, we will obtain some explanation of the problems found by Einstein and others; some of the matters mentioned will be referred to in more detail later in the book. In 1952 Bohr[16] produced his own theory of hidden variables improving upon that of de Broglie's 1927 theory. Though this theory has been controversial, it has long been considered extremely seriously by a comparatively small number of advocates, and more recently has attracted the interest of rather a greater number of physicists, who feel that it may address issues that the standard interpretation cannot consider. From the start it was recognised as being non-local.

The idea of using Brownian motion as a classical stochastic model of quantum theory was taken up, in particular, by Edward Nelson,[17] who worked on this programme for almost 20 years from 1966. In the end, however, he admitted defeat.[18] When his theory tackled separated sub-systems, the behaviour of these sub-systems was not independent; his theory did not respect locality. For Nelson that was a failure to provide a physically realistic picture, which was what his whole programme was aimed at. Just like Einstein almost 40 years before, lack of locality was a sticking-point.

Much of the argument and confusion was in fact cleared up by Bell's work of the early 1960s. He was able to show that von Neumann's theorem was incorrect; hidden variable theories could duplicate the results of quantum theory. After all, Bohm's did perform this task. However such theories could not be local; as we have said, Bohm's wasn't. The eventual failure of Nelson's programme, at least according to his own agenda which insisted on locality, became obvious. Of course stochastic theories along the lines of those of Nelson are by no means ruled out,[19] provided one is willing to accept non-locality. If this is done, the apparent distinction between theories of the hidden variable type and those of the Brownian motion type has effectively disappeared. The theory of Bohm and that of Nelson may both be considered as theories that aim at duplicating the results of quantum theory, but add in extra physical content; they must both be non-local. There is much more about Einstein's attitude to the Bohm theory and locality later in this book.

The Solvay Conference 1927—Main Proceedings and Einstein–Bohr Part I

September and October of 1927 were the pivotal months for the assessment of approaches to quantum theory by the worldwide community of physicists. Bohr[20] made his presentation on complementarity at Como in September and at the Solvay Conference[21] in October. Between 1925 and this date, there had been differences between Bohr on the one hand and Heisenberg and Pauli on the other. Born had been criticised by Pauli for making use of the Schrödinger formalism. Heisenberg and Born had been prepared to gain insights from Einstein, while Bohr had been very keen to come to a common position with Schrödinger and was very concerned about Einstein's position. In other words, the situation seemed progressive; different scientists advocated different positions, but there was a willingness to listen to others and a genuine respect for those who differed from you. All this has been discussed earlier in this chapter

By the end of October positions appeared to have hardened almost irreversibly. During the actual talks it was not clear that this was about to happen. Beller[22] has suggested that, in his lecture, Bohr was having a dialogue with all the leading quantum physicists—Heisenberg, Schrödinger, Einstein, Pauli, Compton and others. In particular he gave a measure of both support and criticism to Heisenberg, and the same to the positions of de Broglie and Schrödinger, behind whom, of course, stood Einstein. Indeed Bohr was extremely hopeful that his positioning between the founders of matrix and wave mechanics would lead to Einstein endorsing the Copenhagen position.

For example, it will be remembered that Schrödinger described his theory solely using waves. Virtually as a means of reply, Heisenberg eschewed waves altogether, and, for example, built up his original uncertainty principle argument using only particles. Bohr was able to demonstrate to Heisenberg that both particles and waves were required for this argument to be carried out successfully, and this became his position at Como. For a rigorous discussion of measurement, as distinct from mere calculation of the results that might be obtained as in Heisenberg's initial approach, he argued that both particle and wave pictures were required and they needed to be handled by the methods of complementarity.

While this might be described as extending an olive branch to Einstein, Schrödinger and de Broglie, or at least responding to their conceptual concerns, it became clear that, to the extent that the three did not come into line with Bohr's argued position, there would almost inevitably be conflict. No further concessions would be made! Pauli was a far more experienced theoretician and debater than de Broglie, and his demolition of de Broglie's hidden variables approach bordered on the brutal. His argument claimed to show that, in a particular scattering experiment, de Broglie's theory would not produce a unique result. As has been said, de Broglie was so shell-shocked that he became a convert to Copenhagen for the next quarter-century. Indeed with Einstein's interest in hidden variable theories having lapsed, it was this further quarter-century before the question of hidden variables was raised again, this time by Bohm; at that later time, Bohm was able to show that Pauli's argument against de Broglie was actually rather insubstantial.[11,23]

At the Solvay Conference, Einstein had been asked to talk on indeterminism in quantum theory and initially agreed, but later he pulled out because he felt his grasp of the details of the theory was not yet strong enough. He did however make a brief but pointed comment at the actual session. He raised the question of an electron or beam of electrons passing through a fairly wide slit so, as a wave, being diffracted through a fairly wide angle towards a film. At the film itself, of course, only a very localised area is exposed. Einstein considered two standpoints, interpretation II corresponding to Bohr's approach. Here the wave represents a single electron, so, just before it hits the film, $|\psi|^2$ represents the probability of the electron reaching that particular point. Einstein declared himself not to be in favour of this interpretation because it seems that, while, just before the electron reaches the film, it is potentially present all over the film, the behaviour at all points once it hits the film is correlated—it is registered on the film at one point, and is instantaneously *not* registered at all other points. There appears to be a problem with at least some aspect of locality.

Einstein said that he preferred interpretation I, according to which the wave represents not a single electron but an ensemble of electrons extended in space. $|\psi|^2$ represents the probability that *any* electron hits the film at that point. Einstein was advocating an ensemble interpretation in which the wave-function represents not a single system but an ensemble of systems. Einstein was often to advocate such an approach, and the part that ensemble interpretations played in his overall position is discussed in detail in Chapter 8. When Bohr replied, he clearly felt that Einstein's approach was far too classical in nature. Bohr certainly did not feel that one should think in terms of effective incompatibility between a classical wave picture just before the particle hits the screen, and a classical particle (or observation of a particle) just afterwards. There seemed to be little common ground between the two men.

It seems that Einstein's actual argument attracted little attention at the time or for a considerable period later. However more recently the problem has been revived by de Broglie,[24,25] and discussed in some detail by Fine,[10] and particularly in a recent paper by Travis Norsen.[26] We discuss these arguments briefly in the final section of this chapter.

Back in 1927, Einstein and Bohr did have more productive meetings informally outside the formal conference sessions. These encounters, and similar ones at the next Solvay Conference in 1930, were described in some detail as part of an account Bohr[27] was to write many years later for a collection of articles on Einstein's work and views. The titanic struggles of these two great minds have become legendary in the history of ideas. For the remainder of this section and the following one, we shall describe the encounter more or less as it is described by Bohr, and then in the final section of the chapter, we will return to consider in greater depth the lessons to be learned.

Einstein's aim in all these discussions was to invent an idealised experiment or thought-experiment that would enable one to measure simultaneously two quantities which quantum theory, or at least the Copenhagen interpretation of quantum theory, said could *not* be measured simultaneously. In 1927 position and momentum were used. Since the operators for these observables have no common eigenfunctions, Copenhagen would indeed say that the two observables cannot be measured simultaneously. Actually something less would be required by Einstein: All that he had to achieve was to produce a scheme in which one could measure the quantities to a greater degree of precision than allowed by the Heisenberg principle, $\Delta x \Delta p \approx \hbar/2$.

Used later in the 1930 discussions were the quantities of energy and time. As discussed in Chapter 3, the analogous principle for these quantities, $\Delta E \Delta t \approx \hbar$, actually has a very different meaning to the previous one, since there is no operator for t in non-relativistic quantum theory, and so Δt cannot be an uncertainty in time. Nevertheless in 1927 the two principles were used in a very analogous way, so we will not, and do not need to, dwell on the difference. We shall analyse what would be involved in such joint measurements, and what concepts were really being tested in these discussions in the last section of this chapter. At this point we just say that, for both sides, the matter was accepted to be extremely important; had Einstein come out victorious, Bohr would have accepted that his conception of quantum theory had suffered a mortal blow.

Before proceeding to the discussion between Einstein and Bohr, let us describe some common ground. In Figure 5.1 is shown the formation of a typical diffraction pattern. A beam of electrons or photons is incident on a slit in a screen A. If the slit-width is a, the magnitude of the initial momentum is p, and the diffraction pattern is of angle θ, then the magnitude of the momentum gained in the y-direction, Δp, is $p \sin \theta$. Elementary diffraction theory tells us that $a\Delta p$ is of order h, exactly as the Heisenberg principle demands.

Included in the diagram is also a shutter C, which may be opened for a only a short time T. In this case only a limited wave-train proceeds through the slit, as shown in Figure 5.1. By thinking of this wave-train in terms of Fourier analysis, it is clear that this means there is a spread of frequencies in the wave-train, or in other words a spread in the energy, ΔE, of the electron or photon. Again elementary analysis shows that the product of T and ΔE is of order h. An alternative means of producing these results is to study the exchange of momentum between the screen and the particle, and the exchange of kinetic energy between the shutter and the particle. For the latter, the shorter T is, the faster the shutter will be moving, and the greater the possible exchange of energy. Again detailed calculation shows that the appropriate Heisenberg relation is obeyed.

So far everything has been non-controversial, but now we start discussing Einstein's attempt to pull out more information than the uncertainty principle would allow. First he suggested that we could measure exactly the momentum or energy transferred to the screen or shutter by studying their movement after the particle has passed through. We would then combine exact knowledge of the

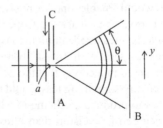

FIGURE 5.1. The figure shows a beam of electrons or photons incident on a slit A of width a. Following diffraction the beam is mostly restricted to an angular width of θ as it travels to plate B. A shutter C is used to open the slit for a limited period T, and so the train of waves is of limited extent.

value of p or E for the particle, with partial knowledge of x or t. More specifically, $\Delta p = \Delta E = 0$, while $\Delta x = a, \Delta t = T$, so $\Delta p \Delta x$ and $\Delta E \Delta t$ are both zero, clearly in violation of the uncertainty principle.

However Bohr replied that the initial analysis assumed that the screen was stationary, and that the speed of the shutter had no uncertainty. It implies that they are able to exchange momentum and energy with the particle, without the screen moving or the speed of the shutter changing as a result of this exchange. This could only be possible if the screen and the shutter were themselves of infinite mass. If, however, as Einstein suggested, we wish to study the motion of the screen and the shutter, we must assume that their masses are finite; but then we must take account of the Heisenberg principle when performing the analysis. There will be uncertainties in the positions of the screen, the slit and the shutter. Since the position of the particle is defined relative to the slit, the uncertainty in the position of the slit is equivalent to an extra uncertainty in the position of the particle. When one takes into account all the uncertainties in detail, including in particular noting that Δp and ΔE of the particle are not now zero, one finds that the usual Heisenberg relations apply. One may say that the two measurements interfere with each other.

Einstein's response to this reply of Bohr was to separate out the parts of the apparatus performing the two measurements, in the hope that they will no longer interfere with each other. As shown in Figure 5.2, he used a screen A_1, with a slit but without a shutter; half way between A_1 and the plate B is a second screen A_2. This screen is supported by a weak spring and has two slits in it, S_1 and S_2. If both screens were rigid, there would be an interference pattern on B. It is important to realise that this is built up by individual particles hitting the particular points on the plate over the duration of the passage of the beam. Large numbers of particles hit the plate in the area around the maxima of the diffraction pattern, far fewer around the minima. Einstein considered the case where the beam is exceptionally weak so that the effect of each particle may be studied individually.

He argued that, with A_2 suspended from the spring, it should be possible to tell which one of slits S_1 and S_2 the particle passed through. This is known as

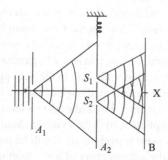

FIGURE 5.2. As for Figure 5.1, but without shutter C and with an additional screen A_2 supported by a weak spring. In A_2 there are two slits, S_1 and S_2.

'Welcherweg' or 'which way' information, and it would indeed appear paradoxical if we knew which slit the particle passed through, while also retaining the interference pattern, which tells us that the 'particle' was actually behaving as a wave and, as such, went through both slits! However Bohr was able to demonstrate that knowledge of the momentum gained by A_2 to the required precision implies an uncertainty in position, and the uncertainties in position of S_1 and S_2 are just sufficient to smudge out the interference pattern. This makes it clear that we may design our experiment to demonstrate *either* particle-like behaviour *or* wave-like behaviour but not both simultaneously.

The Solvay Conference 1930—Einstein–Bohr Part II

We now move to the 1930 Solvay Conference, and we discuss the arguments between Bohr and Einstein outside the main sessions of this conference.[27] There is evidence that Einstein had been thinking about the thought-experiment he was to propose to Bohr on this occasion for a considerable time. In the experiment, he imagined a box containing a certain amount of radiation, with a hole in one of its sides that may be opened and closed by a shutter. Einstein proposed that the shutter be opened for a short period of time T, and that it could be arranged that during this period a single photon would pass through the hole into the box. As already discussed, both Einstein and Bohr were content to regard T as an uncertainty in time, and, although it would not be so regarded today, we shall follow their approach.

The box is weighed before and after this period, and Einstein believed that this weighing could be performed as accurately as one wished, so that the uncertainty of each weighing would be zero. Now the photon that has entered the box has a particular energy equal to hf, where f is its frequency. By Einstein's own relation, $E = mc^2$, it must have an effective mass, and this mass would be the difference between those obtained in the two weighings. Since he assumed that each weighing was known exactly, and the time available for the photon to enter the box was just T, the product of uncertainties in energy and in time was zero, obviously completely contrary to the uncertainty principle.

Einstein felt sure that he had defeated Bohr, and it seems that Bohr was at first half convinced that he was right! After some time, though, the solution came to him; from Einstein's point of view, it was especially ironic that it made use of the great scientific achievement of Einstein himself, general relativity. When Bohr wrote his account of his interactions with Bohr in the Schilpp volume, he illustrated the experiment with a diagram similar to, though actually a lot more elaborate than, Figure 5.3. This was to make it clear that, rather than a term like 'weight' or 'mass' relating just to the box or the photon, in Bohr's view it should relate to an *operation* in which the actual measurement is performed. It will be remembered that such an approach was central to Bohr's conception of how quantum theory must be discussed. In the figure, the box is weighed using a spring balance, and the value of the weight is read off from the position on the scale of the pointer attached to the box.

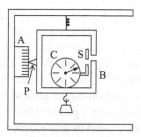

FIGURE 5.3. A diagram based on that used by Bohr in his account of Einstein's photon weighing experiment. A clock C is used to open shutter S for a period of time T. Bohr discusses the weighing by the spring balance of box B and its contents explicitly in terms of the position of pointer P on scale A.

We require the weighings to be performed within an uncertainty Δm. The position of the pointer must have an uncertainty Δq, and this must be related to the uncertainty of the momentum of the box, Δp, by the usual relation $\Delta q \Delta p \approx h$. The uncertainty in momentum must be less than the impulse that would be given by the gravitational field to the mass Δm in time T, which is $Tg\Delta m$, and this in turn is equal to $h/\Delta q$. Now Bohr brought in general relativity. It tells us that when a clock is moved in the direction of a gravitational field, it changes its rate; this causes an uncertainty in t, Δt, given by

$$\frac{\Delta t}{T} = \frac{g\Delta q}{c^2}$$

Putting all the elements of the calculation together, one discovers, as may be expected, that $\Delta E \Delta t \approx h$, so Bohr had again triumphed.

While it was interesting that Bohr used general relativity to defeat Einstein in this instance, the fact has provoked concern. It might be seen to suggest that quantum theory is consistent with general relativity, but not with Newton's theory of gravitation, or in other words that general relativity and quantum theory are intimately connected in a highly unexpected way. The matter has been discussed by Jammer,[28] who argues that in fact Einstein's famous $E - mc^2$ is in fact a formula of relativity, and all proofs of this equation are relativistic in nature. Bohr therefore should be judged not to have introduced relativity into the argument, but rather to have reacted to Einstein having introduced it tacitly. For further comments, though, see Refs. [29, 30]. A more critical approach to Bohr's solution is given in Ref. [31].

Einstein–Bohr: Argument and Concepts in the Early Parts of the Debate

Having described the early parts of the Bohr–Einstein debate much in Bohr's own terms, we now proceed to examine the conceptual factors involved at a deeper level. We ask a number of questions. What was Einstein really trying to show? What would it have meant if he had been successful? Was Bohr's position at all damaged by Einstein's thrusts and his own admissions? Was complementarity a necessary component of Bohr's argument? Why was the argument fought out without use of the quantum formalism, and in fact, except for the calculation of the necessary Δx, Δp, ΔE and Δt, in non-mathematical terms. It is very interesting and significant to compare[32] the answers given to these questions for the particular rounds of the debate that are discussed in this chapter, with the comparable ones for the case of EPR,[33] which is the topic of Chapter 6.

What was Einstein trying to do in these exchanges with Einstein? According to Jammer,[28] his hope was of 'refuting the quantum theory on the grounds of an internal inconsistency'. Defeated by Bohr in these arguments, 'he concentrated on demonstrating the *incompleteness*, rather than the *inconsistency*, of quantum mechanics' in the EPR paper [Jammer's italics]. Fine,[10] however, criticises this as 'wrong on both counts' for 'the issue of completeness was Einstein's concern from the beginning, whereas nowhere do I find him trying to show the inconsistency of the theory'. To analyse the reasons for this disagreement, we shall have to study the meaning of the Heisenberg principle at some length, but first we shall need to explain what Jammer and Fine mean by *completeness*.

According to EPR, a theory is complete if every element in the physical reality has a counterpart in the theory. To see the significance of this criterion, for Einstein in particular, we must note the difference between classical physics, where it is taken for granted that a particle possesses values of, say, momentum and position, at all times, and quantum physics, particularly as argued by Bohr and Heisenberg. In the latter, let us imagine that a particle has a precise value of momentum, in the sense that its wave-function is an eigenfunction of momentum, say $\exp(ik_0x)$ (unnormalised), and so a measurement of momentum is certain to yield the result $\hbar k_0$. With a wave-function of that form, orthodox interpretations will say that the particle just does not have a value of position; a measurement of position may give any value at random. Bohr and Heisenberg have absolutely no problems with this; they would say that, in this circumstance, position admittedly has no value in their theory, but that is because it has no value in physical reality, so the theory is complete according to the above definition. Einstein, however, would argue that, in this circumstance, the momentum *does* have a value in the physical reality though not in the theory, or in other words the theory is incomplete. This was explicitly the aim of EPR in 1935, but now we wish to consider whether it was Einstein's actual aim at the Solvay Conferences in 1927 and 1930.

At these conferences, Einstein's aim was to demonstrate simultaneous measurement of momentum and position more accurately than allowed by the uncertainty principle. (As we have seen earlier, the time—energy uncertainty principle has a different meaning, so we will not discuss that further.) It must be said that this form of words, taken for granted by both Einstein and Bohr, and indeed by everybody involved in the debate at the time, may now be seen to be somewhat confused.

As explained in Chapter 3, Heisenberg's original argument, which was very physical in nature, may in retrospect be seen to be vulnerable to the following charge. It may be said to assume that a particle *has* exact values of both momentum and position, or in other words moves along a defined trajectory, only in order to show that there are limitations on how accurately these values may be measured in combination, or that, in the rather notorious phrase, the measurement disturbs the system. Because both quantities may not be measured simultaneously with certainty, a broadly positivistic argument concludes that, after all, they cannot exist simultaneously. Heisenberg's reasoning, however, is at best suggestive, and certainly far from rigorous. Nevertheless, Jammer was only echoing the opinions of all involved in saying that simultaneous measurement violating the Heisenberg principle would have been reckoned to strike at the heart of the quantum theory as it was understood at the time.

A more rigorous approach to the meaning of the uncertainty principle would start from a particular wave-function. For concreteness, we will make the wave-function of our system equal to the ground-state eigenfunction of the simple harmonic oscillator. It will be remembered that this is the case where in fact the product of Δx and Δp is equal to, rather than greater than, $\hbar/2$. For this state the eigenfunction is given by

$$\varphi_0(x) = (\alpha/\pi)^{1/4} \exp(-\alpha x^2/2) \tag{5.1}$$

where $\alpha = (m\omega/\hbar)$. This is given naturally as a function of x, but of course it may just as well be stated as a function of p:

$$\varphi_0(p) = (1/\alpha\hbar^2\pi)^{1/4} \exp(-p^2/2\alpha\hbar^2) \tag{5.2}$$

Since the expectation values of both x and p, and are both zero, it is easy to see that $(\Delta x)^2$ and $(\Delta p)^2$ are equal to $\langle x^2 \rangle$ and $\langle p^2 \rangle$ respectively, so

$$(\Delta x)^2 = \hbar/2m\omega; \quad (\Delta p)^2 = \hbar m\omega/2; \tag{5.3}$$

and one obtains, as expected, $\Delta x \Delta p = \hbar/2$.

This leads to the appropriate statement of the meaning of the Heisenberg principle. If a measurement is performed of, say, momentum on an ensemble of systems with the same wave-function, Δp is the standard deviation of the measurement results. The standard deviation for results of measurements of position is Δx, and then the product must, in general, be greater than or equal to $\hbar/2$. It is important to note that the argument just given is entirely independent of whether or not there may be hidden variables; it takes into account only the initial wave-function and the distribution of measurement results.

The next comments, though, apply only to the Copenhagen case where there are no hidden variables, which means that, once the form of the wave-function is known, there are no further quantities, known or unknown, relating to each system. Notice that, once this is accepted, the argument of the previous paragraph has nothing to do with uncertainty as such. If one has performed an error-free measurement of position or momentum, the value achieved *is* the measured position or momentum of the particle; the term 'uncertainty' just has no meaning. (If one accepts the projection postulate, then this value is, in addition, the value of the position or momentum after the measurement, again with no uncertainty.)

What one has before any measurement is not a wave-function corresponding to an uncertain momentum or position; this would imply that there exist values of momentum and position which we happen not to know. Rather the values just do not exist to a greater accuracy than the Heisenberg principle allows; the Heisenberg principle should best be called an indeterminacy principle. Neither does the expression 'disturbance of the system' have any meaning in this situation; one has not changed the momentum or position from one value to another by a disturbance; rather it is the measurement that has actually resulted in a realised value for one of these quantities.

We now turn to the case of hidden variables, and we concentrate on the simplest case, where, as well as each system being one of an ensemble with wave-function given by (5.1) or (5.2), each system also possesses, as hidden variables, values of x and p, with the appropriate distributions being those that would be obtained if measurements of x or p were performed on the ensemble. In such theories, and for the case considered above, $\phi_0(x)$ and $\phi_0(p)$ represent not a potential distribution for *each* system, but a distribution over *all* systems. Each system has its own values of x and p.

Now let us turn[32] to the study of simultaneous measurement of x and p, either exact measurements, or combined measurements to a greater accuracy than that allowed by the Heisenberg principle. Again we first consider the standard interpretation. We have said that, according to any standard interpretation, common values cannot *exist* to this degree of accuracy, but that does not necessarily mean that they cannot be *measured* to the same degree of accuracy. This may seem almost contradictory at first reading, but we may consider the case for a single measurement; it should be remembered that, while for a general wave-function for a particular system, values for position and momentum do not, in any sense, exist before a measurement, that certainly does not mean that we cannot measure either of these observables.

To discuss simultaneous measurement within a standard interpretation with no hidden variables, it is profitable to think in terms of the projection postulate. As we have said, following a measurement of position, the system would be left in an eigenstate of position; after a measurement of momentum, it would be left in an eigenstate of momentum. A simultaneous measurement of both observables would imply that the wave-function of the system should be left as a common eigenfunction of the momentum and position operators, but of course there is no such function, so this argument would suggest that there can be no simultaneous measurement.

To discuss what is the best that can be achieved, we may think in terms of a post-measurement state; this will have its own combined indeterminacies in position and momentum, which are equivalent to distributions of possible results to be obtained in a future measurement. These distributions will be governed by the Heisenberg principle, so it is clear that a claim to have performed combined measurements of momentum and position to greater precision than allowed by Heisenberg is indeed contrary to any standard interpretation of quantum theory.

Let us now turn to the question of simultaneous measurement for a hidden variable theory, again considering only the simplest type. We have said that simultaneous values of x and p exist in such a theory, and, at least at first sight, it seems natural to assume that, when a measurement is made of x or p, the appropriate value is obtained as the measurement result, and indeed that combined measurements to any accuracy might be possible.

However, things may not be quite as simple as this. We must note that, though a measurement of momentum may merely 'read off' the previously held value of momentum, such a measurement must in general change the value of position for the particular system. The point is that, while before the measurement the distribution of values of position relates to (5.1), afterwards, if the value of p_0 has been obtained for the momentum, the wave-function has become $\psi = \exp(ip_0x/\hbar)$. We must now consider an ensemble of systems for all of which the momentum has been measured to be p_0 so all have this same wave-function. Then the distribution of values of position must correspond to this distribution . Thus $\psi^*\psi$ is a constant, so in fact all values of x must be represented equally in the distribution. Note that we are assuming here the projection postulate whereby an immediately repeated measurement must give the same result as the initial one. Similarly if a measurement of position is made, in the resultant distribution corresponding to any particular value of x, all values of p must be represented equally, so in general the measurement must have changed the value of x.

The previous paragraph tells us that, as often happens, hidden variable theories turn out in practice to be somewhat complicated when examined in detail. However, and in total contrast to the standard interpretation, since, on the kind of hidden variable theory we have discussed, values of position and momentum do exist for each system, both before and after measurement, it would be entirely natural, from a physical point of view, to hope to measure them simultaneously.

From the point of view of this discussion, then, the aim of Einstein in these early bouts in the Bohr–Einstein debate should be regarded as violation not of quantum theory itself but of its standard or orthodox interpretation, and success for Einstein would have been, as Fine suggests, a demonstration that quantum theory was incomplete, and could be supplemented with hidden variables. (We stressed earlier in the chapter, of course, that Einstein's own solution after about 1927 was not to produce a complete theory by adding simple hidden variables, but to produce an entirely new theory that was complete.) From that point of view, the EPR paper, which unequivocally aimed at demonstrating incompleteness, was a continuation of the same theme rather than a dramatic change. This again supports Fine, though, as said before, one could regard Jammer's contrary opinion as being historically correct, as expressing how things were perceived at the time.

We have argued, for the standard interpretation, from the impossibility of common values of momentum and position, to impossibility of simultaneous measurement. However it must be stressed that one cannot argue the other way. Einstein remained convinced that position and momentum should have simultaneous exact values. However, from 1930 and Bohr's triumph over the photon box, he was forced to admit that it was impossible to violate the Heisenberg relation in a joint measurement. (The photon box argument itself, of course, coupled energy and time, but his admission would have included momentum and position as well.)

However it was quite possible to continue to believe that both had values, but that the two measurements were disturbing each other. Indeed we saw a hint of such an idea in the fact that a measurement of momentum changed the value of position, and vice versa. Note that we have criticised the use of the common expression 'the measurement disturbing the system' as an encapsulation of what is going on in quantum theory. This is because it is often taken to imply that, for example, the system possesses values of momentum and position at all times, which are then disturbed by a measurement. This implication is certainly not what is assumed in orthodox interpretations. However the present discussion of each measurement disturbing the other seems to be quite straightforward, indeed very much along the lines of Bohr's general arguments, though not, of course, his fundamental beliefs. In EPR, Einstein would return with a highly ingenious attempt to consider the results of momentum and position in a context where the measurements did not interfere with each other.

First, though we return to Bohr, and ask whether he was entirely unscathed by these rounds of the debate. In fact, from a rather subtle point of view, he did have to give way. It will be remembered that a corner-stone of his ideas on complementarity was that measurement was classical, that the measuring apparatus should be treated classically. It is this belief that necessitated the existence of the Heisenberg cut between the 'system' to be treated quantum mechanically and the classical apparatus. Yet his winning thrusts against Einstein were centred around the necessity of applying the Heisenberg relations to the measuring device. This may not have been an actual contradiction, but even at the time it must have seen a somewhat awkward blemish. In a recent paper, though, Landsman[34] says that: 'Despite innumerable claims to the contrary (e.g., to the effect that Bohr held that a separate realm of Nature was intrinsically classical), there is no doubt that both Bohr and Heisenberg believed in the fundamental and universal nature of quantum mechanics, and, once more, saw the classical description of the apparatus as a *purely epistemological move*, which expressed the fact that a given *quantum* system is *being used* as a measuring device.' (The italics are in the original.) It is not clear whether Landsman's point would remove the conceptual difficulty discussed in this paragraph.

With the benefit of hindsight we may perhaps see this difficulty as signalling a new stage in the classical limit problem for quantum theory. We may recognise two major problem areas in the orthodox interpretations of quantum theory, both of them picked up by Einstein and dismissed by the main advocates of Copenhagen. The first, which is the better-known one, centred around the questions of measurement, hidden variables and completeness, and later moved onto locality as well. It is the main subject of this chapter and the next, and later Chapter 9 in particular, as well as being the main motivating force behind the non-standard interpretations of Chapter 10. The second problem area is the question of the classical limit, of whether quantum theory moves smoothly into classical physics as systems become large, or in other words macroscopic quantum theory; this is the topic of Chapters 7 and 12.

Other points[32] may be made which are particularly interesting in comparison with the comparable discussion in connection with EPR in Chapter 6. The first point is that, as mentioned, with the exception of the Heisenberg relations themselves and directly related analysis, the arguments were verbal; the quantum formalism was not used at all. This approach, involving the intense study of thought-experiments, was very much to Bohr's taste, and would here have seemed necessary to Einstein also. His aim was to show that he could go beyond what the wave-function could tell him, and it would seem that this could not be achieved by restricting analysis to the use of the wave-function. Detailed analysis of specific physical situations seemed to be required. Einstein's methodology, though not that of Bohr, would be different in EPR.

Finally, an extremely important point is that, in these discussions, Bohr did not need to use complementarity explicitly. One might say, of course, that had he done so, his arguments could not possibly have been found relevant by Einstein, who did not accept complementarity, and indeed whose aim was to disprove it. Rather Bohr used straightforward, though sometimes subtle and surprising, physical arguments to make his points, not conceptual matters of any nature. But it seemed clear that his physical arguments appeared to exemplify in a direct way much of what complementarity demanded—that it was illegitimate even to mention the value of an observable without the presence of an experimental apparatus to measure it. This argument must have given Bohr great confidence that his framework of complementarity faithfully represented physical reality.

But it is important to stress that the arguments do not prove that complementarity is 'correct' in the sense of being the undisputed means of discussing atomic problems. There are, in fact, two arguable positions. The first starts with the so-called quantum postulate, the argument going from the very existence of the quantum to the fact that the interaction between any two bodies must be discontinuous, and must therefore violate determinism. This then leads to complementarity and the completeness of quantum theory, and in the 1930s and for a long time afterwards, this was the orthodox and indeed practically universal position. The second position adopts hidden variables or some analogous approach; hidden variables nullify the quantum postulate since they may specify every detail of the interaction process, and this position leads to quantum theory being incomplete. As we have said, the fact that Einstein accepted that it might not be possible to measure the values of momentum and position simultaneously in no way committed him to accept that they did not exist simultaneously.

Einstein's Boxes

As noted above, Einstein's argument from the 1927 Solvay Conference has more recently been discussed in some detail by several authors. The first seems to have been de Broglie who, in a book of 1964[25] dramatised Einstein's original argument. Rather than the particle impinging at a particular point on a relatively small screen, it is allowed to enter a box. This box is then divided into two separate parts, which are transported to two very distant places—de Broglie suggests Paris and Tokyo. The wave-function of the particle now consists of a linear combination of two parts, one in Paris and the other in Tokyo. De Broglie says that the particle is 'potentially present' in both places. Now we imagine an experiment is carried out in Paris that reveals the particle in that location. The usual interpretation, he says, tells us that the particle has been immediately localised. He is even more exercised by the negative result possibility: Here a measurement in Tokyo that fails to find the particle will be said to localise it in Paris.

De Broglie regards the alleged interpretation as paradoxical. He considers that the particle must have been present in one or other of the boxes from the outset, and concludes that therefore the wave-function does not provide a *complete* description of physical reality, echoing, of course Einstein's expression in EPR arguments.

Twenty years after de Broglie, and independently of him, Fine[10] also discussed Einstein's argument. In fact, he reported that Einstein himself returned to the same basic thought-experiment in a letter to Schrödinger in 1935. This was part of the correspondence between the two men following the publication of the EPR paper. It will be discussed in more detail in the following chapter, and included the first appearance of Schrödinger's notorious cat, and Schrödinger's invention of the word 'entanglement' for the key theoretical feature of the EPR paper. In this letter Einstein expressed his dissatisfaction with the way the paper—in the next chapter we call it EPR35—was written. As we shall see, the published version was produced by Podolsky, and Einstein considered that it was over-formal, and did not make its point clearly.

In this letter, Einstein returned to his argument of 1927, but like de Broglie three decades later, dramatised it somewhat by introducing two boxes. It is known that, at measurement, the particle will be found in one box or the other. From Fine's point of view, while the argument is of some interest in its own right, it would seem that its main importance is in helping Einstein to reshape the EPR analysis into a form which, he thought, made its case in a more convincing manner. Specifically it made the conclusion quite clear; quantum theory must be either incomplete or non-local. Norsen[26] suggests that credit for introduction of the term 'Einstein's boxes' should go to Fine in his discussion of this letter of Einstein.

It was Norsen himself, though, who has argued comprehensively that the Einstein's boxes argument was important not only, and indeed not mainly, as a stage in Einstein's own development, but as an extremely interesting and important thought-experiment in its own right. He claims that it should take 'its rightful place alongside similar but historically better known quantum mechanical thought-experiments such as EPR and Schrödinger's cat'. He claims that the argument makes its point just as convincingly as EPR, and indeed in a simpler and more direct way, and he also argues that it provides a useful staging-ground for discussion of hidden variables, locality and Bell's theorem. So far the only comment on Norsen's paper has come from Abner Shimony.[35]

References

1. Pais, A. (1982). 'Subtle is the Lord ...': The Science and the Life of Albert Einstein. Oxford: Clarendon.
2. Born, M. (2005). The Born-Einstein Letters 1916–1955. 2nd edn., Houndmills: Macmillan.
3. Mehra, J. (1975). Satyendra Nath Bose, Biographical Memoirs of Fellows of the Royal Society 21, 117–53.
4. Heisenberg, W. (1972). Physics and Beyond: Encounters and Conversations. New York: Harper.
5. Forman, P. (1971). Weimar culture, causality, and quantum theory 1918–1927: Adaptation by German physicists and mathematicians to a hostile intellectual environment, Historical Studies in the Physical Sciences 3, 1–117.
6. Hendry, J. (1980). Weimar culture and quantum causality, History of Science 18, 155–80.
7. Moore, W. (1989). Schrödinger: Life and Thought. Cambridge: Cambridge University Press.
8. Przibram, K. (ed.) (1967). Letters on Wave Mechanics. New York: Philosophical Library.
9. Born, M. (1926). Zur Quantenmechanik der Stossvorgänge [On the quantum mechanics of collisions], Zeitschrift für Physik 37, 863–7.
10. Fine, A. (1986). The Shaky Game. Chicago: University of Chicago Press.
11. Cushing, J.T. (1994). Quantum Mechanics: Historical Contingency and the Copenhagen Hegemony. Chicago: University of Chicago Press.
12. Belousek, D.W. (1996). Einstein's 1927 unpublished hidden-variable theory: Its background, content and significance, Studies in the History and Philosophy of Modern Physics 27, 437–61.
13. Holland, P. (2005). What's wrong with Einstein's 1927 hidden-variable interpretation of quantum mechanics? Foundations of Physics 35, 177–96.
14. von Neumann, J. (1955). Mathematical Foundations of Quantum Theory. Princeton: Princeton University Press.
15. Einstein, A. (1949). Autobiographical notes, In: Albert Einstein: Philosopher-Scientist, (Schilpp, P.A., ed.) New York: Tudor, pp. 1–95.
16. Bohm, D. (1952). A suggested interpretation of the quantum theory in terms of 'hidden variables' I and II, Physical Review 85, 166–93.
17. Nelson, E. (1966). Derivation of the Schrödinger equation from Newtonian mechanics, Physical Review 150, 1079–85.
18. Nelson, E. (1985). Quantum Fluctuations. Princeton: Princeton University Press.
19. Bacciagaluppi, G. (1999). Nelsonian mechanics revisited, Foundations of Physics Letters 12, 1–16.
20. Bohr, N. (1927). The quantum postulate and the recent development of atomic theory, Nature 121, 580–90.
21. Mehra, J. (1975). The Solvay Conferences on Physics: Aspects of the Development of Physics since 1911. Dordrecht: Reidel.
22. Beller, M. (1992). The birth of Bohr's complementarity: The context and the dialogues, Studies in the History and Philosophy of Science 23, 147–80.
23. Holland, P.R. (1993). The Quantum Theory of Motion. Cambridge: Cambridge University Press.
24. de Broglie, L. (1959). L'interprétation de la mécanique ondulatoire, Journal of Physics: Radium 20, 963–79.

25. de Broglie, L. (1964). The Current Interpretation of Wave Mechanics: A Critical Study. Amsterdam: Elsevier.
26. Norsen, T. (2005). Einstein's boxes, American Journal of Physics **73**, 164–76.
27. Bohr, N. (1949). Discussion with Einstein on epistemological problems in atomic physics, In: Albert Einstein: Philosopher-Scientist. (Schilpp, P.A., ed.) New York: Tudor, pp. 199–241.
28. Jammer, M. (1974). The Philosophy of Quantum Mechanics. New York: Wiley.
29. Unruh, W.G. and Opat, G.J. (1979). The Bohr-Einstein 'weighing of energy' debate, American Journal of Physics **47**, 743–4.
30. Ghirardi, G.-C. (2005). Sneaking a Look at God's Cards. Princeton: Princeton University Press.
31. de la Torre, A.C., Daleo, A. and Garcia-Mata, I. (2000). The photon-box: Einstein–Bohr debate demythologized, European Journal of Physics **21**, 253–60.
32. Whitaker, M.A.B. (2004). The EPR paper and Bohr's response, Foundations of Physics **34**, 1305–40.
33. Einstein, A., Podolsky, B. and Rosen, N. (1935). Can quantum-mechanical description of physical reality be considered complete? Physical Review **47**, 777–80.
34. Landsman, N.P. (2006). When champions meet: rethinking the Bohr-Einstein debate, Studies in the History and Philosophy of Modern Physics **37**, 212–42.
35. Shimony, A. (2005). Comment on Norsen's defense of Einstein's 'box argument', American Journal of Physics **73**, 177–8.

6
EPR and its Aftermath

Introduction

The Einstein–Podolsky–Rosen paper,[1] usually just called EPR, was written in 1935. Following his discussions with Bohr, which were described in the previous chapter, Einstein was now prepared to admit that it was not possible to measure momentum and position simultaneously to a better accuracy that allowed by the Heisenberg principle. The EPR argument was, at the very least, an ingenious attempt to show that nevertheless both observables do actually have simultaneous values at all times. Bohr fairly quickly produced what he claimed to be rebuttals of the EPR thesis,[2,3] and it is clear that, with very few exceptions, physicists practically at once decided that Bohr's arguments had succeeded yet again in defeating those of Einstein. By this time, of course, Einstein had long since nailed his colours to the mast of opposition to Copenhagen, and was thus almost universally regarded as being completely out of touch with the requirements of modern science. His further criticisms of orthodoxy were taken only as demonstrating yet again his lack of understanding and appreciation; it is doubtful if many physicists of the day bothered to read either EPR or Bohr's response before deciding who was right and who was wrong.

Little changed for several decades. Indeed as late as 1982, Pais[4] totally dismissed the EPR paper; only one phrase, he concluded, was of any interest, and this was interesting only for what it said about Einstein, not for what Einstein actually said! This was striking because Pais at times worked quite closely with Einstein and was a very great admirer of most parts of his work. His biography[4] is, in nearly every way, exceedingly knowledgeable concerning work, life and personality, and includes an exceptionally lucid account of Einstein's contributions to quantum theory up to 1927. His failure even to attempt to understand Einstein's position after 1927 strikes one as utterly extraordinary. Again, even as late as the end of the 1980s, a physicist as eminent as John Wheeler was prepared to agree with Jeremy Bernstein[5] that EPR was a 'nonstarter' and 'just exactly the wrong thing to be asking about'.

Yet by the late 1980s, things were changing, though still perhaps quite slowly. Ten years ago, one of the present authors[6] could still write only that: 'Different as it may have seemed in 1935 and 1965, different as it may largely seem in 1995, I think EPR was a triumph for Einstein'; a decade later one would not have to be quite so hesitant. EPR was the main ingredient of Bell's seminal work in the 1960s. Bell's work also, of course, was slow in gaining attention, but gradually succeeded in doing so, and, at least by the 1980s and 1990s, experimental study of the Bell inequalities was widely recognised as being of considerable importance. Then, from the mid-1990s, both EPR and Bell's work were central in the rise of the hot topic of quantum information theory. Strangely enough, though, there has been a tendency among at least some quantum information theorists to surmise that EPR has been important almost despite Einstein rather than because of his influence, that what has been proved to be significant is not so much the ideas put forward in EPR, as the responses given to them. This particular point is discussed later in this chapter, while a general introduction to quantum information theory is given in Chapter 11.

Even more fundamentally, Bohr's complementarity, which was once accepted in such an unquestioning fashion, has come under increasing criticism. Indeed in 1994, Don Howard[7] suggested that the question now is not whether Bohr's views are pre-eminent; it is the more basic one of whether his philosophy of physics could be given a coherent interpretation at all. Several substantial accounts and clarifications of Bohr's work have been written over the last 25 years.[8-15] The EPR argument and Bohr's response to it are now seen as a third and perhaps crucial round of the Einstein–Bohr debate, following the discussions at the two Solvay conferences discussed in Chapter 5. The EPR paper is fairly generally seen, at the very least, as being of interest and requiring a response, and Bohr's actual response has come under very close scrutiny.[16-19]

While there would still be a range of views on EPR, we have no hesitation in describing it as an exceptionally important and influential paper—influential in helping to clarify many of the fundamental arguments about microscopic processes, influential in bringing attention to the matter of entanglement (to be discussed in detail shortly), influential in leading to the recognition of non-locality as a central aspect of quantum mechanical discussion, and influential in the eventual emergence of quantum information theory. It should rank highly among papers on physics of the twentieth century, and might even scrape into Einstein's top ten!

When we discuss the actual content of EPR, there are tricky historical questions. It seems certain[20,21] that the central idea came from Einstein, who gradually sharpened the basic physical idea in his mind over several years after 1930. He had a deep insight into the conceptual nature of the new quantum theory, but never became particularly fluent at manipulating it for the solution of particular problems. Therefore it must have seemed ideal that, following his move to Princeton in 1933, he established a good working relationship with two younger physicists, Boris Podolsky and Nathan Rosen. The discussions between the three of them were to lead to the EPR paper. While there may have been no clear division of labour, it has been suggested[21] that, with Einstein having provided the general idea, Rosen thought up the specific example used in the paper, while Podolsky worked out the formalism and wrote the paper.

Things did not work out in as satisfactory a way as Einstein might have hoped. The example thought up, presumably by Rosen, was extremely clever, but nonetheless rather awkward from a number of technical points of view. It is no criticism of Rosen to agree with Michael Dickson[18] that the particular example was almost the worst possible for demonstrating clearly what Einstein and his collaborators were interested in, without bringing in irrelevant complications. This is fairly obvious 70 years later, though it certainly could not have been so in 1935. Perhaps even more importantly, Podolsky, who was an expert in formal logic, wrote the paper in a technically rather obscure fashion, introducing a number of complicating issues, notably the infamous 'criterion of physical reality'. Einstein was certainly very disappointed with the result; he complained in a letter to Schrödinger that the matter had been 'buried under learnedness',[21] or in a different translation, 'smothered by the formalism'.[22] Incidentally he was also outraged by Podolsky leaking the publication of the paper to the press, and never interacted with Podolsky after this period, though he remained on close terms with Rosen.

Despite these technical complications, it is still fair to say that, for a sympathetic reader, the paper made its point reasonably clearly. But of course a sympathetic reading was the last thing this paper could expect. Einstein was already regarded by nearly all as unable to appreciate the conceptual steps required to make progress in physics possible. The smothering formalism and technical difficulties merely made it easier for the physicists of the day to disregard the arguments of the paper. Bohr's arguments in reply might well be judged much more obscure still, but people still believed him, not Einstein. Further discussion was very limited, particularly among physicists. As Jammer[20] says, there was rather more discussion by philosophers, but of course this in itself only persuaded further working physicists that the matter was indeed one of armchair philosophy and of no concern at all to them.

Disappointed by this original presentation of EPR, which we shall call EPR35, Einstein returned to the subject at various times, particularly in a paper for a special edition of *Dialectica*[23] (which was reprinted in the *Born–Einstein Letters*[24]) and also in his Autobiographical Notes in the Schilpp volume.[25] His presentations here were very short and clear, and also rather general in approach. Shortly afterwards, in his important book on quantum theory, David Bohm [26] presented a version of the EPR argument involving spin-1/2 particles, which brought out the nature of the argument in a beautifully simple way, avoiding all the technical difficulties of Rosen's original example in EPR35. Bohm's example has been used almost exclusively ever since; sometimes it is called EPR-Bohm, often just EPR. Bohm's example was used by Bell in his important papers based on EPR, and much of the experimental work on Bell's Theorem has used the system of polarised photons, which is mathematically directly analogous to that of spin-1/2 particles. In this book, in an attempt to make clear the central purpose of the EPR argument, we start with Einstein's own presentation, followed by Bohm's example. Only then do we return to consider in detail EPR35, after which we discuss Bohr's reply.

In this introductory section, we make one further point. For many years, the EPR analysis was almost universally given the title of the *EPR paradox*. (In the last few years, this has been perhaps not so common.) In our opinion, this has been highly unfortunate, and helped to make physicists feel even more justified in ignoring the argument actually presented. We would analyse the approach of the original paper as follows. A physical situation is set up and analysed: Various assumptions are made, and the argument comes to a particular conclusion, namely that quantum theory is either incomplete or non-local. These conclusions are at odds with the view of quantum theory held practically universally in 1935 and for many years afterwards. Naturally the assumptions, the argument and the conclusion may all be questioned or criticised from the point of view of complementarity or any other point of view.

Let us contrast this with the image conjured up by the word 'paradox'; it is a proposition leading to an absurd conclusion; so-called paradoxes in recreational books on mathematics might be arguments that $1 = 2$, or that all numbers are even. A claim by EPR to be enunciating a paradox might have been seen as an argument that quantum theory was inherently absurd, or led to incompatible conclusions. This suggestion would have irritated some of those most centrally involved with quantum theory; other physicists may well have felt merely bored, concluding that Einstein was merely constructing a logical conundrum rather than attempting genuine physics. Yet others have felt that EPR had presented a 'puzzle' requiring a 'solution', and that any benefits resulting from the EPR dispute would go to the credit of the solvers rather than EPR themselves. (As we shall see, those giving accounts of quantum information theory often take this last line.)

Where did the word 'paradox' come from in connection with EPR? As discussed by Jammer,[20] it came from a pair of papers by Schrödinger[27,28] in which he discussed and attempted to extend the EPR argument. As Jammer says, Schrödinger often used the word in connection with a result or idea that was merely difficult to understand and appreciate, lying outside our set of general beliefs. Indeed the literal meaning of 'paradox' is 'away from accepted opinion'. With this meaning, the EPR argument, which is that a measurement on one system may have an effect on another sub-system not interacting with the first, could well be called paradoxical. It is quite true that Einstein did refer to the EPR argument as a 'paradox', for example, in his Response to Criticisms in the Schilpp volume;[29] presumably he was using the same definition as Schrödinger, or just following what had become common usage. Rosen[30] much later stressed that, in his opinion, there was no paradox, and we would agree that the use this word was yet another factor that militated against a fair hearing for EPR.

Einstein Locality and Bell Locality

Before discussing Einstein's paper, we discuss the two conceptual elements that are at the centre of the EPR analysis. These are locality, which is explained in this section, and entanglement, which we discuss in the next. It is not argued that Einstein was the first to mention either of these concepts, but it is the case that, for both locality and entanglement, Einstein was the first to draw attention to the implications of what was being discussed. Neither locality nor entanglement was to receive much attention in the period while EPR itself was disregarded, but once EPR was taken more seriously, in Bell's work and in quantum information theory, locality and entanglement were recognised as the crucial aspects of the situation.

First we discuss locality, and for EPR and Einstein's 1948 paper, what is relevant is what we may call Einstein locality. Einstein did not, in fact, use the word 'locality' at all. Rather he took it for granted that one could discuss systems as being situated in different parts of space, and that such systems must behave independently of each other. In the 1948 paper, he discussed what he called the 'principle of contiguity'; this implies that when sub-systems A and B are far apart in space, an influence on one has no effect on the other. This principle, he said, is used throughout field theory; without it, he argued, it would be impossible to postulate and check physical laws, a process which demands separation of the universe into non-interacting sub-systems.

As is well known, Bell sharpened up the concept of Einstein locality to what is usually called Bell locality. Rather than mere spatial separation being enough to guarantee that two sub-systems cannot influence each other, he demanded a space-like separation. In other words, he insisted that, rather than separated sub-systems being considered local, one should talk only of events associated with these sub-systems. Then, if the time between the events and the spatial separation of the sub-systems are Δt and Δx, respectively, Bell insists on the condition $|\Delta x| > c|\Delta t|$. This clearly rules out the possibility that there could be any signal between the two sub-systems and between the two events, and hence any flow of information. (Of course, in the above it is assumed that signals must be luminal or sub-luminal.)

It was the attempt to achieve Bell locality in tests of Bell's inequality that was one of main advances of the work of Alain Aspect in the 1980s over those of John Clauser, Ed Fry and others in the 1970s. Incidentally, it is often implied that what Bell, Aspect and others were concerned about was a perverse and determined attempt by the detection apparatus to frustrate the innocent experimenter. Rather it was the more reasonable worry that, once a measuring system reaches a particular equilibrium, its radiation may act as a signal passing relevant information to the apparatus in the other sub-system.

Entanglement

The concept of entanglement is at the heart of the EPR argument. We have said that Einstein was not the first to think of it, but he was definitely the first to show how large a role it could play in the behaviour of quantum systems, and its importance in arguments concerning the interpretation of quantum theory. Einstein did not name it; this was done by Schrödinger in a famous trilogy of papers he wrote[31,32] in 1935 in response to EPR's argument. Schrödinger famously said of entanglement: 'I would call that not one but the characteristic trait of quantum mechanics, the one that enforces its entire departure from classical lines of thought.'

To explain entanglement, we must consider a system, S_{12}, consisting of two sub-systems, S_1 and S_2 which have interacted in the past but are now separated, so, according to Einstein locality discussed above, can no longer influence each other. Nevertheless, as Schrödinger points out, they do not behave independently; indeed the future behaviour of the two sub-systems is strongly correlated, in the sense that a measurement carried out on the wave-function of either sub-system has an effect on both. To explain entanglement we must first distinguish an entangled state from a product state.

For a product state (which may alternatively be called a separable or non-entangled state), the combined wave-function, ψ_{12} of the total system may be written as a product of a function ψ_1, which depends only on the coordinates of S_1, and a function ψ_2, which depends only on those of S_2, so we have $\psi_{12} = \psi_1 \psi_2$. There is here no mystery over and above what exists for a single system with a particular wave-function. The probabilities of obtaining any particular result in any measurement on S_1 are obtained from ψ_1, those for any measurement on S_2 are obtained from ψ_2, and of course the two sets of probabilities are entirely independent. It is entirely natural just to say that sub-system S_1 has wave-function ψ_1, and sub-system S_2 has wave-function ψ_2.

For an entangled state, on the other hand, the overall wave-function cannot be written as a product, but only as a sum of such products. We may take

$$\psi_{12} = \sum_a c_a \psi_{1a} \psi_{2a} \qquad (6.1)$$

where ψ_{1a} and ψ_{2a} are functions of the coordinates of the particles of S_1 and S_2, respectively. Here we may choose the ψ_{1a} to be a set of eigenfunctions for the operator, $\hat{O}_1$, corresponding to an observable O_1. If we make this choice, we have no further freedom to choose our ψ_{2a}, though, for interesting cases, the ψ_{2a} are likely to be eigenfunctions of an operator for an important observable for S_2, as for the Bohm case below. The same ψ_{12}, however, may be expanded in many other ways. For example, we may choose to replace the ψ_{1a} by a different set of functions ψ'_{1a}, eigenfunctions of an operator, $\hat{O}'_1$ corresponding to a different observable, O'_1, again related to the coordinates of S_1 only. Then we may write

$$\psi_{12} = \sum_a c'_a \psi'_{1a} \psi'_{2a} \qquad (6.2)$$

(This flexibility demonstrated by the passage between Eqs. (6.1) and (6.2) is an important part of most approaches to EPR.)

For an entangled state, it is clear that the results we may obtain in measurements on the two sub-systems are not independent. To take a simple example, suppose we have

$$\psi_{12} = (1/\sqrt{2})(\psi_{1m}\psi_{2m} + \psi_{1n}\psi_{2n}) \qquad (6.3)$$

where ψ_{1m} and ψ_{1n} are eigenfunctions of an operator corresponding to observable O_1 with eigenvalues O_{1m} and O_{1n}, respectively. We suppose that this is a case where ψ_{2m} and ψ_{2n} are eigenfunctions of an operator corresponding to an observable O_2

related to the coordinates of S_2, and that the corresponding eigenfunctions are O_{2m} and O_{2n}, respectively. Let us now imagine we measure O_1 followed by O_2. There is a probability of 1/2 that we will obtain a value of O_{1m} when we measure O_1; in this case, after the first measurement the wave-function collapses to the product state $\psi_{1m}\psi_{2m}$ after the first measurement, and we are certain to get O_{2m} when we measure O_2. However there is also a probability of 1/2 that we will get O_{1n} when we initially measure O_1, and in this case the wave-function collapses to $\psi_{1n}\psi_{2n}$, and we are certain to get O_{2n} when we measure O_2.

Clearly any attempt to form a physical understanding of an entangled state will be far more difficult than for a state of a single system, or a product state of two sub-systems. We shall discuss this point only briefly here, since it will be described in some detail below, as it is the heart of the EPR argument. If we restrict ourselves to a standard interpretation of quantum theory, so the information obtainable from the wave-function may not be supplemented in any way, it seems that the behaviour of the two sub-systems is not independent, as the results obtained in the two sub-systems are correlated, though the state of the system before any measurement cannot explain why. If, however, we allow ourselves the freedom to include hidden variables, it would seem that we may explain the results in a straightforward way by assuming that the experiments on each sub-system merely register the values of quantities which exist before the measurement. (We shall see in Chapter 9 that it is the latter suggestion that Bell shows to be incorrect in his famous work of the 1960s.)

We may ask how easy it is to form entangled states. As implied above, any interaction between two systems is extremely likely to do so. What may be much more difficult will be to maintain the entanglement. The decoherence process, where an entangled wave-function such as that given by Eq. (6.3) decays, so as to behave, not as a superposition, but as a mixture of states, each *either* $\psi_{1m}\psi_{2m}$ *or* $\psi_{1n}\psi_{2n}$, will often occur very rapidly. This is a major problem for quantum computation where maintenance of the superposition is crucial.

Einstein's Version of the EPR Argument

In his paper of 1948, Einstein[23,24] aims to show why he considers any standard interpretation of quantum theory to be fundamentally unsatisfactory, though he admits that the theory itself represents a very important advance in physical knowledge. In the first section of his paper he reviews some of the main ideas discussed in our previous chapter concerning the free particle. Its wave-function, or as Einstein always calls it, its ψ-function, does not provide precise values for momentum or position. Einstein says that there may be two points of view concerning the representation of a single system.

The first is that the free particle actually does have definite values of both momentum and position, even though, as Einstein was rather grudgingly prepared to admit following his bruising encounters with Bohr at the Solvay conferences, they could not be measured exactly and simultaneously. This point of view, which

Einstein says is not one held by the majority of physicists, would imply that the wave-function is incomplete. The concept of *completeness* has been discussed in the previous chapter in particular. To say that the wave-function of a system is incomplete implies that there are elements of reality for the system which are not deducible from the wave-function. For example, it might be the case that a particle does possess a value of position, even though such a value cannot be obtained from the wave-function. If it were discovered that the wave-function was incomplete, there would be an attempt to obtain a complete description, and to find the necessary new physical laws for this description. Einstein adds that the theoretical framework of quantum mechanics would then explode; this was what both he and Bohr had taken for granted in their discussions in the 1920s and 1930s.

The second possibility is that the particle has no definite value of momentum or position; the description by the wave-function is complete. The sharp value of momentum or position obtained at measurement cannot be interpreted as a value held before the measurement. Einstein admitted that this view was held by nearly all physicists at the time he was writing the paper, and rather intriguingly added that it alone did justice in a natural way to the empirical state of affairs expressed in the Heisenberg principle.

He now comes to one of the central points in the paper. He assumes the second possibility, and then considers two wave-functions which differ more than trivially (or, in other words, in more than phase). These must represent two different situations, because the results of measurement of any observable depend only on the wave-function. Even if a measurement of a particular observable *might* give the same result in both cases, the probability distributions for measurements of most or all observables will be different. Einstein insists that this result is also valid for systems containing more than one particle; in the paper discussed, his case is based on its application to systems of two particles, or in general two sub-systems.

In the second section of his paper, Einstein makes a single point, that of Einstein locality, which we have already discussed. We now consider the final section of his paper. Whereas his first section, like the Solvay discussions between Bohr and Einstein, and like the special example considered in EPR35, concentrated on measurement and position, this section, like in fact the main part of EPR35, is far more general in its considerations (except for an example at one point).

In this section Einstein introduces entanglement. As said before, he does not write down explicitly any entangled wave-functions, but makes the point that different measurements on S_1 will result in different forms for ψ_2. For our example above, a measurement of O_1 will leave the overall wave-function in a product state, so we may say that ψ_2 will be either ψ_{2m} or ψ_{2n}. However, had we measured a different observable, O_1', ψ_2 would have been left as a different function; we may say it will be either ψ_{2m}' or ψ_{2n}'.

At first sight, though, this may appear to fall foul of Bohr's argument in his earlier discussions with Einstein. We can, in fact, only perform one measurement; talk of what different measurements will achieve may seem irrelevant. This, though, is where locality comes in. According to Einstein locality, what we do on sub-system S_1 can have no effect on the physically separated sub-system S_2; in particular, whatever measurement we may make on S_1 cannot affect ψ_2. If, after measurement of O_1, S_2 has wave-function ψ_{2m}, then it must have had this wave-function *before* we measured O_1; indeed it must have had this wave-function even if we do not measure O_1, indeed if we make no measurements at all on S_1. But of course, as Einstein stresses, we could make exactly the same argument about any possible result of any conceivable measurement on S_1. It would seem that S_2 has a very large number of distinct wave-functions. But, as he stressed at the outset, different wave-functions must relate to different real situations.

We do not leave the argument here as an awkward conclusion, perhaps almost a 'paradox'. Instead we may note that all the argument really shows is that Einstein's second possibility above, which he used in the above argument, must be incorrect, and so his first possibility must be correct. That, of course, was exactly his aim. Let us see how the argument moves on if we take on board the first possibility. This tells us that the orthodox approach to quantum theory is not correct, and we may assume that all observables have values, even in the absence of any measurement. It is then indeed true that a system may be represented by many different wave-functions.

Let us consider the example of a free particle, and assume it has values of both position and momentum, x_0, and p_0. According to possibility 1, this system may be a member of an ensemble with unnormalised wave-function $\exp(ip_0x/\hbar)$; every member of this ensemble has momentum p_0, but all values of x are represented equally in the ensemble. Equally though, it may be a member of an ensemble with unnormalised wave-function given by the so-called Dirac delta function $\delta(x-x_0)$; in this case every member of the ensemble has position x_0, but all values of p are equally represented.

Einstein's Version of the EPR Argument: Further Consideration

We now discuss Einstein's 1948 argument a little further. It may be suggested that the argument lacks a little conviction because it finishes a little limply with a single system having multiple wave-functions *if* one assumes completeness. EPR35, the Bohm version and most accounts of EPR take a different tack by attempting to show directly that quantum theory as presented in 1925–1926 is incomplete. However they do this by introduction of the EPR *criterion of physical reality*, which certainly requires detailed analysis. We now proceed to adjust Einstein's 1948 conclusion to aim at obtaining his desired result more directly.

This elaboration *requires* that ψ_{12} may be written in two ways, given by Eqs. (6.1) and (6.2). It is required that all four sets of functions included in these equations are eigenfunctions of operators corresponding to simple observables for the particular sub-system to which they relate. We assume that the ψ_{1a} and the ψ'_{1a} are sets of eigenfunctions of operators $\hat{O}_1$ and $\hat{O}'_1$ operating on the coordinates of S_1 and corresponding to observables O_1 and O'_1; and the associated sets of eigenfunctions are O_{1a} and O'_{1a}. Similarly, the ψ_{2a} and the ψ'_{2a} are sets of eigenfunctions of operators $\hat{O}_2$ and $\hat{O}'_2$ operating on the coordinates of S_2 and corresponding to observables O_2 and O'_2; and the associated sets of eigenfunctions are O_{2a} and O'_{2a}. We also assume that there are no common members between ψ_{1a} and ψ'_{1a}, and also between ψ_{2a} and ψ'_{2a}. (We could almost cover this latter point by requiring that neither $\hat{O}_1$ and $\hat{O}'_1$, nor $\hat{O}_2$ and $\hat{O}'_2$ commute, but then we would have to take care to avoid occasional common eigenfunctions which will be possible in some cases.)

We now imagine a measurement of O_1 on sub-system S_1, and suppose that it gives the result O_{1m}. This tells us that the wave-function of the combined system has collapsed to $\psi_{1m}\psi_{2m}$; a measurement of O_2 on sub-system S_2 is now certain to yield the value O_{2m}. However suppose instead that we measure O'_1 on S_1 and that it gives us a result of O'_{1m}. This tells us that the wave-function has collapsed to $\psi'_{1m}\psi'_{2m}$, and a measurement of O'_2 is now certain to give the result O'_{2m}. We may regard each of the possible measurements on S_1 as an indirect measurement of the related observable in S_2.

The fairly obvious riposte of the opponent of EPR-type arguments will probably be that, while we may remark on what will happen if we make either measurement, we are restricted to one or the other. A measurement of O_1 gives us values of both O_1 and O_2, while a measurement of O'_1 gives us values of both O'_1 and O'_2, but we can perform only one of these measurements. This again, though, is where EPR uses the principle of locality. This principle tells us that whatever measurement we perform on S_1 can have no effect on sub-system S_2. Therefore, entirely independently of what measurement we perform on S_1, or indeed whether we perform any measurement at all on S_1, the results of measuring both O_2 and O'_2 are already fixed.

We might be tempted then to say that the next stage of the argument is that, if we know that the result of a measurement of, say, O_2 is certain (although, of course, we will not know what that value is unless we actually measure O_2 itself or O_1), it must already have that value before we perform the measurement. The same would be the case for O'_2, and that would be a direct contradiction with orthodox interpretations, in which it is impossible for O_2 and O'_2 to have simultaneous and exact values. This argument is, indeed, tempting, and would be convincing if it were obvious that, if an observable does have a value before we measure it, the measurement is simply a process of reading off this value. Now it is, of course, the case that in any orthodox interpretation, measurement has a very special role, performing much more than a simple registration of what is already there, and Einstein was obviously highly suspicious and indeed critical of these ideas. But even in the highly realistic hidden variable interpretation of David Bohm,[33] measurement is not in general as simple as a mere registration. It is a deterministic and, in principle, straightforward process in which the measuring device moves to reading a particular value, and the value of the measured observable changes smoothly from an initial value to the one read by the measuring device.

The EPR criterion of reality avoids saying that, just because a measurement of an observable will certainly yield a particular value, the observable must have that value if the measurement is not performed. It says that: 'If, without in any way disturbing a system, we can predict with certainty (i.e., with probability equal to unity) the value of a physical quantity, then there exists an element of physical reality corresponding to this physical quantity.' Indeed one may say that there is an element of reality in the mere fact that the result of a future measurement is known; it is in complete contrast with Copenhagen, where, for example, if the wave-function is not an eigenfunction of momentum, the result of a measurement of momentum is certainly not pre-determined. As so often, Bell[34] puts things very well: 'Since we can predict in advance the result of measuring $[O_2, O'_2 \ldots]$, by previously measuring $[O_1, O'_1 \ldots]$, it follows that the result of any such measurement must actually be predetermined. Since the initial quantum mechanical wave function does not determine the result of an individual measurement, this predetermination implies the possibility of a more complete specification of the state.' This conclusion should be generally satisfactory for Einstein.

Our argument so far has followed the earlier debates between Bohr and Einstein at the Solvay Conferences in that one has considered two observables, O_1 and O'_1, and tried to show that, despite the fact that $\hat{O}_1$ and $\hat{O}'_1$ have no common eigenvalues, the two observables do both possess values at all times. Though there are many differences between the early arguments and the EPR case, EPR subtly attempting to overcome the difficulties of the previous arguments, and aiming to establish simultaneous existence rather than simultaneous measurement of the two observables, there was this element of continuity in EPR35 and in the great majority of presentations of all versions of EPR ever since.

The fact that two measurements take place constitutes, if not exactly a weakness of the EPR case, at least a point easily misunderstood and liable to be denounced as 'the source of the paradox'. Just as an example, Bernstein[5] says that: 'What the EPR paper does not state clearly is that these measurements are separate and *distinct*, and involve different modifications of the apparatus. In no single measurement can one measure *both* the position and momentum with arbitrary precision. If one uses the EPR setup to measure the position, then the momentum becomes completely uncertain, and vice versa. If one insists on using the EPR language one can say that one measurement has conferred "reality" on the position and the other has conferred "reality" on the momentum, but no experiment can confer "reality" on both if one believes in the correctness of quantum mechanics.'

One may remark that this misses the point of the locality condition, which implies that it is the possibility of performing *either* measurement that is essential, rather than the necessity of performing *both*. One might add that the 'scare quotes' around 'reality' seem inappropriate from any philosophical viewpoint; and that what would be disproved by the simultaneous reality of the momentum and position observables would be not quantum mechanics, but its Copenhagen interpretation, which was exactly Einstein's argument. The point being demonstrated here, though, is that the mere mention of two measurements in any EPR argument does cause conceptual confusion. Part of this confusion, in fact, started with Bohr's original arguments against EPR35, as we shall see shortly.

It is interesting, then, that, as has been pointed out by a few people,[19,35–37] that making use, even hypothetically, of two measurements is actually unnecessary. Assuming ψ_{12} is not an eigenfunction of $\hat{O}_2$, Copenhagen would insist that O_2 would not have a value before measurement, and also the value that would be obtained for it if a measurement were made would not be certain. Since EPR denies the latter point, there is a conflict between it and Copenhagen; if EPR is correct, Copenhagen is wrong. We could, of course, rerun the whole argument with $\hat{O}'_2$ and O'_2 replacing $\hat{O}_2$ and O_2, but the point of the argument is that only one of the pairs is required; one does not even need to mention the other. As we have implied, this form of argument is simpler than the more conventional one, and it seems advantageous that it avoids the possibility of criticism of the argument over the two measurements, although, as we have shown, we believe the latter criticism to be broadly misguided.

We now discuss briefly an aspect of Einstein's terminology in the 1948 paper. In his discussion of locality, he expresses his belief in a space–time continuum, and in bodies, fields and so on with a 'real existence' in this space–time continuum in a 'real outside world'. He also discusses which of his points of view (incompleteness or completeness of the wave-function) represents the 'real state of affairs'. One might well wonder to what extent Einstein has imposed his belief in realism on the ideas in the paper, rather than it being the conclusion of his arguments. This question is of some importance because an extremely common view of EPR, particularly among those whose views may have been obtained at second-hand, is that in EPR Einstein was imposing his beliefs of determinism and realism on the physics, and when the physics failed to measure up, he called it a paradox! A mild example might be from the excellent text on quantum information theory by Nielsen and Chuang,[38] who remark that: 'The attempt to impose on Nature *by fiat* properties which she must obey seems a most peculiar way of studying her laws.'

In fact, though, in this particular paper he assumed nothing about reality that would not have been accepted by his Copenhagen opponents. His use of real objects in space-time was effectively restricted to measurement devices being in one or other of the spatially separated sub-systems. Also implicit in the argument is the belief that apparatus settings and measurement results are real and to be considered classically. In saying this we must realise that Einstein's conclusion of different wave-functions for the same system was essentially to be taken as a statement of different sets of future measurement outcomes. All this would be quite allowable to Bohr, for whom measurement was classical, and macroscopic objects could also be treated as classical, albeit, as we saw in the previous chapter, subject to the Heisenberg relations, and existing in space-time. Instead of asking what the 'real state of affairs' is, he might just as easily asked which of the points of view is correct, or even whether position and momentum are both real—where realism is clearly the (possible) deduction of the argument, not the premise.

Lastly in this section, we review briefly what arguments there were concerning Einstein's 1948 version of EPR. The great majority of comment on EPR have, of course, been on EPR35, and we will review them after discussing EPR35 itself. Here we consider comments of Einstein himself at the end of the paper, and of

Born, to whom Einstein sent a copy of the paper, and who replied to it in a letter of 1948 to Einstein. The reply is included in *The Born–Einstein Letters*, together with further discussion from the 1960s.

Einstein's brief remark is to say how orthodox physicists would react to the contents of his paper. He is in no doubt, he says, that they would drop the requirement of locality, which, as he says, quantum theory certainly does not require. This is a fascinating comment of Einstein, because it actually runs completely counter to what one would have thought the orthodox reply *should* have centred on, which would be the possibility even of considering the alternative hypothetical measurements; this is exactly what Bohr did argue about in his reply to EPR35.[2]

Einstein's comment may suggest that even by 1948, it was realised that Bohr's response was a broadly correct application of the Copenhagen approach, but nevertheless it was still felt to be generally unconvincing. It also shows absolutely clearly that, at least by 1948, Einstein was very clear that locality was a key aspect of the EPR argument. (It is also stated in EPR35 although not with the same emphasis.) This is not always realised. Steane,[39] for example, in his excellent fairly early review of quantum information theory, comments that: 'The EPR paper is important because it is carefully argued and the fallacy is hard to unearth.' We note the word 'fallacy' appears to suggest that Einstein had produced a 'paradox' he was unable to resolve, and the resolution had to wait for others. Steane then says that the 'fallacy' can be removed in one of two ways, of which the first is essentially locality, the implication being that Einstein was not aware of the significance of this part of the argument. The second of Steane's ways of removing the 'fallacy' concerns the information content of a quantum system, and this part of Steane's comments will be discussed in our general assessment of EPR later in this chapter.

We now turn to Born's reply to Einstein's article in *Dialectica*. Again, perhaps somewhat surprisingly, Born does not use the kind of argument that for many years was assumed to have won the debate for Bohr, the straightforward use of complementarity. Neither does he pick up the suggestion of Einstein about locality. Rather he says that objects far apart in space may have a common origin so that a measurement on one may give information on both. Born uses an example from optics in which a beam of light is split into two beams by a doubly refracting crystal; a measurement of the polarisation of one beam also tells us that of the second beam. He seems to believe this provides a straightforward victory over Einstein's argument, but in fact it demonstrates only Born's failure to understand it. Bell[40] puts this particularly clearly: 'Misunderstanding could hardly be more complete. Einstein had no difficulty accepting that affairs in different places could be correlated. What he could not accept was that an intervention at one place could *influence* immediately affairs at the other.'

The Bohm Version of EPR

Having studied in some detail Einstein's own approach to the EPR argument, we now turn to the 1951 version of David Bohm.[26] This version may be regarded as a very simple example of Einstein's approach. It is a straightforward argument logically and physically, and it avoids the technical complexities and conceptual difficulties of EPR35, which is discussed in the following section. Bohm's version was the stimulus for Bell's developments of the 1960s, which were themselves the basis for most of the work, theoretical and experimental, on the conceptual foundations of quantum theory carried out subsequently, and also for many of the developments in quantum information theory. All this work has made great use of the Bohm argument, though nearly all has been carried out, not using spins, as in Bohm's original work, but polarisations of photons, the mathematics of which is completely analogous. The Bohm version may be called EPR-Bohm, but it has taken over discussion of the field to such an extent that it is often just called EPR.

In this version, one imagines a spin-0 system which has decayed into two spin-$1/2$ particles travelling in opposite directions. The resulting state may be written as:

$$\psi = \frac{1}{\sqrt{2}} \{s_{z+}(1)s_{z-}(2) - s_{z-}(1)s_{z+}(2)\} \tag{6.4}$$

Here (1) and (2) denote the two spins. It is obvious that the particles are entangled. One will be found to have $s_z = +\hbar/2$ in any subsequent measurement, the other $s_z = -\hbar/2$, but the state-vector does not tell us which will be which.

It should also be noted that the spin part of the state-vector both before and after decay is spherically symmetric. No direction is singled out as special. This is not obvious from the form of Eq. (6.4), where arbitrarily the z-axis has been treated as special, but in fact ψ in Eq. (6.4) may be written in exactly the same mathematical form, but with z replaced by x or y or any other direction. This expresses the physically obvious point that in whatever direction one measures the spin of one particle, if one measures the spin of the second particle in the same direction, one will obtain the opposite result. This part of the argument is a simple example of the flexibility between Eqs. (6.1) and (6.2) discussed earlier in this chapter.

From the ideas in the previous section, it is clear how the argument may be concluded. The most common way is to say that a measurement of s_z on particle 1 would provide information also on the result that would be obtained in a measurement of s_z on particle 2. However since the measurement on particle 1 cannot have had any effect on particle 2, if one assumes locality, the value that would be obtained for a measurement on this particle must have been predetermined. One can make an analogous argument for a measurement of s_x on particle 1. The implication is that, before any measurement, or even in the absence of any measurement, the values of measurements of both s_z and s_x on particle 2 are known. Since there are no common eigenstates of s_z and s_x, this is an argument that quantum theory is not complete. (We could, of course, use any other pair of components of spin.)

All the points we made about the general *Dialectica* argument apply here. First, we should be careful to say merely that the results of measurement are predetermined, not that the observable already has that value, because of the possibility of the disturbance due to the measurement. Secondly, Einstein would prefer to say that different measurements on the first particle would leave the second particle in two different states, represented, for example, by $s_{z-}(2)$ and $s_{x-}(2)$. Thirdly, rather than even discussing two different measurements on particle 1, we may merely say that the result of a single measurement is predetermined, and this itself is an argument against completeness, since ψ in Eq. (6.4) is not an eigenstate of the operator for any component of spin. Again we do not even need to mention more than a single measurement.

The Original EPR Argument of 1935

Having looked at the EPR argument in the form Einstein preferred,[23] and the Bohm version,[26] which we regard as an excellent example of Einstein's approach, we now rejoin the historical route by studying the original EPR paper of 1935, which we refer to as EPR35.[1] It will be remembered that Einstein was unhappy with the presentation of this paper, and that the example chosen and analysed had several awkward technical features. Having discussed EPR35, we then proceed to analyse the responses to it, particularly those of Bohr,[2,3] and also several comments on Bohr's argument itself.[16–19]

EPR35 is a short paper consisting of two sections, each of two pages. The first section starts by discussing the idea of completeness, and the EPR criterion of reality, which we have already met. It also contains some fundamental arguments and assumptions of quantum theory. It explains the eigenfunction–eigenvalue link, the basic point in orthodox quantum theory that, if the wave-function of a system is equal to an eigenfunction of an operator $\hat{O}$ corresponding to an observable O with associated eigenvalue O_n, then a measurement of O will certainly lead to the value O_n. The EPR criterion then tells us that we may speak in this sense of O having a value before measurement, of O being real. It has a value both in the theory and in physical reality. If, however, the wave-function is not equal to one of the eigenfunctions, we cannot predict using quantum theory what value will be obtained in a measurement of O. We may say that O has no value in the theory. If one assumes that quantum theory is complete, one therefore cannot speak of O having a value prior to measurement. We must just say that it does not exist in physical reality.

When we use the phrase 'eigenfunction–eigenvalue link' we include the content of the whole of the previous paragraph; in particular we specifically include the point that, if the wave-function is *not* an eigenfunction of the operator, the corresponding observable does not have a value.

This eigenfunction–eigenvalue link is a central part of what EPR are questioning. The link is violated if one has any kind of hidden variables present. The term 'hidden variable' means exactly that such a variable exists in physical reality though it is not in the theory. We stress this because we shall meet so-called solutions of the EPR 'paradox' which effectively propose the violation of the link; clearly, far from solving any conjectured dilemma on the part of EPR, these arguments are actually supporting the EPR analysis. Again we note how unfortunate the word 'paradox' has been. Convinced that EPR had reached a contradiction in their analysis and had retired baffled, many commentators are keen to show what points they had misunderstood, rather than trying to appreciate what their argument actually was.

The EPR35 argument then continues by saying that, since the operators for measurements of position and momentum do not commute, there can be no common eigenfunctions, and thus it follows that there are no circumstances in which both position and momentum have physical reality. (It will be recognised that the argument misses one point. While the operators for position and momentum do not have any common eigenfunctions, this does not follow directly from their not commuting. Operators for different components of angular momentum, for example $\hat{L}_x$ and $\hat{L}_y$, do not commute, but do have some common eigenfunctions, those with quantum number ℓ, equal to zero. The stipulation is that they cannot have a *complete set* of common eigenfunctions.)

Interestingly, EPR35 admit that the orthodox view is 'entirely reasonable' because what it pronounces real is precisely what may be measured without disturbing the system. Thus Einstein acknowledges Bohr's success in the discussions at the Solvay conferences. Nevertheless EPR35 aim at showing that, even though there may be no simultaneous eigenfunctions of $\hat{x}$ and $\hat{p}$, they can demonstrate that both x and p do exist in the physical reality, and so quantum theory is not complete. To sum up the first section of the paper, one may say that it describes the relevant general quantum background efficiently. By no means is the argument presented obscurely, but it may be said that the criterion of reality and one or two of the supporting statements are made in a rather pedantic way. Compared to Einstein's fairly relaxed account in his *Dialectica* article, the argument does seem more of an exercise in logic, and this is part of what may have disappointed Einstein.

The second section of the EPR35 paper starts with an account of what is now called entanglement, though it is to be remembered that it was only given this name by Schrödinger shortly after the publication of the EPR paper. This is much along the lines of Einstein's later paper, and indeed the conclusion is precisely that different measurements on particle 2 lead to different wave-functions, both though, if one assumes locality, relating to the same physical reality. These might be ψ_{2m} and ψ'_{2m} in our notation earlier in the chapter. As we said earlier, Einstein is entitled to argue that different wave-functions are related to different physical realities.

It is only at this point, very nearly at the last page of the four page paper, that the authors comment that ψ_{2m} and ψ'_{2m} may be eigenfunctions of two non-commuting operators corresponding to different observables for particle 2. More completely we may say that it may be the case that they are members of complete sets of eigenfunctions of the operators. Different members of each set will be the result of collapse of wave-function following a measurement on particle 1. Different results of the measurement of O on particle 1 will lead to different members of the first set of functions; different results of measurement of O' will lead to different members of the second set. This, as we have said, is much the most usual way in which EPR is discussed, the Bohm case being a good example.

However the example given in EPR35 is very much more complicated and in some ways more problematic than, for example, that of Bohm. This point was extremely important in the response to the paper, because discussion of EPR35 has been almost entirely in terms of the particular example, with its admitted problems, rather than the more general considerations that preceded it in EPR35, and constituted the *Dialectica* article. At the beginning of Bohr's main reply[3] to EPR35, for example, after a brief mention of the EPR reality criterion, he wrote that EPR made their point 'by means of an interesting example'. The rest of the paper consisted of Bohr's response only to the example.

We now consider this example, starting with a simple picture of it, which certainly explains the general idea, though, as Dickson[18] points out, cannot be made rigorous. In this picture, a source emits two particles of equal mass; in the quantum mechanical sense there is a distribution of values of the momentum of each particle, but we do know that the momenta are equal and opposite. After some time, a measurement is carried out on particle 1, either of position or momentum. If its momentum is measured as p, we also know that the momentum of particle 2 must be $-p$; if its position is measured as $x_s + x$, where x_s is the position of the source, we also know that the position of particle 2 must be $x_s - x$. But since, from the usual locality argument, our measurement on particle 1 cannot have influenced the state of particle 2, the latter must have had both these properties all along, not merely as a result of the measurement on particle 1. Therefore, since quantum theory cannot provide values for both these quantities simultaneously, the theory cannot be complete.

As Dickson points out, this nice picture cannot be maintained rigorously. If EPR had made this verbal argument, Bohr could have replied that, for the analysis to hold, since the positions and momenta of the particles are defined only relative to the source, one would have to know the positions and momenta of the source precisely. This is what the earlier discussions of Bohr and Einstein had shown to be impossible.

In fact EPR35 provide a mathematical treatment containing the coordinates of the particles though not of the source. Their wave-function is

$$\psi(x_1, x_2) = \int_{-\infty}^{\infty} \exp[(2\pi i/\hbar)(x_1 - x_2 + x_0)p]\mathrm{d}p \qquad (6.5)$$

where x_0 is a constant. Clearly this is an entangled wave-function; the only essential difference from examples we have met before in Eqs. (6.1) to (6.4) is that in this case there is a continuum of entangled wave-functions, each a function of x_1 and x_2, and labelled by a particular value of p; thus the sum that occurs in the earlier cases is replaced by an integral over p. (Because it is a continuum case, EPR35 do not normalise the functions throughout their argument; this has no effect on any conclusions.)

Each term under the integral can be written as a product of a function of x_1 and a function of x_2. The function of x_1, $\exp(2\pi i x_1 p/\hbar)$, is an eigenfunction of $\hat{p}_1$ with eigenvalue p, while the function of x_2, $\exp(-2\pi i x_2 p/\hbar)$ is an eigenfunction of

$\hat{p}_2$ with eigenvalue $-p$, for any value of p. (Again we are ignoring normalisation constants.) Thus a measurement of p_1 with result p_r will cause a collapse of wave-function to $\exp[(2\pi i/\hbar)(x_1 - x_2)p_r]$ (again treating the other component of the term in Eq. (6.5) as part of a normalisation constant which is ignored). This means that a measurement of p_2 is certain to yield the value $-p_r$, or in other words the direct measurement of p_1 may be said to serve also as an indirect measurement of p_2. But, as we have said several times before, if we take into account Einstein locality, the measurement of p_1 has had no influence on particle 2, so its value of momentum must have been an element of reality before the measurement of p_1, or, of course, even if no measurement was made.

However the wave-function of Eq. (6.5) can also be written as

$$\Psi(x_1, x_2) = \delta(x_1 - x_2 + x_0) \tag{6.6}$$

It is a Dirac delta-function, equal to zero except if $x_1 - x_2 + x_0 = 0$, in other words except if the distance between the two particles is x_0. This equation may also be written as

$$\Psi(x_1, x_2) = \int_{-\infty}^{\infty} \delta(x_1 - x)\,\delta(x_2 - x_0 - x)\mathrm{d}x \tag{6.7}$$

(Again in Eqs. (6.6) and (6.7) we are ignoring normalisation.) This way of writing the wave-function is precisely in the usual EPR entangled form we have become used to. Each term in the integral is a product of a function of x_1 and a function of x_2. The function of x_1 is an eigenfunction of $\hat{x}_1$ with eigenvalue x; the function of x_2 is an eigenfunction of $\hat{x}_2$ with eigenvalue $x + x_0$. Thus a measurement of x_1 giving a value of x_r, will collapse the wave-function to a single term in the integral of Eq. (6.7), and is thus also an indirect measurement of x_2 giving the value $x_0 + x_r$.

Again we may say that, according to locality, the measurement on particle 1 can have had no effect on particle 2, so the value of x_2 must have been an element of reality before the measurement, or even in the absence of any measurement. The argument is then that the values of both x_2 and p_2 exist before any measurement is performed, so of course exist even if no measurement is performed. Since quantum theory cannot provide simultaneous values of the two quantities, the theory is not complete.

As we have mentioned in previous sections we will say that, even in its own terms, the argument is unnecessarily complicated. Since $\psi(x_1, x_2)$ is an eigenfunc-tion neither of $\hat{x}_2$ nor of $\hat{p}_2$ if quantum theory is complete, *neither x_2 nor p_2 should have values before any measurement*. The fact that *either* has a value and thus is an element of reality is enough to show that quantum theory is incomplete; one does not require both. Indeed one does not even need to mention the second observable. One may well understand[19] why EPR35 used the argument with two observables as we described it above. It may have seemed a natural development from the previous discussions between Einstein and Bohr, in which, of course, the use of two observables was the whole point of the argument. In the EPR case, use of the two observables may have made the point more dramatically; it perhaps made the point more clearly because it was not necessary to mention the wave-function at all. And certainly since the authors were happy that the argument worked, there was clearly no reason for them to refrain from making it.

Nevertheless, use of the two variables by all versions of EPR turned out in fact to limit even further any positive response that could be expected. We shall see in the following section that Bohr's response, based on complementarity, was totally dependent on measurements of two observables being discussed. Other commentators, arguing from more physical principles than Bohr, were still likely to find the argument confusing, even in the comparatively rare cases where they might be well motivated in principle to Einstein's views. In the majority of cases, where the reader was at the outset prejudiced against Einstein's position, the presence of the two observables in the argument made it complicated enough from a logical point of view that it could be easily dismissed with a minimum of consideration.

We now turn to the specific example discussed in EPR35, and we immediately realise that there are technical difficulties. Neither plane wave states, with wave-functions of the form $\exp(ikx)$, nor states with wave-functions given by Dirac delta-functions are easy to handle rigorously in quantum theory; indeed one may say that, according to the standard mathematical formalism of quantum theory, neither is an acceptable quantum state. (Recently Halvorson[17,41] has extended the formalism by use of $C*$ algebra, and has been able to represent the EPR state in this extended formalism.)

Jammer[20] has provided a lengthy account of published views on EPR up to the publication of his book in 1974. Several criticisms of EPR included here are specifically of the particular mathematical formalism. For example, Lionel Cooper[42] argued that, for Einstein locality to be upheld, each particle must be restricted to half the x-axis; the result is that the whole machinery of eigenfunctions and eigenvalues then fails! Einstein replied that his view of locality did not require a physical division between the two particles, but it must be admitted that it is awkward to conceive of any form of locality between particles whose wave-functions extend over all space.

Another criticism, that of Paul Epstein,[43] pointed out the lack of time-dependence in the EPR35 wave-function, and claimed that including time-dependence would require an assumption that the energy of the system was independent of its momentum. In a rather similar point, Dickson[18] comments that the EPR state would spread under the action of any real Hamiltonian. Jammer[20] reports Rosen's response given to Jammer himself in 1967, to the end that time-dependence is irrelevant—at a certain time a state with the given properties exists, and that is all that is required. While Rosen's point of view is certainly correct, the objection demonstrates a certain physical obscurity in the example, which again may have contributed to unease. Again Dickson[18] points out that, although, from the point of view of Hilbert space, position and momentum are equivalent being Fourier transforms of each other, from the dynamical point of view they are certainly not; momentum is conserved, while position is not. This adds to the conceptual awkwardness of the example, when one compares it, for instance to the later Bohm example.

It would be easy to say that the lesson to be learned from EPR35 was clear, and to accuse its critics of invoking rigour merely to avoid the conclusions of the paper. This would probably be unfair though; EPR felt they were making an important point, and it was only right that their argument should be analysed from different viewpoints. Should it be correct, it would presumably emerge from fair criticism in a stronger state. The real response to criticism of the specific example of EPR35 should presumably have been that one should concentrate on the general arguments of the paper, rather than the admitted problems of a particular example. But as we have said this was not done, not at least until Bohm produced his own example 16 years after EPR35. For nearly everybody, EPR *was* the position-momentum example, and, to the extent that that example was seen to have failed, so was EPR.

Bohr's Response to EPR35

The most important response to EPR35 was bound to be that of Bohr. We have seen that the EPR paper was a continuation of the dialogue between the two men that had taken place at the Solvay conferences. EPR provided a fresh challenge to Bohr, and the question now was whether Bohr was up to it. His published responses[2,3] certainly showed that he was believed he had defeated EPR's argument, and the overwhelming majority of physicists of the time and for several decades afterwards accepted his opinion. His main response[3] is fairly generally reckoned today to be remarkably obscure, so one must surmise that orthodox opinion relied more on a belief in the general superiority of Bohr's position than on detailed study of the various documents.

Indeed even 70 years later, one feels an uncertainty of what Bohr was actually trying to achieve. It will be remembered that, in his previous discussions with Einstein, which were described in Chapter 5, Bohr had not made use of complementarity. Rather his arguments had used detailed physical analysis of how the various combined measurements involved in Einstein's gedanken-experiments would actually be carried out. Thus he was able to show, again from fundamental physical arguments with which Einstein could not disagree, that Einstein's ideas were not successful. The analysis certainly fitted in well with complementarity in that just those experiments which complementarity would not allow were found to be impossible to carry out.

Yet, as we said earlier, in no sense did Bohr's winning of these arguments show that it was impossible to reject complementarity. Any argument involving hidden variables could avoid the conceptual implications of complementarity, though not, of course, making possible what Bohr and Einstein in these debates agreed was impossible, the combined measurement of, for example, momentum and position, to an accuracy greater than that allowed by the Heisenberg principle.

We now wish to consider[19] whether Bohr's argument against EPR was of the same type, that is to say based not on complementarity but on fundamental physics, or whether it was actually based on complementarity. If it was the former, then, as for the earlier debates, and assuming Einstein could not challenge Bohr's arguments, he would have to admit that his ideas in the EPR paper had also failed. However, if Bohr was found actually to have made use of complementarity in his analysis, then, again to the extent that his actual arguments were successful, Bohr himself could claim that complementarity avoided the conclusions of the EPR argument; this could be described as a successful defensive strategy on Bohr's part. But of course his argument could in no sense affect Einstein's position, because Einstein would reject the use of complementarity in the first place. It may be said that this rather obvious argument seems to have largely been overlooked in the 1930s and following decades. In part this may have been because Bohr's insistence that complementarity was the only available approach had been generally believed, so few would have questioned his right to use it if he wished. The von Neumann argument also seemed to undermine Einstein's position, at least in a general sense.

But we must return to the actual question of what Bohr did intend. It may be appropriate to mention that, according to Léon Rosenfeld,[44] who was then his assistant, and became, for the rest of his life, complementarity's strongest advocate: 'This [EPR] onslaught came down upon us like a bolt from the blue. Its effect on Bohr was remarkable.' It was clear that Bohr did feel it seemed possible that his position could be undermined. Yet, in his response to EPR, Bohr[3] was to comment that: 'The special problem treated by Einstein, Podolsky and Rosen . . . does not actually involve any greater intricacies than the simple examples discussed above', the simple examples being the problems discussed at the Solvay conferences. Later in the Schilpp volume,[45] he was to write that: 'We are dealing here with problems of just the same kind as those raised by Einstein in previous discussions.' The clear implication is that, since the Solvay problems did not require complementarity for their solution, neither did the problem raised by EPR.

Yet other evidence would suggest differently. We shall see that Bohr's response[3] commences with a statement that he will 'use the opportunity to explain in somewhat greater detail a general viewpoint, conveniently call "complementarity"', surely an indication that complementarity was to be central in his argument. Rosenfeld,[44] incidentally, says that: 'The refutation of Einstein's criticism does not add any new element to the conception of complementarity', and indeed he characterises the EPR argument as no more than a 'misunderstanding'. This would all suggest that Bohr and Rosenfeld had assimilated the ideas of complementarity to such an extent that they regarded its truth as assured, its use in any argument as beyond criticism or discussion, and any contrary arguments as no more than misunderstanding. As we have already said, such a position was totally unjustified.

There is another interesting remark in Bohr's essay in the Schilpp volume.[45] He comments that, from the viewpoint of complementarity, all the 'apparent inconsistencies' of EPR could be completely removed. The meaning would appear to be that the product of EPR was just confusion, confusion among EPR themselves as much as that of other commentators, and confusion that complementarity could remove. But, of course, EPR by no means saw themselves as raising difficulties and effectively asking others to sort them out, though, as discussed earlier, this idea could have been encouraged by the highly unfortunate use of the word 'paradox'. Rather EPR regarded their paper as making arguments against the whole set of ideas including complementarity. Yet again there seems to have been a lack of understanding on Bohr's part of what a successful argument against EPR would actually have to achieve.

Let us turn to Bohr's actual argument.[3] An interesting remark early in his paper is that an argument of EPR type 'would hardly seem suited to affect the soundness of quantum-mechanical description, which is based on a coherent mathematical formalism covering automatically any procedure of measurement'. The message would seem to be that, as long as one follows the mathematics correctly, all predictions we make must be correct. Yet[19] in fact it is EPR who largely stick to the mathematics. They do have to go beyond it, of course, in bringing in the ideas of locality, and the piece of logic concerning the two possible measurements, but they do not need to discuss any details of carrying out measurements, which was, of course, essential for the earlier discussions. It seems strange that, despite his comment above, it was Bohr who felt it necessary to discuss the physics involved in setting the experiment up and performing the measurements. Indeed, in his paper, apart from some manipulations starting from the Heisenberg principle at the beginning of the paper, Bohr restricts himself, as would be his usual practice, to physical analysis of thought-experiments.

Of course one reason could conceivably have been that Bohr was dubious that the EPR wave-function could be created. We saw in the previous section questions raised about this. However it seems that this was not the case. One of Bohr's first actions is to describe an experimental arrangement by which he suggests this can be achieved, involving two narrow slits in a diaphragm, through which the two particles pass.

Indeed, it is interesting how many even of the more sensitive points of the EPR analysis that Bohr accepts without hesitation. He accepts the general way in which EPR measure an observable belonging to particle 2 by means of a measurement on particle 1; we have called it an indirect measurement. This concept could conceivably be challenged by complementarity, on the grounds that one should not discuss a property of particle 2 without an experimental arrangement in place to measure it on particle 2. Acceptance of the idea of indirect measurement implies that Bohr is also effectively accepting the projection postulate, which EPR use in their analysis of this point. It is certainly true that Bohr could not have found this principle attractive, since he thought of measurement as entirely classical in nature. He did not use the projection postulate himself, but here he was seemingly prepared to accept its use, at least as a procedure which gave correct results.

Incidentally, Bohr also accepted the occurrences of momentum and position as completely symmetrical. He agreed with Bohr that one was faced with 'a completely free choice whether to determine the one or the other [of the two quantities] by a procedure which does not directly interfere with the particle concerned'. Lastly he accepts at least some aspects of Einstein's idea of locality; he says that there is no question of a measurement on particle 1 causing a mechanical disturbance of particle 2 (though he qualifies this statement in an important way, as we shall see).

Having established the virtually total agreement between Bohr and Einstein over single measurements of position or momentum on particle 1, we now reach the area where the agreement stopped – the question of the two measurements, however one may construe the significance of the two measurements and the connection between them. In his paper, Bohr has already discussed at some length the kinds of argument he had made at the Solvay conferences, which demonstrated that simultaneous measurement of position and momentum on a single particle is impossible.

He now repeats them for the EPR case, and says that our 'freedom of choice' is concerned with 'a discrimination between different experimental procedures which allow of the unambiguous use of complementary classical concepts'. Bohr argues that by carrying out a measurement of position of one of our particles 'we have by this procedure cut ourselves off from any future possibility of applying the law of conservation of momentum to the system'. Similarly, he says, that, if we choose to measure the momentum of one of the particles, we will have 'no basis whatsoever for predictions regarding the location of the other particle'.

The argument is indeed exactly of the form that Bohr found successful in his earlier discussions with Einstein. Despite his mention of 'complementary classical concepts', he has not, it would seem, explicitly used the ideas of complementarity; the latter is really just a useful set of ideas to help to explain what he has deduced independently of it. But it seems clear[19] that the argument fails to deal with the subtleties of the EPR argument. The possibility of making a hypothetical measurement of the position of particle 1, according to EPR, establishes the fact that particle 2 has a position after the measurement, therefore had it before the measurement, therefore has it even if we have no intention of making the measurement in the first place! There is no way in which we need to discuss following that measurement with a second measurement of the momentum of particle 1.

Similarly, the mere possibility of performing a measurement of the momentum of particle 1 leads us to accept that, at all times, particle 2 has a value of momentum. Again we have no need to contemplate following this measurement with one of the position of particle 1. Thus the argument is that particle 2 has values of position and momentum, exactly as EPR hoped to show, and it seems that Bohr's argument is powerless against their analysis.

At this point, though, Bohr introduces another condition on what he means by physical reality. He demands that, for 'description of any phenomenon to which the term "physical reality" may be properly attached', an 'inherent element' consists of 'the very conditions which define the possible type of predictions regarding the future behaviour of the system'. These conditions, Bohr says, are influenced by the choice of measurement on particle 1. Actually in his shorter preliminary response,[2] he puts things in a slightly different way, one that perhaps makes his ideas clearer. Here he writes that the measurement influences 'the conditions on which the very definition of the physical quantities in question rests'.

It would seem that, for Bohr, for an entity to have physical reality, indeed even for it to be subject to definition, we must be able to use it to make predictions regarding the future behaviour of the system. Thus Bohr speaks of two experimental procedures permitting the unambiguous *definition* of physical quantities. For Bohr, measuring the position of particle 1 constitutes one experimental procedure which allows the positions of both particles to be defined. The values of the positions of the two particles may then be used in predicting the future behaviour of the system. Measurement of the momentum of particle 1 is a second experimental procedure, which allows the momenta of both particles to be defined, and the values of the momenta may be used in predicting the future behaviour of the system. Clearly the two experimental procedures are mutually exclusive; only one of the two measurements on particle 1 may be performed, as was shown in the earlier discussions between Bohr and Einstein.

Thus if we follow the path of measuring position, the momenta of neither particle is even defined; we may certainly not follow EPR's procedure of discussing the values of the momenta of the two particles, or even of trying to establish that they are elements of physical reality. Such a procedure is ruled out. Similarly if we follow the path of measuring momentum, we may not even discuss the positions of the particles; they are not defined. We thus see that EPR's analysis does not meet this challenge of Bohr; thus, from Bohr's point of view, the kind of argument they make in their paper is rendered illegitimate from the very start.

What though is Bohr's justification for seeking to impose this condition? In the Solvay discussions, we said that complementarity was not imposed on the physics, but something that should dovetail with the physical argument. Here though it is clearly dictating the way forward. Complementarity tells us that we may discuss a physical quantity only in the presence of an experiment to measure it. (Alternatively one may say that one should only discuss the results of measurement, not values the measured quantity may have had in the absence of measurement.) Bohr allows the concept of indirect measurement; in other words he allows one to speak of a measurement apparatus set up to measure the momentum of particle 1 also measuring that of particle 2. But since one cannot measure the momentum and position of particle 1 simultaneously, one must restrict oneself to discussing *either* the position of both particles, *or* their momenta. One cannot discuss the results of both experimental procedures simultaneously.

Indeed Bohr talks of the methodology 'permitting the unambiguous definition of complementary physical quantities', and he argues that it is essential for providing the possibility of the new physical laws of quantum theory. Complementarity, he says, aims at characterising this new situation for the description of physical phenomena. Yet he may have felt that the response rendering EPR out of order on grounds of general principle might be felt by the uncommitted to be a little bald, and for that reason he seems to have mixed in elements from other viewpoints.

In particular we may consider the statement already quoted that an essential element of the attribution of physical reality is the existence of predictions of future behaviour,[3] coupled with the earlier one[2] that essentially identifies these predictions with the actual definition of the physical quantity involved. This would seem to have nothing to do with complementarity, everything in fact to do with positivism; the very existence of a quantity is identified with the predictions that may be made concerning it. Fine[22] indeed calls Bohr's remarks 'virtually textbook neopositivism', giving his opinion that 'while many commentators seem inclined to suppose that Bohr's tendency to obscure language is a token of philosophical depth' he considers that 'where it really matters Bohr invariably lapses into positivist slogans and dogmas'.

We can see why there may be very different views about Bohr's response to EPR. For Henry Folse,[9] for example, who is a keen advocate of complementarity, taken as an *a priori*, the EPR paper required no intrinsic change in Bohr's views, merely some clarification over the term 'disturbance', which we will discuss at the end of this section. Commentators like Fine and Mara Beller, though, who are uncommitted to complementarity, may well feel that Bohr is denying physical reality to one or other of EPR's observables merely because both cannot be measured simultaneously, even indirectly, and this may lead to severe criticism of Bohr, with charges of positivism or instrumentalism being made.[16,22,46]

As to whether Bohr's arguments did, in fact, successfully rebut the EPR argument, it is clear that it was fairly easy to demonstrate that any assumption of complementarity was sufficient to make EPR irrelevant. We may say that Bohr was thus able to defend his own position against this new argument of Einstein. But of course Einstein himself was not obliged to conform to complementarity, so in no sense did Bohr's arguments triumph over Einstein and EPR. It may be that Bohr was uncomfortably aware of the latter point, and thus, perhaps sub-consciously, phrased his argument in such a way that it was not clear exactly what part in it was played by complementarity. It is perhaps this background that made his paper so obscure, a fact that is admitted even by many of his supporters.

We now pick up the point on 'disturbance'. Before 1935, as was briefly discussed in Chapter 4, Bohr was used to buttressing his rather abstract ideas on complementarity with a more physical type of argument, which used the idea of disturbance. It may be remarked that there is nothing inherently wrong with talking about a measurement causing disturbance either in quantum mechanics or in classical mechanics.[9] The projection postulate points us in this direction, and in discussion of the early arguments between Bohr and Einstein, it is at the very worst only a little loose to say that the two measurements of, say, position and momentum interfere with each other or disturb each other.

However, use of the concept of disturbance in quantum mechanics is open to total misunderstanding, and such misunderstanding is exceptionally common. It is easy to assume that the argument is that a general physical observable has a value at any particular time, a value which is unfortunately 'disturbed' by the measurement. It may be natural to assume that the difference between the classical and quantum cases is that in classical mechanics the effect of the disturbance may, in principle, be calculated and allowed for, while in quantum mechanics this is not possible.

The general idea is perhaps seen in the usual name for the Heisenberg principle—the uncertainty principle, which would imply that physical observables do have precise values, of which we are uncertain due to our ignorance: When we measure, say, the value of momentum, we disturb the value of position. In popular discussion of quantum theory, it is often believed that the crucial conceptual point is that the observation disturbs the system. Bohr would not, of course, have supported the more naïve ideas, but did not avoid encouraging the general point of view that there was a physical underpinning to his philosophy.

To return to the early Bohr–Einstein discussions, there was agreement that there could not be simultaneous measurements of position and momentum, and also that if there were successive measurements of position and momentum, they would not be on the same state of the system, because the first measurement would have disturbed the initial state. The implication from Bohr's point of view may have been essentially that it was not even worth considering the possibility that both momentum and position had values, because the disturbance idea would mean that both values could not be measured. (Admittedly this argument does rest on positivism.)

This is why, in EPR35, the idea of locality is expressed by saying that a measurement on particle 1 cannot disturb particle 2. The argument is then that hypothetical measurements on particle 1 show that particle 2 has values for both momentum and position even though these hypothetical measurements would not disturb particle 2. In his reply[3] Bohr agrees that there is no question of a mechanical disturbance. This is the point in his argument where he says that what the measurement on particle 1 achieves is not a mechanical disturbance of particle 2, but a taking up of one or other of the two mutually exclusive experimental procedures. Fine[22] talks of the new position as using 'semantic disturbance' which Bohr could not account for on physical grounds. Semantic disturbance may be said to imply that what we do to particle 1 affects what we can *say* about particle 2.

It may be argued that the removal from Bohr of his use of disturbance was, at least in retrospect, one of the great achievements of EPR35. As we have said, few thought in any depth about either EPR or Bohr's reply for many years. His supporters could, in any case, contend that the disturbance interpretation had never been an intrinsic part of Bohr's argument; it was little more than a physical aside. However, when physicists were persuaded to give some thought to EPR, Bohr's use of complementarity, unsupported by physical argument, increasingly seemed contrived. Eventually physicists would return to examine all the issues involved in EPR, but not for a long time after 1935.

EPR—What did it Mean?

From previous sections, we take the message that the EPR argument was broadly correct, though one may admit that the most famous example, that of EPR35, has technical difficulties. We now discuss what conclusions we should draw from the argument, what actually are its claims.[19] This is important, as many papers claim to 'refute' the EPR paper, or to 'solve' EPR's 'dilemma', yet without really understanding what EPR were arguing, and what might be required for such refutation or solution.

There is a very common belief that EPR merely expressed Einstein's prejudices, or attempted to point towards some sort of internal contradiction in quantum theory itself. We reiterate that in fact it is a clear logical argument leading to a precise and definite conclusion, and should be recognised as such. What EPR had argued was that one could not insist of the truth of *both* of the following propositions:

(A1) Quantum theory is complete.
(A2) Quantum theory is local.

This conclusion is certainly interesting, because, before the publication of EPR35, almost certainly the most influential physicists would have supported both (A1) and (A2). Let us discuss what rejection of (A1) would imply. The easiest answer is that it would imply the acceptance of hidden variables, and most physicists would speak of this as restoring realism. Einstein's views would not be precisely along these lines; we shall mention his beliefs briefly in this section, but in much more detail in the following chapter.

It should not, of course, be suggested that Einstein was undecided about which of (A1) and (A2) he wished to retain, which to get rid of. He was a strong supporter of (A2), a strong opponent of (A1), as even the title of Ref. [1] makes clear. To this extent he has often been criticised for imposing (A2) and (A1′) on to physics. Here we have:

(A1′) Quantum mechanics must be supplemented by hidden variables.

(A2) and (A1′) together may be called local realism, and broadly it may be said that Einstein was in favour of this. To answer these critics, though, one say first that Einstein was well aware that logically it was quite possible that EPR could lead to a rejection of (A2) and acceptance of (A1). Secondly we could say that in 1935 it should have seen immensely sensible to opt for (A2) and (A1′). These were both well-established principles from classical physics, and one would have thought that every effort would have been made to retain them for as long as possible. It may be said that it was only the prejudice of Bohr and his colleagues against hidden variables that made (A1) at all popular, and made Einstein seem at all reactionary for rejecting it.

Bell seems to be making the same point when he told Bernstein[5] that: 'I feel that Einstein's intellectual superiority over Bohr, in this instance, was enormous: a vast gulf between the man who saw clearly what was needed, and the obscurantist.' Admittedly Bell does have to continue: 'So for me, it is a pity that Einstein's idea doesn't work. The reasonable thing just doesn't work.' We shall return to the point of the reasonable thing not working shortly, and then in more detail in Chapter 9.

Let us rewrite the conclusions of EPR as Bell approached them. Bell[40] viewed the statement that quantum theory is not complete as straightforwardly saying that there were hidden variables. Jammer[20] argued that Einstein should not be thought of as a supporter of hidden variables; rather his hopes lay in the production of a new unified field theory, to which he imagined quantum theory would be an excellent approximation, at least in all circumstances studied experimentally up to that date. Bell,[47] though, replied that Einstein's belief in classical field theory, far from being opposed to a hidden variables approach, should be regarded as an example of it. Bell rejected the argument that any hope or expectation that at some point quantum theory would conflict with experiment should exclude one from being considered a proponent of hidden variables.

Einstein's views will be discussed further in Chapter 8. However since we are at the moment following Bell's analysis, we will shortly speak of hidden variables. Recognising the possibility that quantum theory may not give correct predictions in all circumstances, we may then rewrite the results of EPR as follows. Not all the following may be true:

(B1) The physical Universe is not real (as defined above).
(B2) Physics is local.
(B3) Quantum theory is correct in all circumstances.

This conclusion is important, because, before EPR, Bohr and his supporters would almost certainly have argued, at least in effect, that all three statements *were* true. They would have insisted on the truth of (B1) and (B3); on (B2), it may perhaps be best said that the possibility of it requiring consideration would scarcely have occurred to them. It may be argued that Bohr's response to EPR constituted a tacit rejection of (B2), although this would not have been admitted explicitly; we shall return to this point shortly.

Bell's initial approach to the EPR argument would then have been to replace (B1) with (B1′) given by

(B1′) Physics has hidden variables

and to expect that (B1′) and (B2) would be upheld—Bell's conception of local realism. However it was precisely this that his great result, discussed in detail in Chapter 9, seriously threatened. One could not retain (A1′) and (A2); quantum theory is *not* a local realist theory. Clearly, though, one could hope to retain (B1′) and (B2), provided one is willing to reject (B3); the Universe may possibly respect local realism although quantum theory does not. This is what the experimental tests of the Bell's inequality are designed to check.

While that is the subject of Chapter 9, here we are studying some typical responses to the EPR argument, some of which do relate to Bell's work. We have noted that, much as Einstein would have wished to retain local realism, he was well aware that the logical conclusion of the EPR argument was that *either* quantum theory was not complete *or* that there was non-locality. It should therefore be obvious that no argument involving explicitly or implicitly the use of hidden variables or non-locality (in whatever form of words thought appropriate) may be regarded as a refutation of the EPR argument, or a solution of Einstein's (supposed) dilemma. Unfortunately history has often judged differently. Encouraged by the 'paradox' terminology, several analysts have thought it appropriate to present 'solutions' to his difficulties, or alternatively to present 'rebuttals' of the EPR argument that are actually no more than examples of his own conclusions.

Much of Bohr's long-term response to Einstein's ideas may be seen, at least according to one reading, in this way. Folse[9] argues at some length that after 1935 Bohr took great pains to stress that the object of scientific study should be not an individual particle, but a 'phenomenal object', a physical system as it appears in particular experimental conditions. This may be described as an extension of the idea of *wholeness*, the idea that measured and measuring sub-systems cannot be considered separately, to the idea that the entire EPR-system comprising the two particles and any measurements carried out on them, must be considered as a unity. The question immediately raised is: Does this not violate locality? Dugald Murdoch,[10] though, argues that it violates only the *principle of independent existence*, which says that the real states of spatially separated objects are independent of each other.

Einstein appears to recognise the distinction between violation of locality and violation of independent existence, though he approves of neither. In his auto-biographical notes in the Schilpp volume,[25] he writes: 'On one supposition we should, in my opinion, absolutely hold fast: the real factual situation of the system S_2 is independent of the system S_1, which is spatially separated from the former.' He adds that: 'One can escape from this conclusion only by either assuming that the measurement of S_1 (telepathically) changes the real situation of S_2 or by denying independent real situations as such to things which are spatially separated from each other. Both alternatives appear to me entirely unacceptable.' Einstein's possibilities are denial of either locality or independent existence.

In his responses to criticisms[29] in the same volume, and specifically discussing the EPR debate, Einstein says that: 'Of the "orthodox" quantum theoreticians whose position I know, Niels Bohr's seems to me to come nearest to doing justice to the problem'; he describes Bohr as believing that 'if the partial systems A and B form a total system, there is no reason why any mutually independent existence (state of reality) should be ascribed to the partial systems A and B viewed separately, *not even if the partial systems are spatially separated from each other at the particular time under consideration.*'

It would seem clear, though, that loss of independent existence actually implies loss of locality. It may be a particularly meaningful and helpful way of discussing non-locality, but nevertheless it *is* just a way of thinking about non-locality! It follows that, to the extent that Bohr's answer to EPR relies on loss of independent existence, it is certainly *not* a refutation of EPR or a solution of EPR's supposed dilemma; rather it is very much in line with EPR's analysis.

We now return to Steane's comments[39] on EPR, which were introduced above. Having claimed to have unearthed the 'fallacy' in the EPR analysis, Steane claims that this fallacy 'can be removed in one of two ways: one can either say that Alice's measurement does influence Bob's particle [where, as always in quantum information theory, Alice and Bob are the participants in the two wings of the EPR set-up], or (which I prefer) that the quantum state vector $|\Phi\rangle$ is not an intrinsic property of a quantum system, but an expression for the information content of a quantum variable. In the singlet state there is mutual information between A and B, so the information content of B changes when we learn something about A.' Steane's first method of solving the difficulty is clearly non-locality. For his 'mutual information' to be successful, either one must have non-locality—the state of affairs at B, about which we have information, changes when we learn something about the state of affairs at A; or one must have hidden variables—what one discovers about B was already in existence before we did any measurement at all.

Thus Steane's arguments certainly do not unearth any fallacy in EPR; rather its suggestions are precisely along the lines already pointed out by Einstein. This is not, of course, to deny that Steane's use of the concept of quantum information is interesting, informative and potent, as it stands at the gateway to the important subject of quantum information theory. But it must be stressed that there is nothing in Steane's discussion that could in any sense explain or rebut the EPR argument. Steane's use of quantum information, far from being a refutation of the EPR argument, or somehow solving a problem left by EPR, is rather a precise and interesting example of precisely the type of thing that EPR said must exist. In contrast to those who claim to point out elementary ways of explaining away Einstein's arguments, we would assert that there are no elements of the basic physics of the situation that were not already totally obvious to Einstein.

Another generally excellent account of quantum information theory has been the comprehensive text of Nielsen and Chuang.[38] Unfortunately, though these authors explain the importance of EPR in this area, like Steane they do not appreciate the role of the original authors. We have already quoted them as saying that: 'The attempt to impose on Nature *by fiat* properties which she must obey seems a most peculiar way of studying her laws.' But, of course, far from trying to force properties onto nature, EPR argued in a sustained and effective way for these properties.

Nelson and Chuang add, in reference to the experiments on the Bell's inequalities that 'Nature has had the last laugh on EPR.' Now of course the EPR conclusions, as stated in the form that not all of (B1), (B2) and (B3) can be true, are certainly not contradicted by the results of the Bell experiments, which, as discussed in Chapter 9, say that, subject to closure of loopholes, *either* (B1') *or* (B2) is false, *or* both of them are false, (B3) not being disproved. If it is assumed that quantum theory is correct, we may say that the result of the experiments is to rule out the combination of (A1') and (A2); this is in no way in contradiction to the conclusion of EPR. It is, of course, against, what Einstein *hoped* would be the eventual conclusion of study of the topic, but he certainly never imposed this wish on his own analysis. As will be argued in detail in Chapter 11, Einstein's work, particularly that centred around the EPR analysis, is one of the great contributors to quantum information theory. It is entirely false to imply that he left only misunderstandings and mistakes, and that others cleared up the conceptual confusion and deserve much or even all of the eventual credit.

Einstein and Schrödinger: The 1935 Correspondence

In this section, we discuss an important and influential interaction between Einstein and Schrödinger that followed the publication of EPR35. This correspondence[22,48] touched on several topics discussed in various chapters of this book. We have already mentioned in the previous chapter that it was in this correspondence that Einstein reintroduced his remarks to the 1927 Solvay conference in the form that was to become known as 'Einstein's boxes'. Earlier in this chapter we said that, also in this correspondence that Schrödinger extracted the central element of EPR and christened it 'entanglement', a term which now permeates all today's work on the

Bell inequalities and quantum information theory. In Chapter 8, we shall describe how Schrödinger also used this interaction to criticise one of Einstein's preferred options for solving the problems of quantum theory, the method of ensembles. In this section, we mention some of the other aspects of this correspondence, including the famous Schrödinger's cat.

It was clear that the EPR paper of 1935 immediately struck a chord with Schrödinger, who was temporarily in Oxford, having left his prestigious post in Berlin with the coming of Hitler. Only 3 weeks after the publication of the EPR paper, he wrote to Einstein with a detailed analysis[48] of its argument, which he claimed had 'caught dogmatic quantum mechanics by the coat-tails'. In this letter, Schrödinger stressed the significance of locality for the argument, and it was in a reply written only 12 days after Schrödinger's letter, that Einstein outlined the idea of Einstein's boxes.

In a later letter, Einstein confided in Schrödinger that he was the only person with whom he was 'willing to come to terms'. All the other physicists, he said, argued from the theory to the facts rather than the other way round, and they had thus become trapped in a 'conceptual net'. In this letter, he outlined his idea of ensembles to Schrödinger, and also suggested yet another thought-experiment, in which there is a certain probability that a mass of gunpowder would explode in the course of a year. By the end of the year, Einstein suggested, the wave-function of the system would be a linear combination of two totally different types of state, the first corresponding to no explosion and the maintenance of order, and the second to an explosion and the associated large-scale chaos it would create. Such a superposition of two entirely different macroscopic situations appeared to Einstein to be totally contradictory to his idea of macroscopic realism.

Schrödinger addressed both issues in further letters to Einstein, but also in publications. His three-part paper[31] in Die Naturwissenschaften titled 'The present situation in quantum mechanics' was a direct response to EPR, while he also published a pair of more formal papers[27,28] on the same general set of themes. His response to Einstein's idea of ensembles was negative. As will be discussed in some detail in Chapter 8, he presented detailed arguments against at least the most obvious way of understanding Einstein's suggestion.

However his response to Einstein's gunpowder thought-experiment was to become celebrated. He made the magnification from microscopic superposition to macroscopic superposition implicit in Einstein's idea more vivid by use of the following 'diabolical device'. A radioactive atom is left in a container for a period of time such that there is a 50% probability of survival, 50% probability of decay. The wave-function of the atom at the end of the period will be a superposition of states corresponding to survival and decay. Of course such a situation at the microscopic level is very familiar in quantum theory, and, at least by 1935, was very much taken for granted.

But Schrödinger's thought-experiment went considerably further. It was arranged that if the atom decayed it caused the death by poison of a cat imprisoned in the vessel containing the radioactive atom. If the atom survived, so did the cat. Thus at the end of the period the wave-function of the full system would be a superposition of a state consisting of surviving atom and cat, and one consisting of decayed atom and dead cat. Such a linear combination of a dead cat and a living one seemed contrary to any conception of macroscopic reality. So did the idea that the person who opened the box could be regarded as performing a measurement which would collapse the macabre wave-function, and actually kill or spare the cat.

The cat argument occupied only a small fraction of Schrödinger's trilogy of papers, which incidentally he suggested might be called either a 'report' or a 'general confession'. From the strictly conceptual point of view, probably the most important part was his extracting from the EPR paper the idea of entanglement, christening it, and describing in some detail how an experimenter in one wing of an EPR-type experiment can draw the sub-system in the other wing into a range of different states, each with its own wave-function. Entanglement was to play a great part in the work of John Bell, and then in the development of quantum information theory, as we shall see in Chapters 9 and 11. But it was the Schrödinger's cat concept that caught the imagination, and was eventually to play a part, alongside EPR, in persuading at least some physicists to question Copenhagen.

Einstein had a close, usually positive, intellectual relationship with Schrödinger, who had, of course, admitted Einstein's part in stimulating his own great creation of wave mechanics. Both men were in the small minority of physicists who could not accept complementarity and the Copenhagen interpretation, and they took some comfort in sharing their negative views. While their ideas on what should replace Copenhagen differed, they were able to debate these in a relaxed and positive way. It might be mentioned that another point of disagreement was that Schrödinger was quite positive about giving up determinism, much in contrast, of course, to Einstein. From the 1930s on, Schrödinger was one of the few physicists who supported Einstein in attempting to develop a unified field theory. It is true that this topic did lead to a major debacle[48] when Einstein clearly felt that Schrödinger had extracted enormous publicity for what was, at best, a very minor advance; Schrödinger aptly called this affair *Die Einstein Schweinerei*. Overall, though, their interaction was mutually stimulating, and its results were to bear fruit, though not for a quarter of a century after the publication of EPR35 and Schrödinger's own trilogy.

Note added in proof

Arthur Fine has recently written an interesting paper 'Bohr's response to EPR: Criticism and defense', referring to some of the points mentioned in our section 'Bohr's response to EPR35'. It was presented at the *Workshop in Memory of Mara Beller* and will be published in *Iyyun*.

References

1. Einstein, A., Podolsky, B., and Rosen, N. (1935). Can quantum-mechanical description of physical reality be considered complete? Physical Review **47**, 777–80.

2. Bohr, N. (1935). Quantum mechanics and physical reality, Nature **136**, 65.
3. Bohr, N. (1935). Can quantum-mechanical description of physical reality be considered complete? Physical Review **48**, 696–702.
4. Pais, A. (1982). 'Subtle is the Lord ...': The Science and Life of Albert Einstein. Oxford: Clarendon.
5. Bernstein, J. (1991). Quantum Profiles. Princeton: Princeton University Press.
6. Whitaker, A. (2006). Einstein, Bohr and the Quantum Dilemma. (1st edn. 1996, 2nd edn. 2006) Cambridge: Cambridge University Press.
7. Howard, D. (1994). What makes a classical concept classical? In: Niels Bohr and Contemporary Philosophy (Faye, J. and Folse, H.J., eds.) Dordrecht: Kluwer, pp. 3–18.
8. Howard, D. (1979). Complementarity and ontology: Niels Bohr and the problem of scientific realism in quantum physics, Ph.D. dissertation, Boston University.
9. Folse, H.J. (1985). The Philosophy of Niels Bohr: The Framework of Complementarity. Amsterdam: North-Holland.
10. Murdoch, D. (1987). Niels Bohr's Philosophy of Physics. Cambridge: Cambridge University Press.
11. Honner, J. (1987). The Philosophy of Nature: Niels Bohr and the Philosophy of Quantum Physics. Oxford: Clarendon.
12. Krips, H. (1987). The Metaphysics of Quantum Theory. Oxford: Clarendon.
13. Faye, J. (1991). Niels Bohr: His Heritage and Legacy: An Anti-Realist View of Quantum Mechanics. Dordrecht: Kluwer.
14. Petruccioli, S. (1993). Atoms, Metaphors and Paradoxes: Niels Bohr and the Construction of a New Physics. Cambridge: Cambridge University Press.
15. Ottaviani, J. (2004). Suspended in Language: Niels Bohr's Life and the Century he Shaped. Ann Arbor: G.T. Labs.
16. Beller, M. and Fine, A. (1994). Bohr's response to EPR, In: Niels Bohr and Contemporary Philosophy. (Faye, J. and Folse, H.J., eds.) Dordrecht: Kluwer.
17. Halvorson, H. and Clifton, R. (2003). Reconsidering Bohr's response to EPR, In: Non-Locality and Modality. (Placek, T. and Butterfield, J., eds.) Dordrecht: Kluwer, pp. 3–18.
18. Dickson, M. (2003). Bohr on Bell: A proposed reading of Bohr and its implications for Bell's theorem, In: Non-Locality and Modality. (Placek, T. and Butterfield, J., eds.) Dordrecht: Kluwer, pp. 19–36.
19. Whitaker, M.A.B. (2004). The EPR paper and Bohr's response, Foundations of Physics **34**, 1305–40.
20. Jammer, M. (1974). The Philosophy of Quantum Mechanics. New York: Wiley.
21. Jammer, M. (1985). The EPR problem in its historical development, In: Symposium on the Foundation of Modern Physic: 50 Years of the Einstein-Podolsky-Rosen Gedankenexperiment. (Lahti, P. and Mittelstaedt, P., eds.) Singapore: World Scientific, pp. 129–49.
22. Fine, A. (1986). The Shaky Game. Chicago: University of Chicago Press.
23. Einstein, A. (1948). Quantum mechanics and reality, Dialectica **2**, 320–4.
24. Born, M. (2005). The Born-Einstein Letters 1916–1955. 2nd edn., Houndmills: Macmillan.
25. Einstein, A. (1949). Autobiographical notes, In: Albert-Einstein: Philosopher-Scientist. (Schilpp, P.A., ed.) New York: Tudor, pp. 1–95.
26. Bohm, D. (1951). Quantum Theory. Englewood Cliffs: Prentice-Hall, pp. 614–23.
27. Schrödinger, E. (1935). Probability relations between separated systems, Proceedings of the Cambridge Philosophical Society 31, 555–62.

28. Schrödinger, E. (1936). Probability relations between separated systems, Proceedings of the Cambridge Philosophical Society **32**, 446–52.

29. Einstein, A. (1949). Response to criticisms, In: Albert-Einstein: Philosopher-Scientist. (Schilpp, P.A., ed.) New York: Tudor, pp. 665–88.

30. Rosen, N. (1985). Quantum mechanics and reality, In: Symposium on the Foundation of Modern Physic: 50 Years of the Einstein-Podolsky-Rosen Gedankenexperiment. (Lahti, P. and Mittelstaedt, P., eds.) Singapore: World Scientific, pp. 17–33.

31. Schrödinger, E. (1935). Die gegenwartige Situation in der Quantenmechanik [The present situation in quantum mechanics], Naturwissenschaften **23**, 807–12, 823–8, 844–9. Translation by J.D. Trimmer in Ref. [32].

32. Wheeler, J.A. and Zurek, W.H. (eds.) (1983). Quantum Theory and Measurement. Princeton: Princeton University Press, pp. 152–67.

33. Bohm, D. (1952). A suggested interpretation of the quantum theory in terms of 'hidden variables', I and II, Physical Review **85**, 166–79, 180–93.

34. Bell, J.S. (1964). On the Einstein-Podolsky-Rosen paradox, Physics **1**, 195–200 (1964); reprinted in Speakable and Unspeakable in Quantum Mechanics. (1st edn. 1987, 2nd edn. 2004) Cambridge: Cambridge University Press, pp. 14–21.

35. Redhead, M. (1987). Incompleteness, Nonlocality and Realism: A Prolegomenon to the Philosophy of Quantum Mechanics. Oxford: Oxford University Press.

36. Maudlin, T. (1994). Quantum Non-Locality: Metaphysical Intimations of Modern Physics. Oxford: Blackwell.

37. Ghirardi, G.-C. (2005). Sneaking a Look at God's Cards: Unraveling the Mysteries of Quantum Mechanics. Princeton: Princeton University Press.

38. Nielsen, M.A. and Chuang, I.L.(2000). Quantum Computation and Quantum Information. Cambridge: Cambridge University Press.

39. Steane, A. (1998). Quantum computing, Reports on Progress in Physics **61**, 117–74.

40. Bell, J.S. (1981). Bertlmann's socks and the nature of reality, Journal de Physique **42**, C241–61 (1981); reprinted in Speakable and Unspeakable in Quantum Mechanics. Cambridge: Cambridge University Press (1987), pp. 139–58.

41. Halvorson, H. (2000). The Einstein-Podolsky-Rosen state maximally violates Bell's inequalities, Letters in Mathematical Physics **53**, 321–9.

42. Cooper, J.L.B. (1950). The paradox of separated systems in quantum theory, Proceedings of the Cambridge Philosophical Society **46**, 620–5.

43. Epstein, P.S. (1945). The reality problem in quantum mechanics, American Journal of Physics **13**, 127–36.

44. Rosenfeld, L. (1967). Niels Bohr in the thirties: consolidation and extension of the concept of complementarity, In: Niels Bohr: His Life and Work as Seen by his Friends and Colleagues. (Rozental, S., ed.) Amsterdam: North-Holland, pp. 14–36.

45. Bohr, N. (1949). Discussions with Einstein, In: Albert-Einstein: Philosopher-Scientist. (Schilpp, P.A., ed.) New York: Tudor, pp. 199–241.

46. Beller, M. (1999). Quantum Dialogue. Chicago: University of Chicago Press.

47. Bell, J.S. (1976). Einstein-Podolsky-Rosen experiments, In: Proceedings of the Symposium on Frontier Problems in High Energy Physics. Pisa: Scuola Normale, pp. 33–45.

48. Moore, W. (1989). Schrödinger: Life and Thought. Cambridge: Cambridge University Press, pp. 302–14.

7
Einstein and the Macroscopic Limit of Quantum Mechanics

Introduction

In a letter to David Bohm which was written on 24th November 1954 and has been quoted by Jammer,[1] Einstein wrote: 'I do not believe in micro and macro laws, but only in (structural) laws of general validity.' This remark reflects one of the central tenets of his belief, and Einstein used it to substantiate his arguments which attempted to demonstrate the incompleteness of quantum mechanics. Einstein put forward various versions of these arguments, but there were two main themes. The first, which has certainly received the most attention in books and articles, concerns the question of hidden variables, entanglement and so on. We have discussed it in the previous two chapters, and, of course, this theme led to the work of John Bell, which is discussed in Chapter 9, and much that is in later chapters of this book.

The second theme, which has received, in general, far less attention, concerns the problem of consistency between the macroscopic limit of quantum mechanics, and classical mechanics. In this area too, Einstein felt that the failings of the Copenhagen interpretation were manifest. During his lifetime, there was considerably less discussion over this area of dispute than the first. Einstein did use arguments concerning the macroscopic region to emphasise his general position on realism in a few major publications, as we shall see, but most of his more general discussion of the classical limit occurred in his correspondence with Max Born, and it remained unpublished at the time. We discuss the correspondence in this chapter. Since Einstein's death, many of these topics have been investigated in considerable detail, confirming Einstein's opinion that there were many subtle and awkward issues that required much study. We discuss this more recent work in Chapter 12.

Einstein's line of reasoning stemmed from the fact that, in the macroscopic domain, classical mechanics provided a treatment which was *objective*, in the sense of being intersubjective, and *realist*, in the sense of being observer-independent. Moreover it treated the behaviour of an individual system in terms of an *event-by-event description*. Hence Einstein felt that one of the necessary requirements for the completeness of quantum mechanics should be the emergence of such a realist individual description from quantum mechanics in the macroscopic limit.

141

Einstein used a number of examples to highlight his view that quantum mechanics failed to satisfy this requirement of completeness, and we shall now discuss some of these examples.

Macrorealism: Examples

In 1949, the volume *Albert Einstein: Philosopher-Scientist* (the so-called Schilpp volume) was published to commemorate the seventieth birthday of Einstein. It contained a number of articles by several luminaries including Bohr, Pauli, Born and Heitler explaining their disapproval of Einstein's attitude towards quantum theory. As a rejoinder, Einstein wrote a piece titled *Reply to Criticisms*[2] in which he analysed the following example.

He considered a single radioactive atom, the decay product of which can be detected by a Geiger-counter connected to a chart-recorder (an automatic registration-mechanism) which records a 'mark' on paper. This mark corresponds to the triggering of the Geiger-counter caused by the occurrence of the decay. In this way the fact that a decay has occurred is registered in a discernible form. Here the key features in Einstein's argument are that the Geiger-counter connected to a chart-recorder is a macroscopic arrangement, that the mark on the paper has an objective reality in the sense that it is 'out there' independent of whether one is looking at it, and that its location on the paper corresponds to a particular instant at which the decay occurs.

However the quantum mechanical formalism does not describe the actual instant of the decay of an individual radioactive atom, but only the probability of a decay occurring within a certain time. The quantum mechanical description of the above macroscopic arrangement for detecting the decay of a radioactive atom does not demonstrate the objective existence of an individual mark at a definite spot on the chart-recorder. Rather the final wave-function of the entire arrangement corresponds to a superposition of various possibilities of the mark being localised at different positions on the chart-recorder. Thus quantum mechanics does not ensure an event-by-event objective description of the registration of an individual mark at a definite spot on a macroscopic chart-recorder.

Einstein interpreted the above feature as showing that the quantum mechanical wave-function failed to specify *completely* the state of an individual system, since it was not consistent with the realist description for an individual macrosystem. Hence he asserted: 'The attempt to conceive the quantum theoretical description as the complete description of an individual system leads to unnatural theoretical interpretations, which become immediately unnecessary if one accepts the interpretation that the description refers to an ensemble of systems and not to an individual system'.[3]

The phrase 'unnatural theoretical interpretations' was used in the above quote in the context of the *incompatibility* between the wave-function description of an individual system, and *realism* at the macroscopic level. It is important to stress here that lack of determinism in quantum mechanics was not the central point which

Einstein wanted to highlight through such examples. Rather it was the persistence of the lack of realism in quantum mechanics when extrapolated from the micro to macro level that Einstein sought to use as a key ingredient in his examples to illustrate his critique of quantum mechanics. This particular example is discussed further in the following chapter, together with a much more detailed and general discussion of the importance for Einstein of realism and determinism.

In a letter to Born[4] written in September 1950, Einstein came up with a variant of his earlier example. This involves a macroscopic body which can rotate freely about an axis whose state is determined by an angle of rotation. We may imagine that the initial conditions (angle and angular momentum) may be defined as precisely as permitted by quantum theory. This accordingly fixes the initial wave-function. The Schrödinger equation then determines the wave-function at any subsequent instant. Because of the initial uncertainties, if this time evolution occurs for a sufficiently long time, states corresponding to various angles of rotation will become equally probable.

Now, if an observation is made, for example, by flashing a torch, the macro-rotator will be found in a definite state corresponding to a definite angle of rotation. Einstein argued that, although this in itself does not prove that the angle had a definite value before it was observed, this ought to be the case, as he reasoned, 'because we are committed to the requirement of reality on the macroscopic scale'. Thereafter he concluded: 'Thus the ψ-function does not express the real state of affairs perfectly in this case. This is what I call "incomplete description".'

Macrorealism: Discussion

It is evident that these examples of Einstein were specifically designed to justify his thesis of the incompleteness of quantum mechanics, on the basis of what may be called the incompatibility between quantum mechanics and the concept of macrorealism. In the same letter, Einstein stressed once again that he differed from physicists like Born primarily on how they look at this incompleteness of quantum mechanics; while Einstein believed it was a flaw which required to be removed, most physicists, like Born, were prepared to live with it. As Einstein put it: 'For the time being, I am alone in my views as Leinbiz was with respect to the absolute space of Newton's theory.'

Interestingly, in a note written later commenting on this letter of Einstein, Born[5] justified his position on the ground that 'an exact description of the state of a physical system presupposes that one can make statements of infinite precision about it, and this seems absurd to me'. This indicates the crux of the confusion about Einstein's viewpoint. Born and many of his contemporaries refused to recognise the conceptual distinction between the impossibility of *determining* an initial state with infinite accuracy, and the question of whether the state actually *has* an objective reality independent of observation. It is this latter issue on which Einstein had focused, particularly in the examples we have discussed, in the context of applying quantum mechanics at the level of macroscopic systems. It is indeed surprising that, in spite of Einstein's repeated clarifications, a number of physicists like Born continued to miss this point.

However, Pauli did see this point after discussions with Einstein. In a letter to Born written in March 1954,[6] while agreeing with Einstein that 'quantum mechanics must, in principle, be able to claim validity for macroscopic spheres; their finer structure (atomic constitution) clearly does not come in to play', Pauli summarised Einstein's position by saying: 'Now, from my conversations with Einstein I have seen that he takes exception to the assumption, essential to quantum mechanics, that the state of a system is defined only by specification of an experimental arrangement. (By the way, Einstein says instead of "specification of the experimental arrangement": "that the state of a system depends on the way one looks at it". But it boils down to the same thing). Einstein wants to know nothing of this.'

After the above remarks Pauli made a curious observation as regards the validity of the 'assumption' he had mentioned as 'essential to quantum mechanics'. He tried to justify this 'assumption' by a somewhat vague remark: 'If one were able to measure with sufficient accuracy, this would of course be as true for small macroscopic sphere, as for electrons . . . But Einstein has the philosophical prejudice that (for macroscopic bodies) a state (termed "real") can be defined "objectively" under *any* circumstances, that is *without* specification of the experimental arrangement used to examine the system (of the macro-bodies), or to which the system is being "subjected". It seems to me that the discussion with Einstein can be reduced to this hypothesis of his, which I've called the idea (or the "ideal") of the "detached observer". But to me and other representatives of quantum mechanics, it seems that there is sufficient experimental and theoretical evidence against the practicability of this ideal.'

In essence, what Pauli claimed in the above statement was that the concept of *macrorealism* which Einstein regarded to be a sacrosanct notion (and which he believed to be *incompatible* with quantum mechanics) was already experimentally shown to be untenable. However, this claim was not justified. To see this, we first note that both Einstein's examples discussed above crucially involved *superpositions* of *macroscopically distinguishable states*. In the radioactive atom example, the quantum mechanical evolution resulted in a superposition of the mark being localised in different regions on the chart-recorder, while the macroscopic rotator example was based on the feature that, after a certain time, the quantum mechanical evolution resulted in a superposition of macroscopically distinguishable states corresponding to the macro-rotator being oriented along different directions.

It is important to stress that, in contrast to the views of Pauli, the question of the validity of the quantum mechanical superposition principle for the macroscopically distinguishable states of a macrosystem was not carefully studied until 1980. It was Tony Leggett[7] who first argued persuasively that decisive experiments were lacking which critically tested the quantum mechanical superposition principle for macroscopically distinguishable states of a macrosystem. Subsequently, Leggett and Garg[8] succeeded in formulating a scheme seeking to test specifically the hypothesis of macrorealism—the Einsteinian assumption that whenever a macro-scopic system has available to it a number of macroscopically distinct states, then at 'almost all' times it is definitely *in* one or the other, independent of whether the state of the system is observed, and the compatibility of this assumption with the quantum mechanical superposition of macroscopically distinct states of a macrosystem. (In the above, the phrase 'almost all' recognises that there may be a time of transition between the various macroscopically distinct states.)

The most significant experimental studies to date on this issue, motivated by Leggett's proposal, use superconducting devices based on the Josephson effect (called SQUIDs or superconducting quantum interference devices), and involve the quantum superposition of states in which the collective electron current, which has a magnitude of the order of a few micro-amperes, circulates clockwise and anticlockwise respectively in a macroscopic superconducting loop.[9-12] We will discuss the basic ideas and the present status of the studies along this direction in Chapter 12, where we deal with the contemporary investigations which are connected with these questions that were raised by Einstein, and also in Chapter 13.

The necessity and considerable interest of these recent research studies clearly indicate the illegitimacy of the claim, made by Einstein's contemporaries like Pauli, of the empirical invalidity of Einstein's assumption of macrorealism. In fact, Einstein posed his questions about the tenability of the notion of macrorealism when quantum mechanics is extrapolated to the macroscopic level by clever use of the superposition of macroscopically distinct states; this line of study was actually far ahead of his time. The non-triviality of this whole issue has become increasingly clear only in recent times due to investigations over the last two decades. For a comprehensive review, see Leggett.[13-15]

Particle in a Box

Subsequent to the exchanges with Born following the publication of the Schilpp volume, Einstein contributed a provocative article in a collection of essays pre-sented to Born in 1953[16] on the occasion of his retirement from the Tait Chair of Natural Philosophy in the University of Edinburgh. In this piece Einstein gave another example to substantiate his thesis that quantum mechanics did not provide an objective realist description of an individual system, even in the macroscopic limit. We now consider this example.

Einstein considered a single macroscopic particle with a specified energy con-fined in a box consisting of an infinite potential well, say along the x-axis between two perfectly reflecting and smooth parallel walls located at $x = 0$ and $x = L$. The potential is then given by $V = 0$ for $0 < x < L$ and $V \to \infty$ for $x \leq 0, x \geq L$.

The quantum mechanical description for such a system bound between $x = 0$ and $x = L$ is provided by an energy eigenfunction of the form

$$\psi_N(x,t) = \left(\frac{2}{L}\right)^{1/2} \sin\left(\frac{N\pi x}{L}\right) \exp\left(\frac{-iE_N t}{\hbar}\right) \tag{7.1}$$

where N is an integer and $E_N = \hbar^2 k_N^2/2m$, where $k_N = N\pi/L$; $\psi_N(x, t)$ vanishes at $x = 0$ and $x = L$.

If the stationary state wave-function given by Eq. (7.1) is considered to describe the state of an individual particle with a definite energy, this implies that the probability of finding the particle at any given position within the box does not change with time. Hence the above wave-function does not correspond to the to-and-fro motion of a single particle within the box, where the probability of finding the particle at any position changes with time. This feature remains true under any limiting condition.

Furthermore, using Eq. (7.1), the Fourier transform of $\psi_N(x, t)$ yields the probability distribution function of observed momentum values given by

$$|\phi_N(p)|^2 = \frac{\hbar}{2\pi L} \left\{ \frac{\sin[(p_N + p)L/\hbar]}{(p_N + p)} - \frac{\sin[(p_N - p)L/\hbar]}{(p_N - p)} \right\}^2 \quad (7.2)$$

where $p_N = \hbar k_N$, $k_N = N\pi/L$.

Then if one passes to the macroscopic limit as characterised by Einstein by the requirement that the dimension of the box should be much larger than the de Broglie wavelength ($L \gg 2\pi/k_N$), this requires $N \gg 2$. For large values of N in the macroscopic limit, it can be shown[17,18] that the momentum distribution given by Eq. (7.2) becomes a sum of two effectively non-overlapping distributions peaked around the classical values of momentum $\pm p_N$:

$$Lt_{N\to\infty}|\phi_N(p)|^2 = \left(\frac{1}{2}\right)[\delta(p + p_N) + \delta(p - p_N)] \quad (7.3)$$

This apparently corresponds to the case of a classical particle moving to-and-fro uniformly within the box with a definite energy.

The point that Einstein stressed was that even if the above limiting condition given by Eq. (7.3) was satisfied, all that one could say within the standard interpretation of quantum mechanics was that if one *measured* the momentum of a single individual particle described by the wave-function given by Eq. (7.1), the probability of obtaining the value either $+\hbar k_N$ or $-\hbar k_N$ was the same. Consistent with the quantum mechanical formalism and the wave-function of Eq. (7.1), one is *not* entitled to assume, even in the macroscopic limit, that an individual particle confined within a box *has* a definite value of momentum (either $+\hbar k_N$ or $-\hbar k_N$) *before* any measurement is made. This is because the Fourier transform of Eq. (7.1), which gives rise to the momentum distribution function corresponding to Eq. (7.3) in the macroscopic limit, represents a *superposition* of momentum eigenfunctions corresponding to the eigenvalues $+\hbar k_N$ and $-\hbar k_N$, and *not* a mixed state of these momentum eigenfunctions, however large the mass of the particle may be for a fixed energy.

From the preceding considerations, Einstein concluded that the example again showed that the interpretation of the ψ-function as a complete description of the state of an individual system was untenable. This was because of its failure, in the macroscopic limit, to provide an appropriate description of the motion of a macroscopic object with a fixed energy trapped between perfectly reflecting walls separated by a macroscopic distance. Such problems, Einstein reasoned, would go away if one regarded the ψ-function as a statistical description of the state of an ensemble of systems, all of which have been prepared by an identical procedure.

That is why Einstein wrote to Born about this work,[19] 'For the presentation volume to be dedicated to you, I have written a little nursery song about physics ... It is meant to demonstrate the indispensability of your statistical interpretation of quantum mechanics.' Here we stress once again that by the term 'statistical interpretation of quantum mechanics', Einstein always meant the interpretation of a ψ-function in terms of an ensemble of identical systems it represents. (In attributing this concept to Born, he was maybe being a little provocative.)

However, Einstein did not note that even if one adopted the statistical or the ensemble interpretation of quantum mechanics, the problem of the macroscopic limit of the quantum mechanics did not necessarily disappear. For example, in this particular case, in order to assess the macroscopic limit problem pertaining to an ensemble, one needs to compare the preceding quantum mechanical evolution of, say, the position probability density, with that of an ensemble of classical particles confined within the box potential, all having the same magnitude of the initial momentum $p_N = \hbar k_N$, energy $E_N = p_N^2/2m$ and an initial position distribution function $\rho_{cl}(x, t = 0)$ which is the *same* as that given by $|\psi_N(x, t = 0)|^2$. The time evolution of the classical ensemble in this case can be obtained by solving the relevant Liouville's equation in the box potential.[20] The final result gives the time evolved position probability density at any instant t.

As discussed in Ref. [20], the classical probability distribution corresponds to a flow of probability within the box, resulting from the underlying to-and-fro uniform motion of the particles with momenta $\pm p_N$. This classical probability distribution exhibits nodes, and their locations change with time. Hence the nonclassical nature of $|\psi_N(x, t = 0)|^2$ given by Eq. (7.1) does not lie in its exhibiting nodes, but essentially in the feature that the quantum distribution of nodes does *not* change with time. This particular difference as well as the overall difference between the classical probability distribution and $|\psi_N(x, t = 0)|^2$ obtained from Eq. (7.1) cannot be bridged under any limiting condition imposed on $|\psi_N(x, t = 0)|^2$. Einstein was therefore *not* correct in believing that by taking recourse to an ensemble interpretation of the ψ-function, it could be possible to ensure complete consistency between the macroscopic limit of quantum mechanics and classical dynamics. (A full account of Einstein's apparent advocacy of ensembles is given in Chapter 8.)

Interestingly, Born's response to Einstein's example of the particle in a box was as follows. In a letter to Einstein written in November 1953,[21] just the day after the volume of essays containing Einstein's article was presented to him, Born wrote: 'For the time being I have read only a few of the articles—yours was the first, of course, ... I must take the liberty of asserting that your treatment of the example (that of a ball rebounding between two walls) does not prove what you say it does; namely, that in the limiting case of macroscopic dimensions, the wave mechanical solution does not become the classical motion. This is due to the fact that—forgive my cheek—you have chosen an incorrect solution which is inappropriate to the problem.'

By the phrase 'incorrect solution', what Born meant was that, for studying the passage to the macroscopic limit of quantum mechanics, he considered that the energy eigenfunction was *not* the appropriate one to be used. In the same letter, Born pointed out that, if the ψ-function was taken to be a wave-packet localized in position space, formed by superposing a sufficient number of energy eigenfunctions, one could show that the wave packet moves between the walls 'to-and-fro in exactly the same way as a particle'. As he said, there must be, of course, the quantum mechanical spreading of a wave-packet, but it becomes negligible in the macroscopic limit as mass becomes very large. Hence Born concluded: 'I am convinced that in this sense quantum mechanics represents the motion of a macroscopic single system according to deterministic laws.'

In fact, Born later carried out a detailed calculation of this example using a Gaussian wave packet[22] to illustrate his contention. However, in his entire analysis, Born overlooked the basic point that, even by ignoring the spreading of a wave packet, one cannot literally *identify* the motion of a wave packet with that of a particle. This is simply because, within the standard framework of quantum mechanics, the coordinates appearing in the argument of a wave-function represent only the potential values of the dynamical variable concerned (say, position in the position representation, or momentum in the momentum representations), one of which is actually realized with a probability $|\psi|^2$ when a measurement is performed. In the absence of any measurement, the coordinates in the argument of the wave-function do not correspond to the actual values of, say, position or momentum of a particle in any ontological sense. Hence the time evolution of a wave-packet *per se* does *not* denote the actual motion of a particle. Thus Born's claim about obtaining from quantum mechanics 'deterministic laws' of the 'motion of a macroscopic single system' did not have adequate justification. As Holland[23] put it: 'How can one pass from a theory in which ψ merely represents statistically knowledge of the state of a system to one in which matter has substance and form independently of our knowledge of it?'

The fallacy in literally identifying a wave-packet with a single particle is also obvious from the example such as that of a particle going through a beam splitter. Then its corresponding wave-function is split into two parts with different probability amplitudes corresponding to transmission and reflection. Obviously this could make sense only if one was to interpret the wave-function as corresponding to an ensemble of particles, either a bunch of particles, or a series of single particles incident on the beam splitter.

Interestingly, Einstein's response to Born's treatment of his example in terms of a localised wave-packet was as follows. Einstein[24] mentioned his objection to the basis premise of Born, that the discussion concerning the macroscopic limit of quantum mechanics should be restricted to localised wave-packets. Einstein remarked: 'But when one looks at it in this way, one could come to the conclusion that macro-mechanics cannot claim to describe, even approximately, most of the events in macro-systems that are conceivable on the quantum theory.' Here by the phrase 'most of the events in macro-systems that are conceivable on the quantum theory', Einstein was alluding to the macroscopic limit of quantum mechanical wave-functions other than localised wave-packets.

Dephasing of the Wave-Packet

Even as regards the macroscopic limit of a localised wave-packet, Einstein was not prepared to admit that the mere fact that the spreading or dephasing of a wave-packet became negligible for large mass, entitled one to regard it as able to represent the corresponding classical situation. Referring specifically to the argument that the time required for a wave-packet to spread 'becomes appreciably large on a "cosmic" scale as the mass is increased', Einstein made a cryptic remark in the above mentioned letter: 'But one could easily quote some quite pedestrian examples where the divergence time is not all that long. I consider it too cheap a way of calming down one's scientific conscience.' However, Einstein did not elaborate in detail, either in that letter on in any later writing, what type of 'pedestrian examples' he was considering.

Before proceeding further, we will make a few remarks on the above point. One indirect indication of what type of 'pedestrian examples' Einstein probably had in mind is provided by a brief remark he made in the same letter. While discussing Born's criterion for restricting the discussions concerning the macroscopic limit of quantum mechanics to localised wave-packets, Einstein commented that: if only situations compatible with localised wave-packets were permitted in the macroscopic limit, '. . . one would then be very surprised if a star, or a fly, seen for the first time, appeared even to be quasi-localized'. For an object like a star, for example, the time elapsed since its birth is, in general, astronomically large. Hence if it is represented by a wave packet, Einstein probably thought that its spreading could be appreciable over such a large time. However, a simple-minded calculation shows that this belief was *not* justified. If one considers, for example, a star of the size ($\approx 10^9$ m) and of the same mass ($\approx 10^{30}$ kg) as the Sun, and it is represented by a localised Gaussian wave-packet, the amount of spreading over a time scale of, say, 10^{18} s, which is the age of the Universe, is still found to be vanishingly small, because of its large initial width.

However it is interesting to consider other cases. An instructive example from the contemporary point of view is to consider large molecules in the *mesoscopic regime* (intermediate between the microscopic and macroscopic regimes). Consider, for example, a C_{60} molecule with mass $= 720 m_H$ where m_H denotes the mass of a hydrogen atom, often known as a buckyball, whose quantum mechanical wave property has been experimentally verified using the double-slit interference effect.[25] If it is represented by a Gaussian wave-packet (in terms of the center-of-mass coordinates) of width, say, 10^{-7} m, it can be shown by simple calculation that the spreading of the wave-packet becomes appreciable (the change of width becomes larger than the initial width) within a time scale as short as 10^{-3} s. If one considers a *macromolecule*, say a typical protein molecule of mass of the order of $10^4 m_H$, represented by a Gaussian wave packet of the same width, the comparable time is of the order of 10^{-2} s, and for a macromolecule of mass $10^6 m_H$, the time is around 1 s.

It is thus seen that for a typical macromolecule, which can be seen directly by an electron microscope, the quantum mechanical spreading of a wave-packet is not negligibly small. Hence the very concept of *quasi-localisation* that Einstein repeatedly questioned is indeed an ill-defined notion in the *mesoscopic regime*. This gives rise to the conceptually vexed question as regards at which precise stage during the transition from the micro to macro domain, the concept of quasi-localisation of a macrosystem becomes a well-defined notion. It is this problematic issue which lay at the core of Einstein's contentions discussed earlier.

Localisation of the Wave-Packet

We now turn to further exchanges between Einstein and Born, following Born's assertion that the macroscopic limit of quantum mechanics is meaningful only in terms of localised wave-packets. In a letter to Born written at the beginning of 1954,[26] Einstein pointed out that, if one superposes two or more wave-functions which are individually localised in position or momentum space, the resulting wave-function is, in general, *not* localised. Hence Einstein asserted Born's demand that the ψ-function of a macrosystem should necessarily be 'narrow' with respect to the position or momentum coordinates 'is irreconcilable with the superposition principle for ψ-functions'.

Commenting on this contention by Einstein, Pauli wrote to Born in March,[27] expressing his agreement with Einstein on the point that, if quantum mechanics is applied consistently all the way from the micro to macro level, one *cannot* restrict the allowed wave-functions of a macrosystem to being localised in position or momentum space. In fact, Pauli made a strong statement that *all* mathematically possible solutions of the Schrödinger equation, including superpositions of localised wave-functions, should occur in nature under appropriate conditions, even for macrosystems, whatever be the nature of the macroscopic limit of these wave-functions. Thus Pauli implied that, if appropriate conditions are realised, the quantum mechanical interference effects of superpositions of localised wave-functions should, in principle, be observable, even for macrosystems. As Pauli put it: '... the difficulties are only going to be *technical* because of the small size of the wavelength'.

Then in the same letter Pauli identified the point of his disagreement with Einstein to be Einstein's argument that 'a macro-body must always have a quasi-sharply-defined position in the objective description of reality', and hence that, for this reason, the class of wave-functions which are *not* localised in the position space cannot correspond to the description of individual macrosystems. Pauli's criticism of this contention was interesting. While remarking: 'I believe it to be *untrue* that a "macrobody" always has a quasi-sharply-defined position,' Pauli's point of view was that the fact that a macrobody *is* observed to have a definite position should be regarded as 'a "creation" existing outside the laws of nature . . . the natural laws only say something about the *statistics* of these acts of observation'.

Thus Pauli was effectively endorsing Einstein's contention that the 'reality' pertaining to *individual* macrosystems or *individual* events at the macrolevel was not described by a quantum mechanical wave-function. However, Pauli felt no discomfort with this feature. He remarked, 'As O. Stern said recently, one should no more rack one's brain about the problem of whether something one cannot know anything about exists all the same, than about the ancient question of how many angels are able to sit on the point of a needle. But it seems to me that Einstein's questions are ultimately always of this kind.' Born later commented[28] 'Pauli's analysis of our fundamental difference of opinion was the correct answer to Einstein's paper.'

Conclusions

Advocates of the Copenhagen interpretation were completely sure that they were in possession of absolute truth in the area of determinism, hidden variables, entanglement and so on. When their views were questioned, in particular by EPR, it seemed they did not even consider that they might have something to learn; all they felt required to do was to demonstrate what they thought were the misunderstandings of their critics.

Their attitude to questions regarding macroscopic quantum theory and the classical limit apppeared to be much the same. Since they assumed that the Copenhagen interpretation was the final solution to all the conceptual problems of quantum theory, it followed that it *must* be able to explain the classical limit. Since Einstein questioned this claim, it followed that he *must* be wrong. This they seemed sure of, even before they had any reasons for their belief concerning any particular argument, and sometimes it must be said that the reasons they did come up with seemed a little contrived.

For example, one finds Born making what seems to be an arbitrary choice about which quantum mechanical wave-functions should be expected to behave suitably in the macroscopic limit. One finds Pauli making the unjustifiable assumption that Einstein's idea of macrorealism had already been experimentally proved to be untenable. One finds a determination to ignore the fact that the *ontologies* of quantum and classical physics are totally different; the first uses a wave-function giving only statistical information about the result of any conceivable measurement, while in the second, a particle has substance and dynamical variables independent of measurement. How, then, can it be possible to move from one to the other by a simple mathematical limiting procedure? In addition, the idea of dephasing was scarcely taken seriously.

Looking back after half a century or more, Einstein's criticisms and probing seem well motivated and carefully constructed. Many of his points have been found to be essentially correct. The dephasing argument is now taken very seriously, and the response has been the substantial mathematical theory of decoherence, which has developed over the last three decades or so, and is described in Chapter 12. Overall it has become clear that the theory of the classical limit of quantum mechanics is vastly more complicated, interesting and demanding than imagined by the advocates of Copenhagen, who, it might be said, hoped to get by with a little bit of mathematics when it worked, and a little bit of hand-waving when it didn't! There are now many aspects to study of the classical limit and macroscopic quantum mechanics. A few of these are sketched in Chapter 12, while the work of Leggett is described in a little more detail in Chapter 13.

References

1. Jammer, M. (1974). The Philosophy of Quantum Mechanics. New York: John Wiley, p. 219.
2. Einstein, A. (1949). Remarks to the essays appearing in this collective volume, In: Albert Einstein: Philosopher-Scientist. (Schilpp, P.A., ed.) Evanston: Library of the Living Philosophers, pp. 665–88.
3. Einstein, A. in Ref. [2], p. 671.
4. Einstein, A. (1971). In: The Born-Einstein Letters. (Born, M., ed.) London: Macmillan, pp. 187–9.
5. Born, M. in Ref. [4], p. 189.
6. Pauli, W. in Ref. [4], pp. 217–9.
7. Leggett, A.J. (1980). Macroscopic quantum systems and the quantum theory of measurement, Progress of Theoretical Physics. (Supplement) **69**, 80–100.
8. Leggett, A.J. and Garg, A. (1985). Quantum mechanics versus macroscopic realism – is the flux there when nobody looks? Physical Review Letters **54**, 857–60.
9. Van der Wal, C.H., ter Haar, A.C.J., Wilhelm, F.K., Schouten, R.N., Harmans, C.J.P.M., Orlando, T.P., Lloyd, S., and Mooji, J.E. (2000). Quantum superposition of macroscopic persistent-current states, Science, **290**, 773–7.
10. Friedman, J.R., Patel, V., Chen, W., Tolpygo, S.K., and Lukens, J.E. (2000). Quantum superposition of distinct macroscopic states, Nature, **406**, 43–6.
11. Vion, D., Aassime, A., Cottet, A., Joyez, P., Pothier, H., Urbina, C., Esteve, D., and Devoret, M.H. (2002). Manipulating the quantum state of an electrical circuit, Science, **296**, 886–9.
12. Yu, Y., Han, S., Chu, X., Chu, S., and Wang, Z. (2002). Coherent temporal oscillation of macroscopic quantum states in a Josephson junction, Science, **296**, 889–92.
13. Leggett, A.J. (2002). Testing the limits of quantum mechanics: Motivation, state of play, prospects, Journal of Physics: Condensed Matter **14**, R415–51.
14. Leggett, A.J. (2002). Probing quantum mechanics towards the everyday world: Where do we stand? In: Proceedings of the Nobel Symposium 2001, Physica Scripta **T102**, 69–73.
15. Leggett, A.J. (2002). Physics—Superconducting qubits—a major roadblock dissolved? Science, **296**, 861–2.
16. Einstein, A. (1953). Elementaire Überlegungen zur Interpretation der Grundlagen der Quanten-Mechanik [Elementary considerations on the interpretation of the fundamentals of quantum mechanics], In: Scientific Papers Presented to Max Born. Edinburgh: Oliver and Boyd, pp. 33–40.
17. Holland, P. (1993). The Quantum Theory of Motion. Cambridge: Cambridge University Press, p. 245.

18. Robinett, R.W. (2000). Visualizing the collapse and revival of wave packets in the infinite square well using expectation values, American Journal of Physics **68**, 410–20.
19. Einstein, A. in Ref. [4], p. 199.
20. Holland, P. in Ref. [17], pp. 240–2.
21. Born, M. in Ref. [4], pp. 205–7.
22. Born, M. (1955). Continuity, determinism and reality, Danske Videnskabernes Selskab, Matematisk-fysikie Meddelelser **2**, 1–26.
23. Holland, P. in Ref. [17], p. 222.
24. Einstein, A. in Ref. [4], pp. 208–9.
25. Arndt, M., Nairz, O., Vos-Andreae, J., Keller, C., van der Zouw, G. and Zeilinger, A. (1999). Wave-particle duality of C^{60} molecules, Nature **401**, 680–2.
26. Einstein, A. in Ref. [4], p. 212.
27. Pauli, W. in Ref. [4], pp. 221–5.
28. Born, M. in Ref. [4], p. 227.

REFERENCES

8
Summary of Einstein's Views

Did Einstein 'Reject' Quantum Theory?

Jeremy Bernstein[1] has written that 'In 1926, even before Heisenberg's paper [on the uncertainty or indeterminacy principle], Einstein abandoned the quantum theory'. Bernstein also writes of the depth of Einstein's feelings about the theory, that, at least at one period: 'He wished to destroy it.' And indeed it is extremely commonly implied that Einstein 'did not accept' quantum theory; faced with a shelf of popular physics books, or a series of relevant websites, it would take a very short time to gain the overwhelming impression that Einstein 'rejected quantum theory', usually stated without any qualification.

It must he stressed that, at the very most, this is a tremendous over-simplification; much qualification is certainly required. In Chapter 4, it was stressed that Einstein was left 'breathless' by Heisenberg's initial matrix mechanics paper in 1925, and he must have been at least as excited by Schrödinger's first ideas a few months later. His nomination of the two for Nobel Prizes (one each, he suggested, since their work was too important for them to share a single prize) continued long after his initial enthusiasm had been tempered, long after he had reached his final views of quantum theory.

Of his later views, Heisenberg was to write, in his foreword to the Born–Einstein letters,[2] that 'Einstein agreed with Born that the mathematical formulation of quantum mechanics ... correctly described the phenomcna within the atom.' Heisenberg added that 'He may also have been willing to admit, for the time being at least, that the statistical interpretation of Schrödinger's wave function, as formulated by Born, would have to be accepted as a working hypothesis.' Towards the end of his life, Einstein himself wrote [3] 'It is my opinion that the contemporary quantum theory by means of certain definitely laid down basic concepts, which on the whole have been taken over from classical mechanics, constitutes an optimal formulation of the connections.'

Somewhat earlier he had written[4]: 'There is no doubt that quantum mechanics has seized hold of a good deal of the truth, and that it will be a touchstone for any future theoretical basis, in that it must be deducible from that basis, just as electrostatics is deducible from the Maxwell equations of the electromagnetic field or as thermodynamics is deducible from classical mechanics.' And much earlier still, just after the 1927 Solvay Conference, he had written[5] to Arnold Sommerfeld that: 'On "Quantum Mechanics" I think that, with respect to ponderable matter, it contains as much truth as the theory of light without quanta. It may be a correct theory of statistical laws, but an inadequate conception of individual elementary processes.'

Thus it may be said that Einstein 'accepted' quantum theory in much the same way that physicists of today would say that they 'accept' classical or Newtonian mechanics, or Maxwell's electromagnetism. They would regard the latter theories as highly respectable and important contributions to physics, which have given a very great deal of genuine understanding of an immense range of physical phenomena. It may be said that classical mechanics always, in fact, yields essentially correct answers in its very wide range of applicability, and would routinely be used by today's physicists when discussing problems within this range. However it is not successful outside this range, the appropriate limits being discovered towards the beginning of the last century as elements of the theories of special and general relativity, and of quantum theory. (The relationship between classical physics and quantum theory is a little over-simplified in the above, as discussed in Chapter 12 in particular, but certainly the broad thrust of the remarks is appropriate.) To put this in Einstein's own words[3]: 'Newton ... found the only way which in [his] age was just about possible for a man of the highest thought and creative power. The concepts created are even today still guiding our thinking in physics, although we know that they will have to be replaced by others farther removed from the sphere of immediate experience, if we aim at a profounder understanding of relationships.'

It has been stressed that classical mechanics gives the correct results in the appropriate region, but it is equally important to recognise that its ontology, the picture or description of the world that it provides, so convincing and final as it must have seemed to physicists and others in the nineteenth century, has been superseded by the very different ontologies of relativity and quantum theory. In special relativity, for example, Newtonian absolute space and time have been succeeded by the relativisation of space and time to become spacetime, and the recognition of the speed of light as an absolute and as a maximum speed. In general relativity, perhaps even more surprisingly, the concept of gravitational attraction, surely, one might have thought, the most permanent legacy of Newtonian physics, has given way to the idea of curvature of spacetime. And in quantum theory, the time-honoured picture of independent particles with definite values of position and momentum interacting according to the precepts of locality has also disappeared, leaving the conceptual problems discussed in this book.

It is clear that classical physics, though an exceptionally useful and important theory, is not a final theory. Many, perhaps even most, contemporary physicists may have a similar expectation about today's fundamental theories—quantum theory, quantum field theory, special and general relativity. While these theories are obviously highly successful in very many regions of physics, there are still major clear-cut difficulties, in particular the failure so far to obtain a convincing theory of quantum gravity, and the incompatibility of quantum theory and relativity. There are unresolved issues in cosmology—the role of inflation, the presence or otherwise of dark matter, and in elementary particle physics—the lack of substantial empirically confirmed success in moving beyond the standard model by introduction of Grand Unified Theories or supersymmetry.

Perhaps these problems may be resolved by relatively minor adjustments within the overall constraints of our present theories; perhaps indeed the issues in cosmology and particle physics may be handled by manoeuvres restricted to these areas of physics, such as the emergence of string theory. But alternatively these difficulties may contribute to a growing 'crisis' in physicists' views of today's fundamental theories, and hence to the search for new and deeper theories and concepts produced by another 'scientific revolution'. (The terminology, of course, is that of Thomas Kuhn[6] in his famous approach to the development of science.) Such a revolution would almost certainly require a further change of ontology, the picture, limited as it is, provided by quantum theory and perhaps relativity, becoming as dated as that of classical physics today.

These views, as we have said, are probably quite typical of the general expectations, often not strongly felt, of many physicists. It should be said that others, perhaps notably Stephen Hawking[7,8] have put forward an entirely different point of view. They believe that our present theories are quite close to providing a final 'theory of everything'.[9] Incidentally, while Hawking himself may see this as 'the end of physics', others, especially Philip Anderson,[10] winner of the Nobel prize for work in condensed particle physics, disagree. They argue that a comprehensive understanding of fundamentals can only encourage and support a search for the genuinely new, important and technologically useful phenomena that may unfold as individual atoms and molecules are built up into complex structures, such as life itself. Superconductors, superfluids and Bose–Einstein condensates are among many examples already discovered.

The general point we are making is that Einstein's views on the quantum theory of Heisenberg and Schrödinger, that its mathematical content is used in a highly successful way to make predictions agreeing with experiment, but that it is probably not a final theory, and that the picture it gives of the world may be changed utterly in any replacement theory, are not out of line with those of many other physicists who remain well within scientific orthodoxy, what may be called the scientific 'establishment'.

In a 1976 paper, Bell[11-13] made some interesting related points about hidden variable theorists. Max Jammer[14] had criticised Bell himself for expressing the view that Einstein supported the idea of hidden variables. In reply, Bell argued that Einstein's ideas, centred on fields, should be regarded as a particular conception of a hidden variable approach, not as opposed to such an approach. The point we are stressing here is that Jammer considered the fact that Einstein expected quantum theory to come eventually into conflict with experiment prevented him being considered as a proponent of hidden variables. For Jammer, it seemed that to have 'orthodox' views on quantum theory implied assuming that it was a final theory, and that, even though supporting hidden variables at all certainly made one 'unorthodox', nevertheless even hidden variable theorists would be expected to accept that, once hidden variables were added to quantum theory, it would require no further change and be a final theory. One might say that hidden variable theorists regarded their hidden variables as rescuing quantum theory, not as undermining it fundamentally.

Bell however disagreed. He replied that, if a requirement for being considered a hidden variable theorist is a belief that quantum theory will never conflict with experiment, there would be no hidden variable theorists, certainly not, he implied, himself. Hidden variable theorists also, he suggested, expect an eventual failure of quantum theory. In this paper Bell explicitly made the point that, from this perspective, Einstein could actually be regarded as rather a strong *supporter* of quantum theory. Few supporters of hidden variables, he said, would expect quantum theory to be so comprehensively vindicated at the statistical level. He quoted Einstein, again from the Schilpp volume,[15] as saying of quantum theory that: 'The formal relations which are given in this theory—that is, its entire mathematical formalism—will probably have to be contained, in the form of logical inferences, in every useful future theory.'

So Einstein should certainly not be regarded as having rejected quantum theory just because he believed it would eventually be replaced. At least, if he is to be so regarded, he would be far from alone. Where Einstein did stand practically alone, though, was in believing that one should not wait for a Kuhnian crisis to develop before beginning the search for a replacement for quantum theory. In contrast to the overwhelming majority of the physicists of his time, he was not prepared to accept, even tentatively, that Bohr's Copenhagen interpretation of quantum theory was the final word on how one should approach the conceptual problems of quantum theory. As we shall see in the remainder of this chapter, he could not accept the type of argument put forward by Bohr and his supporters to justify their programme. Equally he found several of the consequences of complementarity, dilemmas in the area of determinism, realism, locality and classicality, intolerable.

The central point of his opposition to Copenhagen was its inability, as he saw it, to handle individual systems convincingly. Satisfied as he was with its effectiveness at the statistical level, he[15] felt it necessary to add: 'What does not satisfy me in that theory, from the standpoint of principle, is its attitude towards that which appears to me to be the programmatic aim of all physics: the complete description of any (individual) real situation.' He also wrote that the theory 'is apt to beguile us into error in our search for a uniform basis for physics, because, in [his] belief, it is an incomplete representation of real things, although it is the only one which can be built out of the fundamental concepts of force and material points (quantum corrections to classical mechanics).' Heisenberg[2] followed his comments on Einstein's acceptance of the mathematical formulation and, at least tentatively, the statistical interpretation, by remarking that Einstein 'did not want to acknowledge that quantum mechanics represented a final, and even less a complete, description of these phenomena'.

Before we discuss Einstein's views in more detail, we return to the remarks of Bell just quoted. As explained in Chapter 5, we would actually broadly agree with Jammer rather than Bell, since, after his own early attempt turned out to be a failure, Einstein showed no interest in simple hidden variable theories, theories where a number of simple extra elements are introduced to smooth over conceptual problems. He clearly did not believe they were capable of doing what was hoped for them. We must remember Bell's argument that Einstein's real aim of producing a unified field theory could be described as a hidden variable theory, though presumably not a simple one, since elements would be added to quantum theory, though again not simple ones.

Jammer's point, though, is presumably that Einstein was far from committed, and far from restricted, merely to adding elements to the quantum formalism as it stood. Rather he intended a totally new theory, and allowed himself the freedom to introduce new concepts as required. Only once the theory was constructed would it become clear that it reduced, from the observational point of view, to today's quantum theory, at least in those regions where the latter has already proved successful. The conceptual nature of the new theory, though, Einstein hoped, would relate to an entirely novel ontology, in which what he regarded as the unacceptable elements of complementarity would be absent. To put things another way, Einstein believed that quantum theory was not complete; rather than completing it by addition of other elements, he aimed at producing an entirely different theory which would be complete.

It will be admitted that Bell did produce yet another quotation from Einstein's response[15] to the papers in the Schilpp volume, which does appear to lend support to Bell. Einstein wrote: 'Assuming the success of efforts to accomplish a complete physical description, the statistical quantum theory would, within the framework of future physics, take an approximately analogous position to the statistical mechanics within the framework of classical mechanics. I am rather of the opinion that the development of theoretical physics will be of this type, but the path will be long and difficult.' As Bell remarks, this does seem a moderately clear commitment to hidden variables. One might seek to downplay the strength of the argument by drawing attention to the word 'approximately'. Einstein, it may be said, is attempting no more than to give a general idea of the relationship he sees between modern quantum theory and the new theory at which he is aiming, not to give a full description. While there may be merit in this argument, it must be admitted that, in many points of his writings, Einstein appears to be backing precisely the point of view Bell suggests—that quantum theory is an ensemble or Gibbs ensemble theory. This argument will be discussed in detail later in this chapter.

Einstein's Philosophical Position—General Remarks

In Chapter 1, we reviewed Einstein's philosophical journey, which might simply be described as one from positivism to realism. In Chapter 3, we described the obvious conceptual difficulties that appeared to emerge from quantum theory (at least, it might be said, if quantum theory as initially presented by Heisenberg and Schrödinger is deemed to be complete): loss of determinism and realism, the Heisenberg principle, the measurement problem, the problem of the classical limit and so on. In Chapter 6, the problem of locality was added to these. These difficulties would be worrying to most people who studied the theory, but Einstein's general views made it natural that he would be particularly disturbed.

In Chapter 4, we studied the various so-called orthodox positions, those of Bohr and von Neumann in particular, and their suggested solutions of the difficulties. Broadly these were rejected by Einstein, and in Chapter 5, we saw his initial attempts to demonstrate their weaknesses. However these were generally unsuccessful. He was forced to admit that his attempt to construct a simple hidden variables theory was a failure, and Bohr was able to dispute his attack on the uncertainty principle with great effect. In 1949, incidentally, Einstein[15] was to state of what he called 'Heisenberg's indeterminacy-relation' that its correctness should be 'rightfully regarded as finally demonstrated'. However he continued to discuss and criticise the orthodox positions. (Abraham Pais[16] reports a remark made by Einstein to Otto Stern that he had thought a hundred times as much about quantum theory as about general relativity. The quantum, Pais says, was Einstein's demon.) As described in Chapter 6, with the EPR paper he was able to focus attention on locality to add to his arguments on determinism and realism, while in Chapter 7 we described his criticisms of the Copenhagen approach to the classical limit of quantum theory. In this chapter we commence by describing his concerns directly with the methods of Copenhagen, then moving to his views more specifically on determinism, realism and locality, and then to his ideas on how to solve the riddles of quantum theory.

First we will say a little more about Einstein's basic philosophical or epistemological standpoint. We have described his eventual position as one of realism, and will certainly continue to recognise this as expressing a major part of the truth. Yet Einstein himself was sometimes less specific. In his 'Reply to criticisms'[15] in the Schilpp volume, he wrote that, while the scientist accepts gratefully epistemological conceptual analysis, he cannot afford to be too much restricted in constructing a conceptual world by adherence to a particular epistemological system. The facts of experience do not allow it! To the systematic epistemologist, Einstein admits that the scientist must seem an 'unscrupulous opportunist'. He appears to be a realist as he attempts to provide a description of the world which is independent of perception; he appears as idealist as he looks on concepts and theories as free inventions of the human spirit, in no way derivable from the empirically given; he appears as positivist as he regards his concepts and theories as justified only to the extent that they provide a logical representation of relations among sensory experiences. He may even, Einstein says, appear as a Platonist or Pythagorean, since his research will be guided by the aim of logical simplicity.

Indeed it may be said that Einstein has, at least in some regards, a rather pragmatic approach to theoretical physics, in particular to the generation of new physical concepts and theories. Jagdish Mehra[17] describes him simply as 'one of the most pragmatic scientists who ever performed research'. This remark may be greeted with considerable scepticism by anybody who cares to remember his resistance over several decades to the Copenhagen interpretation; his advocacy of the idea of the photon again over several decades and against practically universal opinion; his steadfast pursuit of general relativity, ultimately successful, over a considerable number of years; and his even more steadfast pursuit of a unified field theory, unfortunately remaining unsuccessful, for much longer still. Certainly pragmatism did not imply lack of resolution, unwillingness to come to a conclusion about the acceptability of a particular theoretical or conceptual standpoint, or uncertainty about the necessity of following a particular path. It did imply, though, especially towards the beginning of his career, a willingness to follow paths of interest even when their philosophical implications appeared uncertain or confusing, or even frankly uncongenial given Einstein's general views.

For example, as stressed particularly by Roger Penrose,[18] Einstein's two most important papers of 1905 made exceptionally uneasy bedfellows; they were clearly inconsistent with each other. The special relativity paper was a testament to the overarching power of Maxwell and his equations. It might be said that Michelson, Morley, Fitzgerald, Lorentz, even to an extent Poincaré were arguing for or studying the consequences of a limitation in the scope of Maxwell's equations to a particular frame of reference; they were effectively giving primacy to Newton's laws. In the special relativity paper, Einstein argued for the position established ever since for classical (non quantum) physics—that Maxwell's laws are universal, and to allow for this, it is Newton's laws that must become a non-relativistic approximation.

Yet Einstein's light particle or photon paper did quite the opposite. If it were accepted, it would imply that Maxwell's equations would be at best an approximation for systems of many photons where their individual nature and properties would become irrelevant. Einstein's pragmatism allowed him to feel it possible to promote the two incompatible theoretical structures simultaneously, though making the gesture of calling the photon paper a 'heuristic point of view' rather than a formal theory. Despite his unease about the implications for the nature of radiation, he was clear-minded enough to recognise the photon concept as an essential part of the fundamental truth, and determined enough to continue to advocate it as practically a lone voice for almost two decades.

Again he was pragmatic enough in 1916 to write his paper on the A and B coefficients, even though he was aware that this work would encourage the very views against determinism to which he was so opposed. In the following decade he again saw the significance of the ideas of de Broglie and Bose; pragmatism enabled him to be excited by these ideas, and to promote them as a necessary part of any final solution, again irrespective of the conceptual problems they were certain to raise.

Pragmatism allowed him to recognise rather disparate arguments as portions of the truth, elements that would have to be taken account of in any final outcome. However it definitely did not allow him to accept as this final outcome itself any position that he felt was incoherent or based on unsound dogma. He was equally unable to endorse what he might have felt was the shallow positivism of Heisenberg's early position, and the more sophisticated, but he felt equally unfounded, complementarity of Bohr. It might be said that he had been the progenitor of wave-particle duality, and he was not prepared to see the problem side-stepped, or 'solved' by what he considered verbal subterfuge. We shall proceed to discuss this in more detail.

First though we mention that pragmatism appeared to desert Einstein in his later years. Pragmatism might have encouraged him to move in the directions taken by David Bohm and John Bell in the 1950s and 1960s. It would certainly have caused him to be much more enthusiastic about Bohm's hidden variable theory when it did emerge in 1952. From Einstein's perspective, Bohm's theory had very great drawbacks; it was non-local, and in any case was exactly the simple kind of approach to hidden variables that Einstein had already rejected, as explained in Chapter 5. Nevertheless pragmatism would have caused him to be delighted that Bohm's work did show a deep contradiction in, and a possibility beyond, the Copenhagen edifice. Pragmatism was however lacking for Einstein on this occasion, and it was left to Bell, in his role as follower of Einstein,[19] to come to the necessary conclusion, and to draw as much attention to Bohm's work as possible.

In his later years, Einstein's pragmatism appears to have been replaced by what he called, in the argument discussed above, Platonism, or, one might say, he became absorbed in this type of approach to physics above all the others he mentioned in that paragraph. It would seem that a major cause of this change was his highly successful work on general relativity. His approach to general relativity was based on a few general precepts, in particular the principle of equivalence, and involved setting up and solving complicated sets of equations, his guiding principle being the logical simplicity of the mathematics. While it is true that obtaining the correct behaviour for the perihelion of mercury acted as a lodestone for his theory, it is clear that, while it could provide a check on his theories, it certainly could not have provided assistance in their actual construction. The great success of general relativity meant that, when he moved on to unified field theories, he took the message that what was required was an even more elaborate set of equations, chosen according to very general and rather abstract principles. This procedure, we know, did not have the success of Einstein's work on general relativity.

If one wishes to discuss *why* Einstein's years of success were followed by years of disappointment, the cause will not be found, as is often suggested, for example by Heisenberg,[2] in a simple inability of Einstein to follow the new conceptual demands imposed by Bohr and complementarity. (Einstein did state[15] that he had failed to achieve a sharp formulation of complementarity despite much effort on his part, but we must suspect that, in his opinion, this is much more a criticism of complementarity than an admission of his own failing powers.) Neither will the cause be blamed on a movement of Einstein's philosophical position from positivism to realism. It seems much more likely that it was a result of a movement from a virile pragmatism to an almost total reliance on Platonism. We will return to this point towards the end of this chapter.

Einstein's Approach to the Copenhagen Interpretation

While Einstein was clearly unhappy about many aspects of the Copenhagen interpretation, and the conclusions that might be drawn from it, in the Schilpp volume
he concentrated[15] on two aspects that were very much connected. So far in this
chapter we have said very little *explicitly* about realism. We shall say much more
about it later in the chapter. As we have seen in Chapters 5 and 6, Einstein definitely
believed that, for example, the position and momentum of an individual particle
had sharp values at all times, even though he came to admit that both could not
be measured simultaneously. However any argument for this position, which by
1935 he recognised had to be an argument of EPR-type, was subtle and, as such,
open to criticism of all types.

For a case where, he felt, the orthodox position was analogous, but its weaknesses were more obvious, he chose the decay of a radioactive atom; this case
was discussed briefly in the previous chapter. The description of the system given
by quantum theory, after a period has been allowed for decay, will be of a wave-
function for the decay-particle extending out from the parent atom in all directions.
The orthodox position takes it for granted that the wave-function relates to an individual system, not to an ensemble of such systems. In the absence of measurement
(which will be discussed shortly), the decay-particle then cannot be said to be at a
particular point in space, but more significantly it cannot be said to have originated
from a decay at a particular instant of time. (We are assuming, of course, that the
speed of the decay-particle is at least approximately the same for any decay.)

If the period allowed for decay is considerably more than the lifetime of the
radioactive atom, we may say that it is all but certain that the atom has decayed,
but that it does not have a definite decay-time. It must be noted that this is a
much stronger statement than the one that we do not know the decay-time; the
latter statement will obviously be true if we have made no attempt to discover
the decay-time by arranging to intercept the decay-particle, and it is clearly not
problematic in any way.

Einstein now describes the approach of the orthodox quantum theorist as follows. Against his own assertion that the atom must, in fact, have decayed at a
particular time, the reply is that, since there has been no measurement of the time
of decay, such an assertion is 'not merely arbitrary, but actually meaningless'. If
one did attempt to obtain experimentally a value for the decay-time, one would
be disturbing the system, and so such a measurement could tell us nothing about
the unobserved system. Any difficulty Einstein might claim to experience in the
orthodox account, it may be said, thus arises from his illegitimate postulation of
an entity, in this case the decay-time, as 'real', which is not observable.

Einstein was clearly totally unsatisfied by such an argument, which he described
as displaying the 'basic positivistic attitude', in essence being prepared to consider
properties as 'existing' only if they are observed. Indeed, as Einstein describes the
argument, particularly with his use of the word 'meaningless', it is indeed virtually textbook positivism. Complementarity may come to the same conclusions
as positivism by a marginally more sophisticated route, talking of the necessity,

when discussing the property of decay-time, to restrict oneself to situations where measurement apparatus is in position to observe the decay itself or the decay-particle. However, when discussing this particular physical situation Einstein was as equally unimpressed by complementarity as by the more naked positivism. Complementarity makes itself logically quite impregnable, since it renders discussion of problematic situations out of order by *fiat*. Nevertheless Einstein had chosen an example where not only he himself, but, one suspects, many physicists and others thinking about the problem, would find the argument of complementarity rather contrived.

Einstein's other argument directly against the Copenhagen interpretation, though it may well be thought of as a different aspect of the first, concerns the central nature for Copenhagen of the measurement process. If Einstein was not impressed by the positivistic argument that, if an entity or property were not observed, it should not be considered real, he was perhaps even less impressed by the fact that, for any strand of orthodox interpretation, the act of observation (or detection or measurement) has an effect far transcending its role in classical physics.

In classical physics, measurement is a straightforward physical process, governed totally by the usual laws of physics. It is a special type of process only in the extremely limited sense that its role is to determine the value, within a particular range of accuracy, of a property of the system before the measurement. It is obvious that the act of measurement will almost inevitably have an effect on the system being observed, though not necessarily on the actual property being measured. However in classical physics, this effect of measurement, or, to put things another way, the interaction between measuring and measured systems, may itself be studied in detail, and it again strictly follows the laws of classical physics. There will be particular measurements for which the effects of this interaction on the measured system may be vanishingly small.

In orthodox versions of quantum theory, as discussed in Chapter 4, in particular, measurement plays a much more important part, and in particular the measurement process transcends the laws, not only of classical physics, but also of quantum laws for processes other than measurement. The most clear-cut position is that of von Neumann.[20] For von Neumann, it is 'measurement' that achieves the transformation or collapse of the wave-function of the measured system from, in general, a linear combination of eigenfunctions related to the observable being measured, to the single eigenfunction corresponding to the value of the observable actually obtained in the measurement. The transformation is a non-unitary irreversible process, and so certainly not governed by the Schrödinger equation. This is, of course, the measurement problem of quantum theory.

How (and where) does collapse take place? Let us consider, as an example considered by von Neumann, a measurement of temperature. It will consist of a thermometer in contact with the body whose temperature is being measured, and some photons travelling from thermometer to the retina, then to the optic nerve tract and the brain; finally there is what von Neumann calls the 'abstract ego'. One must assume collapse takes place at some point in the macroscopic range of the measurement chain. It was easy for von Neumann to show that, from the point of view of the analysis, it made no difference at what point within this range the collapse occurs; it does not affect the result of the mathematics, for example, on which side of the optic nerve tract one imagines the collapse.

Though it does not affect the working of the collapse idea, it is still an important question to discuss where it occurs, and von Neumann gives a very clear answer: it is at the final stage, which must be entirely different in nature from all the previous stages. He writes that: 'It is inherently entirely correct that the measurement or the related process of the subjective perception is a new entity relative to the physical environment and is not reducible to the latter. Indeed, subjective perception leads us into the intellectual inner life of the individual which is extra-observational by its very nature.' He also states that 'experience only makes statements of this type: an observer has made a certain (subjective) observation; and never like this: a physical quantity has a certain value'.

So, for von Neumann, collapse is performed by the observer, by, in fact, the *abstract ego*. It may be said that measurement does not *just* measure the value of the appropriate property of the observed system; it effectively creates it, since, before the measurement, in most cases the property does not have a definite value. Von Neumann's ideas are somewhat vague, and one may well understand why some of those who have discussed them have articulated them in more concrete terms. Eugene Wigner[21] interpreted von Neumann's *abstract ego* directly as consciousness. Others such as Rudolf Peierls,[22] and more recently David Mermin,[23] and Anton Zeilinger and Časlav Brukner[24] have suggested that what is most easily changed in the mind of an observer is the knowledge of, or information concerning, the physical state of the universe, and not the physical state itself. Thus they have suggested a knowledge or information interpretation of quantum theory. There is a little discussion of these ideas in Chapter 10.

None of this would cut any ice at all with Einstein. For him[3]: 'Physics is an attempt conceptually to grasp reality as it is thought *independently of its being observed*'; and (completing the quotation given above)[15]: 'What does not satisfy me in [the statistical quantum theory], from the standpoint of principle, is its attitude towards that which appears to me to be the programmatic aim of all physics: the complete description of any (individual) real situation (*as it supposedly exists irrespective of any act of observation or substantiation*).' (In both quotations the italics are ours.) Both quotations refer to the 'real' and we shall study Einstein's views on 'realism' later in this chapter. Here we dwell on the fact that, for Einstein, an 'observer-independent realm', as Arthur Fine[5] calls it, is perhaps the most important requirement in any physical theory. He would be extremely reluctant to admit that, rather than the observer being free to do exactly that—just observe some behaviour or events occurring independently of himself, the observation is an intrinsic part of creating what may be observed.

Let us now turn from the ideas of von Neumann to the Bohrian version of orthodoxy, complementarity.[25,26] Though the two aproaches are very different, for Bohr also measurement plays a crucial role in any account of quantum theory. If one wishes to discuss, for example, the momentum of a particle, one must do so in terms of an apparatus to measure its momentum. Indeed it may be said that what is open to discussion is not the momentum itself, but the result of the measurement, or in other words a combined property of measuring and measured systems. It is clear that this set of ideas could be no more to Einstein's taste than those of von Neumann. In fact in Bohr's case we are not permitted to describe the behaviour of the system under measurement even in mathematical terms. We do not have any equivalent to the collapse process of von Neumann. We are simply forbidden to consider anything apart from the results of measurements.

Thus we do not have a measurement-independent realm in Bohr's approach. It is fair to ask, though, whether we have an *observer*-independent realm. In other words, though we are not allowed to discuss the properties of a particle in the absence of measurement, may we describe that measurement itself as a physical process independent of the brain, mind or *abstract ego* of any human observer? Certainly Bohr's writings would give the answer that no observer is required. From his point of view, the measurement process is regarded as classical, and the macroscopic measuring apparatus is also to be treated as classical. For example, in the early rounds of the Bohr–Einstein debate,[27] as described in Chapter 5, this is exactly how measurement is described; no human observer is required, but every effort is made in text and diagrams to present the measuring apparatus as definitely macroscopic, in fact as massive. Thus one is definitely led to the belief that, for a system to be used as a measuring device, a requirement is that it is macroscopic. For Bohr, it might be said, extremely (perhaps even inexcusably) loosely, that the equivalent to von Neumann's collapse is caused, not by a human observer, but by a suitable macroscopic system.

As Jammer[14] in particular has pointed out, and as mentioned in Chapter 5, there is a problem with this set of ideas. Bohr's refutation of Einstein in the early rounds of the debate relied on application of the Heisenberg relations to the measuring system. It is not easy to reconcile this with the description of this system as, in principle, classical. It should be noted that, to say that something is classical, one is not merely saying that it is large enough that any quantum properties may be ignored in nearly all situations, though perhaps not in absolutely all. In the context of complementarity, to call something classical should imply it has no quantum properties at all, so there is, for example, no questions of forming linear combinations of different classical states. It appears that Bohr's effective answer to what ensures classicality, the macroscopic nature of the system, may not be definitive.

Tony Leggett[28] has for several decades investigated the topic of macoscopic quantum theory, the essentially experimental question of whether it is possible to form linear superpositions of macroscopically distinguishable states. A related step has been the demonstration[29] by Anton Zeilinger's group of the diffraction of the Buckminster fullerene molecule, C_{60} (the buckyball). While this is, of course, an early stage in the move from microscopic to macroscopic, it again puts in question the idea that macroscopic necessarily means classical (and thus effectively assumes collapse of wave-function). The macroscopic limit of quantum theory is discussed in Chapters 7 and 12 in this book, while Leggett's work is described in Chapter 13.

Einstein took as the orthodox position on quantum theory that it was the (human) observer who caused collapse and thus enabled measurement to be performed or completed, rather than just the onset of classicality. He regarded this as one of the major defects of the orthodox position. (Of course he would also have found gravely unsatisfying the idea that measurement, even considered more generally as a physical operation, was required somehow to create values for physical quantities, but it was the assumed necessity for an actual observer that particularly affronted him.) Half a century after Einstein's death, it is probably true that, equally among the orthodox and the less orthodox, the consensus is to downplay the actual observer when interpreting quantum theory. For the orthodox, even in discussing von Neumann's ideas, there is a tendency to draw a veil over the fact that the *abstract ego* is the ultimate cause of collapse. Rather the idea is stressed that it is the macroscopic nature of the measuring apparatus that allows a measurement result to be obtained, though, as we have seen, Leggett's work and the experimental advance to macroscopic quantum theory may put that approach under threat. And many of the more modern non-standard interpretations, such as ensemble and many world interpretations, have the specific aim of removing, not just the special role of measurement, but especially the observer, from quantum theory.

To this extent it might be suggested that Einstein's arguments on this point have succeeded, or at least that his criticisms of the interpretation of quantum theory are not as relevant today as when he made them. However it is interesting to read in Bernstein's book[1] of the views on quantum theory held by John Wheeler, a physicist of massive achievement in many areas of the subject. Wheeler is prepared to admit himself a follower of Bohr, though he sees Bohr as more flexible than many others would. The essence of quantum mechanics, Wheeler says, is that: 'No elementary quantum phenomeon is a phenomenon until it is a registered phenomenon, that is to say brought to a close by an irreversible act of amplification.' While this certainly follows Bohr in not definitively requiring an actual observer, Wheeler confesses himself 'puzzled' by, for example, measurements of photons consisting of a flash on a zinc sulphide screen on a faraway planet with no life. He also likens the Bohr approach to measurement to his 'quantum mechanical umpire' for whom 'They ain't nothing till I calls them,' rather than 'I calls 'em the way they are.'

Wheeler may be said to promote with vigour the idea of measurement as a very special event, and also one of some mystery. He admits he has no answer to the question of Eugene Wigner—what does 'irreversible' mean in the 'irreversible act of amplification'? It is all, he says, a puzzle, though an immensely stimulating one. With his mention of the 'umpire', and his concern about the point of unobserved measurements, he also seems to be giving at least some credence to the idea of the 'observer' playing a role in quantum measurement, and thus in quantum theory.

Wheeler is definitely an opponent of Einstein's views on quantum theory, and thus of Bell's also—he regards EPR as exactly the wrong thing to be worrying about, and considers Bell's inequality, to be discussed in the next chapter, as simply a part of ordinary quantum mechanics. He considers quantum theory absolutely correct, but still criticises physicists who take it for granted. He believes the correct question is not the one Einstein or Bell might ask—What is wrong with quantum theory?, but one that searches for a deeper foundation for the theory—Where does it come from? Wheeler's views are interesting, and come from a lifetime of thought at the conceptual frontiers of physics, but they certainly leave considerable scope for disagreement and counter-argument along the lines of Einstein.

Einstein on Determinism

In Chapter 3 we sketched the apparent problems quantum theory, if regarded as complete, caused for determinsm and realism, while in Chapter 5 we discussed Einstein's initial attempts to solve these problems. In Chapter 6 we explained his later ideas, and described how he attempted to use the EPR argument to demonstrate the necessity to complete quantum theory, and thus tackle the difficulties for determinism and realism, and also that for locality. In the previous section we examined Einstein's particular criticisms of Copenhagen; it will be noted that there we did not make explicit use of the terms—determinism, realism, locality, though it could well be argued that at least realism played a considerable though largely unannounced role in the argument. We now attempt to elucidate his final approach to these matters, following, for example the response of Bohr and others to EPR, the arguments of Born,[30] particularly on the classical limit of quantum theory, the 1952 hidden-variable theory of Bohm,[31] and of course the constant thought on his own part that was stressed by Bernstein[1] and Pais.[16] It may be said that we are now analysing Einstein's concerns not so much about Copenhagen itself, but about its implications for physics.

Let us start with determinism (or causality, as Einstein calls it), and of course the one thing that everybody knows, or at least thinks they know, about Einstein and quantum theory is that his supposed rejection of it was largely or entirely a result of his unwiillingness to accept its lack of determinism. His famous comment is 'God does not play dice.' As early as December 1926, he[30] wrote to Born that: 'Quantum mechanics is certainly imposing. But an inner voice tells me that it is not yet the real thing. The theory says a lot, but does not really bring us any close to the secret of the 'old one'. I, at any rate, am convinced that *He* is not playing at dice.' As late as 1953, he refers in another letter to Born to ' "that same non-dice-playing God" who has caused so much bitter resentment against me ... amongst the quantum theoreticians'.

Certainly it is true that Einstein was extremely positive towards determinism; his preference would undoubtedly have been for an entirely deterministic universe. Yet there are reasons for doubting whether it was as sacred for him as is often imagined; in particular it is misleading to suggest that lack of determinism was, as he saw it, the main defect of the Copenhagen interpretation of quantum theory, the one that made him ultimately reject it.

The clearest evidence for this occurs in an intervention by Wolfgang Pauli into a discussion on quantum theory between Einstein and Born carried out in a series of letters[30] written in 1953 and 1954. This was mentioned briefly in the previous chapter. In fact it was more than a discussion, practically an argument, even a moderately heated one, not so much because their views on quantum theory differed—the two men had been great friends for many years and accepted this, but because Born found it completely impossible even to understand Einstein's position concerning a particular physical problem. (Born writes in one letter of 'the anxiety which your last letter had caused me. Its tone was irritable and angry, as if you had regarded the difference of opinion between us as a personal attack.')

Fortunately Pauli then made a visit to Princeton and was able to report back to Born. His letters are included with the Born–Einstein letters.[30] Pauli suggested that Born had chosen to erect a dummy Einstein, and then knocked this dummy down with great pomp; Pauli himself was unable to recognise the Einstein that Born wrote about. In particular, Pauli said that Einstein had emphatically told him many times that he did not consider the concept of determinism to be as fundamental as is often believed; he certainly did not use as a criterion for admissibility of a theory the question—'Is it rigorously deterministic?' Pauli went through Einstein's analysis of the physical situation, discussion of which had caused the clash with Born. He agreed with Born that Einstein had got 'stuck in his metaphysics', but said that his argument did not contain the concept of determinism at all. Rather than his point of departure being determinism, Pauli suggests, it is realism.

We will discuss Einstein's views on realism shortly, but for now we appear to have reached a conundrum concerning his views on determinism. Not for the only time when discussing Einstein's approach to an aspect of quantum theory, we seem to find two very different positions, one very commonly accepted and not without some evidence, but also another position altogether, again with seemingly good evidential support. To move forward we shall review a number of quotations from Einstein's writings over quite a long period of time.

We start from an early letter to Born, well before the arrival of modern quantum theory, but 4 years after his 1916 work on the A and B coefficients which, taken in the most natural way, did appear to threaten determinism. He wrote:[30] 'The business about causality causes me a lot of trouble, too. Can the quantum absorption and emission of light ever be understood in the sense of the complete causality requirement, or would a statistical residue remain? I must admit that there I lack the courage of my convictions. But I would be very unhappy to renounce *complete* causality. … (The question whether strict cauality exists has a definite meaning, even though there can probably never be a definite answer to it.)'

The next quotation is from a 1924 letter[30] to Born. It followed the publication of the Bohr–Kramers–Slater (BKS) paper discussed in Chapter 2, in which, in what seems at least in retrospect to be a last desperate attempt to avoid acknowledging the existence of the photon, Bohr relinquished determinism, as well as conservation of energy and momentum. (His continuous 'ghost field' had to interact with atoms with quantised energy-levels, which made these renunciations inevitable.) Einstein wrote: 'Bohr's opinion about radiation is of great interest. But I should not want to be forced into abandoning strict causality without defending it more strongly than I have so far. I find the idea quite intolerable that an electron exposed to radiation should choose *of its own free will*, not only its moment to jump off, but also its direction. In that case, I would rather be a cobbler, or even an employee in a gaming-house, than a physicist.'

We now move forwards to March 1927, a fairly short period after the epoch-making work of Heisenberg and Schrödinger. Einstein was asked to write an article[32] on the two hundredth anniversary of Newton's death. Naturally he took the opportunity to praise not only Newton's physical insight but in particular his development of the fundamental mathematical techniques which 'formed the program of every worker in the field of theoretical physics up to the end of the nineteenth century'. Einstein stressed the importance for Newton of causality. He wrote '[B]efore Newton there existed no self-consistent system of physical causality which was somehow capable of representing any of the deeper features of the empirical world'; and 'The differential law is the only form which completely satisfies the modern physicists' demand for causality. The clear conception of the differential law is one of Newton's greatest intellectual achievements.' Einstein admits that the specific theoretical structure built by Newton was threatened by the necessity to develop the idea of the physical field, but argues that, in a more general sense, Newton's ideas were a crucial step in the conception of field theories, including the theory of general relativity.

Only in the last paragraph of this paper does Einstein turn to the quantum theory and the problems it presents for any conceptual system related even in a general way to that of Newton. Many physicists, he says, believe that not just the differential law, but causality and even the idea of the space–time representation of physics must be abandoned. Einstein finishes the article with a question—'Who would presume today to question whether the law of causation and the differential law, those ultimate premises of the Newtonian view of nature, must definitely be abandoned?' In the translation of the recent biography of Einstein by Albrecht Fölsing,[33] the wording is a little more aggressive: 'Who would have the temerity today …?' Nevertheless Einstein does not appear to be saying that determinism will never have to be abandoned, merely that it would be brave or reckless scientist to say that that decision must already be made.

If that conclusion seems at least marginally open-minded, Fine[5] points out that a remark on the same issue in a letter to the Royal Society[34] on the same anniversary appears less so. Here Einstein writes: 'May the spirit of Newton's method give us the power to restore unison between physical reality and the profoundest characteristic of Newton's teaching—strict causality.'

In 1933, at the conclusion of his Herbert Spencer lecture, Einstein[35] noted that the probabilistic interpretation of the wave-function is logically unobjectionable and has important successes to its credit, but nevertheless concludes that: 'I cannot but confess that I attach only a transitory importance to this interpretation. I still believe in the possibility of a model of reality—that is to say, of a theory which represents things themselves and not merely the probability of their occurrence.' It is important to recognise that this statement contains references to both determinism and realism, and we shall return to this point. In fact when we analyse Einstein's most important discussions of quantum theory over the next decades, his 1936 article[36] on 'Physics and Reality' and his contributions[3,15] to the Schilpp volume, there is much quite directly about realism, comparatively little at least explicitly about determinism.

We end this series of quotations with a rather discordant remark of Einstein's in a letter[37] of December 1950 to Schrödinger. He says that: '[I]t seems certain to me that the fundamentally statistical character of [quantum] theory is simply a consequence of the incompleteness of the description', which had been his belief for at least the previous 15 years. However he continues by saying that: 'This says nothing about the deterministic character of the theory; this is a thoroughly nebulous concept anyway, so long as one does not know how much has to be given to determine the initial state ("cut").'

We now try to sum up what has gone before, leaving to the following section the question of whether Einstein actually conflated determinism and realism, or, as Fine[5] suggests, whether realism, for Einstein, was inevitably causal realism, so any distinction between the importance of the two concepts for Einstein is practically meaningless.

From his early thoughts on the apparently indeterministic nature of radioactive decay, but particularly from the publication of his 1916 paper and right through the 1920s, before and after the coming of modern quantum theory, determinism was obviously exceptionally important for Einstein. He was extremely uneasy that his own 1916 paper might have played a part in its downfall, at times tremendously concerned about the future of theoretical physics, as demonstrated in his 1924 letter to Born. He certainly had no intention of even questioning determinism on the basis of a theory such as BKS, which he must have considered somewhat shoddy.

So he hoped for, worked for, argued for, and expected the triumph of determinism. Yet he was in the main open-minded enough to recognise that eventually it might be necessary to admit defeat. He argued rather that this admission should not be made on the basis of fragments of theory that were little more than suggestive such as BKS. It should not be made even on the basis of modern quantum theory, which was certainly exceptionally successful in practice, but, Einstein felt, had been supported by untenable conceptual dogmas to produce the Copenhagen interpretation, which did renounce determinism. The position of this paragraph is not out of line with Pauli's report to Born; for all that Einstein himself favoured determinism, it was not (perhaps, one might say, not quite) a criterion for his even being prepared to consider a theory, that it be absolutely deterministic.

Even before fuller consideration of Pauli's remarks, one must recognise the apparent shift of focus of Einstein's concerns from determinism to realism that seems to have occurred in the 1930s, a shift actually denied by Fine, as we shall see in the following section. It could be argued that Einstein's remark in the letter to Schrödinger is a continuation of this process. It seemed to imply that he had lost at least some belief in the conceptual significance of determinism, perhaps again in comparison with his lasting belief in realism.

Einstein on Realism

We now turn to realism. It may scarcely be denied that Einstein was, in some sense, a realist, that realism was extremely important to him. However that remark can only be useful if we are able to come to a conclusion on what he meant by realism, and what kind of realist he was. We have, in fact, already met aspects of his realism in the section of this chapter on his response specifically to Copenhagen; these were his desire for an observer-independent realm, and his opposition to what he felt was the most stubborn form of positivism, for example the refusal to acknowledge that a decayed atom had decayed at a particular time, unless its decay had actually been observed.

For Einstein, oberver independence was perhaps the most important aspect of realism. It seemed intolerable that the observer should be intrinsically and decisively a central performer in the content of the physics being observed. Equally he could not accept that the properties of atomic quantities could be discussed only at the act of measurement. The refusal to accept that the atom had a definite time of decay must have seemed to him the worst form of positivism, taking much too far the slogan that the job of physics was only to correlate directly experienced sensations. Einstein[3] says that 'one is driven to the conviction that a complete description of a single quantum system should, after all, be possible; but for such complete description there is no room in the conceptual world of statistical quantum theory'.

As with determinism, it should not be said that Einstein *demanded* of the physical universe that it satisfied his strongly held intellectual desires. Fine[5] quotes a 1955 letter to M. Laserna, where having said 'It is basic for physics that one assumes a real world exisitng independently from any act of perception,' Einstein adds 'But this we do not *know*.' However, at least in the case of observer-independence, it is difficult to imagine Einstein actually *accepting* a theory which did not satisfy this requirement; one feels he would always have demanded the freedom to work towards an alternative theory.

Having established Einstein's bare criteria for realism, we must inquire whether he would agree with broader tenets often associated with the term—that there is an objective physical reality; that our theories attempt to describe or correspond to this reality, and may, in an always partial and corrigible way, succeed in doing so; indeed that our successive theories are a closer and closer approach to this reality, and that the realist aim is that there should eventually be complete convergence. To what extent did Einstein accept this set of beliefs?

Let us start to answer the question by following Einstein's well-known advice[36] that, if you want to find anything out from theoretical physicists about the methods they use, you should look at their deeds rather than their words. Much of Einsteins's labours on quantum theory, described in Chapters 5 and 6, were aimed at showing that, contrary to any standard approach to quantum theory, both the position and momentum of a particle had specific values at any particular time, even though Einstein came to accept that both could not be measured simultaneously. This was a definite denial of positivism on this particular point, and a call for at least a measured amount of realism.

To turn to words rather than deeds, in connection with EPR Einstein's requirement of completeness was explicitly a demand that every element of reality should be matched by a corresponding one in the theory. In his *Dialectica* article,[38] he wrote in connection with EPR that: 'If one asks what, irrespective of quantum mechanics, is characteristic of the world of ideas of physics, one is first of all struck by the following: the concepts of physics relate to a real outside world, that is, ideas are established relating to things such as bodies, fields, etc., which claim a "real existence" that is independent of the perceiving object.' We already mentioned that the quotation from his Herbert Spencer lecture was at least as much about realism as determinism. One is forced to conclude that Einstein spoke of an objective reality, the idea of which he used, to a considerable extent, for his own conceptual analysis of quantum theory, and his criticism of the Copenhagen interpretation.

However the question of whether successive theories would become closer to and gradually approach an ultimate reality is much more difficult to answer positively. Einstein wrote several articles (e.g., Refs. [32], [36], [39]), which are effectively testaments to realism. He describes how the mechanical approach of Newton was succeeded by the field-theory of Maxwell, Lorentz and others, and then by general relativity, all broadly realistic theories. For Einstein this is all excellent progress. He emphasises what connections there are between successive theories, but in no way feels it necessary to hide, or apologise for, the massive conceptual shift between, for example, Newtonian gravitation and general relativity. Newtonian gravitation, one is forced to conclude, is an important theory because of its empirical success, not because any picture it gives of reality is close to the ultimate. And general relativity has replaced Newtonian gravitation because of its greater empirical success, perhaps because its foundations were more coherent, but not because it moved closer to this same ultimate reality.

Before following this point up, we comment briefly on one aspect of Einstein's support of realism, his determination to consider individual systems. At first sight this may seem a totally obvious aspect of any realist theory, and indeed a point that would be conceded *in principle* even by non-realists, even though the latter might regard it as unnecessary and unhelpful. However it must be remembered that in the 1920s and 30s it was commonly thought to be impossible to deal with individual atoms, and also that this state of affairs would almost certainly be permanent. For example as late as 1952, Schrödinger[40] wrote that: '[W]e never experiment with just *one* electron or atom … We are not *experimenting* with single particles, any more than we can raise Icthyosauria in the zoo.' In what seems to have been a lazy, at most semi-articulated positivism, supporters of orthodox positions appear to have taken the view that, if we can never experience individual systems outside a measurement context, that is an even stronger reason to be satisfied that little attention is paid to them in the theory.

Einstein was never impressed by any aspect of this argument. He[36] asked: 'Is there really any physicist who believes that we shall never get any insight into these important changes in the single systems, in their structure and their causal connection, regardless of the fact that these single events have been brought so close to us, thanks to the marvellous inventions of the Wilson chamber and the Geiger counter? To believe this is logically possible is without contradiction; but, it is so very contrary to my scientific instinct that I cannot forego the search for a more complete conception.' From the time of very early discussions with Heisenberg,[41] Einstein insisted on the right and necessity to discuss the individual particle and its properties. This was the basis of what we call his ensemble interpretation, to be discussed later in this chapter.

The assumption that study of an individual atomic particle over a reasonable length of time would forever be impossible has been shown to be quite wrong, and Einstein, on this matter, quite right, by the construction of the ion trap. From the 1950s on, physicists have developed techniques to trap single ions, and as experimental sophistication has developed, and in almost deliberate defiance of the orthodox theoreticians of previous decades, it has become possible to trap a single ion for periods of weeks;[42,43] some researchers become so attached to a particular ion that they give it a pet name! This experimental success has resulted in Hans Dehmelt and Wolfgang Paul being awarded the Nobel prize for Physics in 1989. The ion trap is one of the most important techniques for the development of quantum computation, and is discussed a little further in later chapters.

Another important aspect of Einstein's approach to realism is the continuity between microscopic and macroscopic. This is in strong contrast with orthodox approaches where two tactics are used. Particularly when measurement is being discussed, orthodox quantum physicists may assume a 'cut' or 'Heisenberg cut' between microscopic and macroscopic. That is to say that a *lack of* continuity is assumed without argument. However when measurement is not being discussed, a variety of mathematical techniques are used to argue that, in an appropriate limit, quantum theory could represent a macroscopic system in a satisfactory way; the particle would be localised, at least to a very good approximation. Einstein felt that these arguments were designed selectively to give the result required, and this is why he and Born[30] had intense and somewhat detailed arguments about the way in which the limiting procedures should be used, as discussed in Chapter 7 of this book.

At the macroscopic level, Einstein felt it was clear that objects were localised. There were two aspects to this statement. First, there should be no appreciable fuzziness in the wave-function at any particular time, and the wave-function should not disperse with time. Secondly the wave-function should not consist of two or more components, each component perhaps being individually localised, but the various components being separated by a considerable distance.

His intention, though, was to argue that lack of realism at the microscopic level could not be made to disappear at the macroscopic level, at least not without the arbitrary cut, or the equally arbitrary restriction of the type of wave-function that should be subject to the appropriate limiting procedure. If this were the case, then clearly the wave-function could not be of the form required to represent a macroscopic object. The macroscopic particle itself must be localised to a much greater extent than the wave-function representing it, or in other words the wave-function must be incomplete. But again, if the macroscopic is to be regarded merely as the limiting case of the microscopic, the wave-function corresponding to a microscopic particle must also be incomplete, which, of course, was exactly what Einstein had wanted to demonstrate.

Einstein's concern with macroscopic systems, and his desire to use them to explore what he considered to be the weaknesses of the Copenhagen position, were at the basis of his famous question—'Is the moon still there when you are not looking at it?' This question, often used to encapsulate Einstein's naivety concerning quantum theory, is in fact quite subtle. It not only raises the fundamental question as to whether the existence of even a macroscopic object can be discussed without a measurement (or observation) being involved. It also hints at the question of classicality—how a macroscopic system becomes classical, and why its wave-function does not spread in time, causing it to become delocalised and lose its classicality. Einstein raised this question in a letter to Born[30] written in December 1953.[44] It was a question that the orthodox felt at the time could be best answered by hand-waving discussions of the effects of observation. Only comparatively recently has it become clear that the question was extremely pertinent and important, and can be given a detailed and convincing answer in the theory of decoherence, for which see Chapter 12.

Einstein used discussion of the macroscopic skilfully in his discussion of the decaying atoms in the Schilpp volume.[15] (This discussion was mentioned in the previous chapter; here we review some further points.) He dismissed the positivistic belief that, just because the decay of an individual atom has not been observed, one should not consider that it has a definite time of decay. Recognition of reality, he said, is not derived directly from the senses. Rather we should think of something as real if it makes intelligible what *is* given by the senses. Thus the significant question is not whether a definite time of decay exists, but: 'Is it, within the framework of our theoretical total construction, reasonable to posit the existence of a definite point of time for the transformation of a single atom?' This, Einstein admitted, is a question that may well be answered in the negative by a supporter of Copenhagen.

However he now used an argument of the type of Schrödinger's cat, as discussed in Chapter 6, to make this response, as he puts it, impractical. (Einstein acknowledges Schrödinger at this point.) Einstein included a Geiger counter, and a registration mechanism with a registration strip moved by a timer. Einstein now said that the registration strip will show a single blackened portion at a definite location, since it is a macroscopic object, but theory still predicts only relative probabilities for the blackened region being at any location. While, as he said, it is not logically impossible that the theory is correct, and it is only the observation of the registration strip that has created a localised blackened region, it is unlikely that anybody would accept this argument. Rather we believe the strip, being macroscopic, will behave in a realist fashion, even in the absence of observation. For the case where the macroscopic object is included, Einstein believed that the Copenhagen interpretation clearly cannot provide a complete description of the system, and he concluded that such must therefore be the case even for a microscopic system.

This is precisely the kind of argument made by Einstein that Pauli analysed for Born, as mentioned above; Pauli[30] explained that the argument was based on realism rather than determinism, though Pauli agreed with Born that Einstein was 'stuck in his metaphysics'. It is interesting to examine Pauli's response to Einstein's points. Pauli agreed with Einstein that all mathematically possible solutions of the Schrödinger equation occur in nature, even for a macroscopic object, not just those that correspond to a sharply defined position for the object. Pauli calls the latter class of solutions K^0, and he acknowledges that if ϕ_1 and ϕ_2 are wave-functions in K^0 with widely separated mean positions, then the wave-function $c_1\phi_1 + c_2\phi_2$ will *not* be in class K^0. Also, if $\phi(x, t_0)$ is a wave-function in class K^0 at time t_0, then at a later time t, $\phi(x, t)$ will no longer be in K^0 for $|t - t_0|$ large enough. (Both these types of functions correspond to examples mentioned above which Einstein considered gave problems for classicality of macroscopic systems.) So it is impossible to restrict consideration even for macroscopic bodies to functions of type K^0.

Pauli also agreed with Einstein that when a macroscopic object is observed, it has a (very nearly) sharply defined position; also that it is not reasonable to invent a causal mechanism that might ensure that this is the case. (Pauli specifically says that he does not believe the appearance of this definition can be deduced from natural laws.) Yet at this point the opinions of the two men diverge. Einstein believed the argument shows that the object must have had the sharply defined position even before the observation; in other words, the wave-function is not complete. Pauli however did not believe that a macroscopic body always has a (nearly) sharp position, since he can see no fundamental difference between microscopic and macroscopic bodies. (This would seem to be missing Einstein's point; Einstein was using the argument to show completeness for both macroscopic and microscopic objects, since he too saw no fundamental distinction.) Pauli commented that one must always have some indeterminacy in position to allow the wave-aspect of the object to manifest itself; here he was treating the Heisenberg principle in the way dictated by Copenhagen, but a way which, as shown in Chapter 3, is arbitrary, and so which Einstein is by no means obliged to follow.

Pauli himself believed that the appearance of the definite position at observation must be regarded as 'a "creation" existing outside the laws of nature, even though it cannot be influenced by the observer'. He commented that the natural laws only tell us about the statistics of these acts of observation. Paulii's analysis certainly avoided considering the problem that worried Einstein, but that can be regarded as a good move only if one agrees that Einstein's question was indeed irrelevant and unnecessary, and that must be a matter of opinion. One might equally contend that it was Einstein who is raising respectable if debatable arguments, and Pauli, with his talk of 'a "creation" existing outside the laws of nature', whose analysis was shallow.

Realism as a Programme

We have said a great deal about what may be called Einstein's commitment to realism, and his various strategies to make realism appear natural and attractive. Yet when one asks what his realism consists of at basis, it is much more difficult to get a convincing answer. He will occasionally speak of an objective world. We saw above that in his *Dialectica* article he talks of 'a real outside world ... things such as bodies, fields, etc. which claim a "real existence"'; however we note that the effect of this is diminished by the fact that the quotation is preceded not by 'I believe in ...' but by '[T]he concepts of physics relate to ...' As often, when Einstein appears to be about to commit himself over what realism actually means, he usually falls tantalisingly short of doing so. In 1944, though, he had written to Born[30] that: 'You believe in the God who plays dice, and I in complete law and order in a world which objectively exists, and which I, in a wildly speculative way, am trying to capture.'

Einstein said a little about an objective world, but never said at all explicitly that the theories of physicists were intended to mirror or correspond to this objective world, and certainly not, as mentioned above, that successive theories would be expected to approach this world ever closer. Even in his paper titled *Physics and Reality*,[36] he says a certain amount about how scientists come to create and accept physical concepts, and much about the history of physical theories—the mechanistic point of view, field theories, relativity, quantum theory and so on, but practically nothing about any objective realistic world. The nearest he comes to a clear statement about reality is that 'the concept of the "real external world" of everyday thinking rests exclusively on sense impressions'. Certain commonly repeated sense impressions are correlated, by a free creation of the mind, to relate to bodily objects. Elementary concepts of everyday thinking may also be built up intuitively from sense experiences. Propositions about these concepts may be called 'statements about reality' or 'laws of nature'.

If these discussions of what Einstein means by realism have appeared inconclusive and even sterile, we may get much closer to his genuine interest in realism by noting that, in the Schilpp volume,[15] he says that 'the "real" in physics is to be taken as a type of programme, to which we are, however, not forced to cling *a priori*'. He continues by remarking that nobody is likely to attempt to give up this programme for the macroscopic region, but that, since macroscopic and microscopic are so inter-related, it seems impractical to give it up for the microscopic region as well. He certainly sees no reason based on quantum theory for giving it up, unless one 'clings *a priori*' to the belief that the interpretation of Bohr and Heisenberg is final.

The point to be stressed is the idea of the programme. Realism is not a desire to produce a theory relating, to as large extent as possible, to a hypothetical objective physical world. We are certainly not aiming at an ultimate theory in one-to-one correspondence with this world. Indeed, inasmuch as there is an 'ultimate', it has rather a different nature. In 1940, Einstein[45] wrote that: '[F]rom the very beginning there has always been present the attempt to find a unifying theoretical basis for all these single [branches of physics], consisting of a minimum of concepts and fundamental relationships, from which all the concepts and relationships of the single [branch] might be derived by logical process. This is what we mean by the search for the foundation of the whole of physics. The confident belief that this ultimate goal may be reached is the chief source of the passionate devotion which has always animated the researcher.' Thus the 'ultimate' is not a perfect mapping of theory to an objective physical reality, but a theory with as few independent conceptual elements as possible, empirically successful across the whole of physics.

The means of approaching this basis is by a 'free creation of the mind', but in a realist theory the data from experiment are coordinated by means of some conceptual structure or model. The mental process of the construction of this model will always be a matter of ingenuity; scientists will regard it as an act of creativity, and, particularly for models which encompass substantial areas of experience, such as mechanics or the study of light, those constructing the models will be regarded as the most important figures in the history of science. While the model may contain elements and concepts of different types, particles, waves, fields and so on, and indeed alighting upon the most useful, appropriate and perhaps novel type of element may be the most difficult and creative aspect of the creation of the model, there are barriers that must not be broken. In particular the set of concepts of the model must be observer independent. An important component of the idea of a realist theory is that it should stand apart from the observer, who is indeed able to observe the posited real world externally.

While the construction of a model may be a laborious and stressful process for the scientists performing the task, the great advantage once a successful model has been obtained is that it is itself the centre of all subsequent progress in the field, at least for a period of time, often a substantial period. The model will suggest new properties of the physical world which may be analysed and studied experimentally; it will raise questions and difficulties which must be answered by theoretical and experimental study; it may of itself suggest applications to technology or to completely different areas of knowledge, perhaps even removed from what would strictly be called science.

Einstein contrasted the fruitfulness of the realist approach with the sterility of the Machian system. Mach was able to relate the various data of experience to form what Einstein called a 'catalogue', as we saw in Chapter 1. However for Mach there was no hypothesised centre of realism, which could be used as an intellectual resource even by scientists of moderate ability to produce new ideas and discoveries. Mach himself did produce important results, which suggests that, in the hands of a scientist of the greatest ability, his approach did not prevent the production of useful progress (assuming, of course, that his strictly scientific work followed his methodological precepts). However, as mentioned in Chapter 1, his devoted followers themselves produced little if any science that should be remembered.

It will be noted that we have said nothing about whether this model or construct of reality necessarily bears any relation to some objective real physical world. We may take it, of course, that, particularly in the case of a well-established realist model for a major area of science, Newtonian mechanics or the wave theory of light, for example, practitioners of the particular area of science will almost certainly come to think of the central ideas of the construct as directly corresponding to aspects of ultimate reality. We may say we are close to Thomas Kuhn's idea of the paradigm.[6] Kuhn spoke of the paradigm as a totally accepted set of ideas in a particular area of science, for which not only were the central concepts and laws not open to question, but also the the kinds of problems that should be tackled, the range of methods available and even the background knowledge and understanding that were required to be a practitioner in the area. Such immersion in the particular model or the paradigm may in fact be useful for such a practitioner; it will enable the maximum use to be made of the associated conceptual ideas.

Yet the model, of course, remains just that—a model. There will almost inevitably come a time when it outlives its usefulness. To use Kuhn's term, anomalies will occur which cannot be removed by judicious use of secondary assumptions; in retrospect at least, these anomalies will be seen to have struck deeply into the fundamental structure of the model. From the point of view of the expert practioner, totally at one with the paradigm, this period will of course be one of great stress and intellectual turmoil. Eventually a new paradigm may emerge, separated from the old by a scientific revolution, again using Kuhn's terminology. Experts on the old paradigm may well never be able even to accept the new ideas, let alone to work productively with them.

This revolution may well cause stress and even dismay to the scientists of the day. From the point of view of the realist programme, however, the move from one theory to another is perfectly natural and very much to be expected. No commitment should be made to the ultimate truth of a particular realist theory, so no surprise should be felt that one theory has past its usefulness, and that a successor is required. There are, of course, issues of scientific methodology, such as the fact that a successor theory must be able to include the successes of its predecessor, while also being able to remove whatever anomalies that caused its downfall; these issues have been discussed in Chapter 3.

We see the reason why, for Einstein, an account of physics and realism must contain some ideas about how realistic theories are produced, but much more describing the various important realist theories through history—Newtonian mechanics, field theory, special and general relativity. There need be no dismay or embarrassment that Newtonian mechanics, for example, has had to be replaced by special and general relativity. What is important is not the fate of any individual realist theory, but that of the realist programme, consisting maybe of a succession of realist theories, and intended to produce increasing empirical content. For Einstein the realist programme had been exceptionally successful; for several centuries realism had been the automatic belief system of the scientist, and had produced theories of steadily increasing scope. This is why Einstein was disappointed but not discouraged by the practically universal acceptance of complementarity. His articles on the topic practically invariably remark that he believed this state of affairs to be temporary. Over many years, the realist programme had been sufficiently strong and successful that in the long term Einstein was sure that it would again prevail.

Broad support for this position is provided by the content of a book called *The Evolution of Physics*[46] published in 1938 under the joint authorship of Einstein and Leopold Infeld. The genesis of this work is as follows.[33] Infeld had corresponded with Einstein since 1927, and, after a few months working with Born in Edinburgh in 1936, came to Princeton to work with Einstein, supported by a scholarship provided by the Institute for Advanced Study. Einstein and Infeld, together with Banesh Hoffman, published an important paper on the extremely interesting topic of motion in general relativity, and Einstein was not pleased when the Institute refused to renew Infeld's scholarship. He wrote to Born[30] in 1937 that 'Infeld is a splendid chap. We have done a very fine thing together. Problem of astronomical movement with treatment of celestial bodies as singularities of the field. The institute has treated him badly. But I will soon help him through it.'

Einstein's initial idea was that he would support Infeld from his own pocket, but Infeld, embarrassed by that idea, suggested rather that they should write a history of physics for a general readership. With Einstein one of the authors, the book, of course, sold well and Infeld's career was rescued.[16] He was to become professor at the University of Toronto from 1938 to 1950, during which time he published two further papers with Einstein, and he then moved to the University of Warsaw. In 1949, he was invited to contribute to the Schilpp volume on Einstein.[47]

With this background for the preparation of the book, one could suspect that Einstein's role in the writing might have been rather perfunctory, and that it would be unwise to attribute whatever methodological position might be apparent to Einstein himself. Indeed Pais[16] suggests that Einstein was not enthusiastic about the book; the quotation that Pais gives suggests rather that the relationship between the two authors may have suffered somewhat during its writing. However, on the basis of Einstein's correspondence with Maurice Solovine,[48] his very great friend almost since the beginning of the century, who translated the book into French, Fine[5] stresses that Einstein took a great interest in the book, and particularly the picture it provides of the philosophy of physics working throughout its history. He planned the book as an extended account of the realist approach in the history of physics, and as a argument against the positivistic approach. Regarded in this light, it helps to provide an enhanced understanding of how Einstein viewed the way in which physics had developed.

In the Preface the authors state that: 'We have tried to show the active forces which compel science to invent ideas corresponding to the reality of our world.' In the final section titled 'Physics and Reality', they write that 'Science ... is a creation of the human mind, with its freely invented ideas and concepts ... Physical theories try to form a picture of reality and to establish its connection with the wide world of sense impressions.' Also: 'We want the observed facts to follow logically from our concept of reality. Without the belief that it is possible to grasp the reality with our theoretical constructions, without the belief in the harmony of our world, there could be no science. This belief is and always will remain the fundamental motive for all scientific creation.'

While such remarks confirm much that we have said regarding Einstein's views, the book also contains exceptionally interesting comments about our theories, and whether they relate to an underlying ultimate reality. Having explained the theory of the planets moving around the sun, and commenting that it works splendidly, the authors say that another systerm based on different assumptions might work just as well. It is the whole system of assumptions, they say, which is proved or disproved by experiment, not an individual assumption. Physical concepts, they explain, are not uniquely determined by the external world; rather they are free creations of the human mind.

They say that the situation is like a man who tries to undertstand the mechanism of a watch. He can see the face and moving hands, he can hear the ticking, but he cannot open the case. If he is ingenious enough, he may be able to form a picture of a mechanism which could explain what he observes of the watch, but he can never be sure that his picture is the only one that could do this, since he can never compare his picture with the actual mechanism. What he does believe is that, as his knowledge increases, his picture of reality will become simpler and simpler, and will explain a wider and wider range of his sense impressions. The authors conclude that 'He may also believe in the existence of the ideal limit of knowledge and that it is approached by the human mind. He may call this ideal limit the objective truth.' It should be noted that, even in this last extremely cautious statement, buttressed by the two uses of 'may', there is, in any case, no suggestion that any ultimate theory corresponds to an ultimate reality.

It may be said that Einstein's support of the realist position, the realist programme, was mainly pragmatic. Of course it is true that he came to be irritated equally by what he may have considered to be the evasiveness of Mach and Copenhagen, and part of his commitment to realism may have been merely a riposte to these arguments. Equally as a scientist himself, he may naturally have been attracted to the scientists' automatic belief that they were exploring the properties of a real world. However his genuinely intellectual commitment to realism was the pragmatic one that it was the system of beliefs that led to empirical progress, to the production of concepts, theories and laws that gave increasingly good agreement with experience and experiment, and maybe to totally new and extremely interesting developments. In the following part of this book, which examines events after Einstein's death, we shall argue that his approach has indeed stimulated a range of exciting conceptual advances, and also potentially major technical achievements.

It is interesting that his pragmatism actually made him a little cavalier about how theories actually were produced. In this he seems to have differed somewhat from Planck. It will be remembered from Chapter 1 that Planck had broken decisively with Mach following his own seminal work which initiated the quantum theory, and particularly after Boltzmann's suicide in 1906. Planck was an embittered opponent of Mach at least from 1908, and at that stage it was Einstein who defended Mach, but Einstein's own switch of views around 1921 left him and Planck as allies.

However, Planck came to feel that his views were, at least to an extent, undermined by the great success of Heisenberg's work of 1925,[49] and he was to express himself in much more moderate tones about Mach's ideas from then on. For it seemed to him that Heisenberg had deliberately and systematically followed Mach's prescription for scientific progress. He had left aside realist constructs in the form of the Bohr orbits, and argued directly from quantities derived from experiment. Planck could not help but feel that his strong criticisms of Mach's position must have been misplaced.

It does not appear that Einstein was ever troubled in that way, even momentarily. It will be remembered that he was excited by Heisenberg's work and his results, but it does not seem to have improved his view of Mach's approach to physics at all. It is interesting to speculate on the reason for this. It may be that Einstein regarded Heisenberg's approach to the problem not as representative of a philosophical position that Einstein would have opposed, but as little more than a ploy, and an extremely fruitful one. Heisenberg, it might be said, faced with a major challenge, tried a new and interesting approach, which worked out even more successfully than he might have hoped for.

It will be remembered that, following a scientific revolution between one paradigm and another, it is highly likely that the new paradigm may contain, or even be centred around, new and strange theoretical structures. An example could be the physical field introduced in the revolution in which Newtonian mechanics was replaced as the basis of electromagnetism by field theories. It might have been doubted by conservative elements that the new concepts were meaningful, in the sense of providing a real model as basis of the new paradigm. Work would clearly have to be done to demonstrate the strength of these new conceptual elements, and also their ability to provide some physical understanding, which would be essential if future workers were to feel confident to develop the new paradigm.

Einstein may initially have felt that the new concept of the matrix produced by Heisenberg, though initially forbidding, might well be developed to provide direct physical understanding. (He would have been considerably more sure that the wave-function from Schrödinger's formalism could be interpreted fairly directly to give physical understanding.) And it may only have been when it was clear that neither matrix nor, in fact, wave-function could be interpreted in such a reassuring way that he became suspicious of these developments, and their interpretation by orthodox interpretations.

Whatever Einstein's feelings, it seems that pragmatism allowed him to be excited by Heisenberg's work, without giving credence to Heisenberg's stated philosophical methodology. It seems that mathematical and conceptual structures which gave good results were acceptable, whatever the methods used to obtain them, and the fact the results were good did not necessarily put the methods beyond criticism. The later developments in quantum interpretation, where philosophical ideas which Einstein considered unacceptable were used, not to produce new concepts or to explore ways forward, but, as Einstein saw it, to close down argument and discussion, was an entirely different matter. In having *these* thoughts, Einstein and Planck were united.

Einstein on Locality

Having studied Einstein's approaches to determinsm and realism, we now turn to his approach to locality. Locality was of course at the heart of the EPR paper, discussed in detail in Chapter 6. However before we can even discuss the question of locality, we must confirm that a space–time representation is to be used. It is fair to say that Einstein favoured a space–time description, which would be essential, of course, for his desired replacement of quantum theory by a theory using continuous fields. In 1940, for example, he[45] summed up many of his beliefs as follows: 'Some physicists, among them myself, cannot believe that we must abandon, actually and forever, the idea of direct representation of physical reality in space and time; or that we must accept the view that events in nature are analogous to a game of chance.'

However he did discuss the possibility of having to renounce the possibility of a space-time description. In 1936, he[36] wrote: 'To be sure, it has been pointed out that the introduction of a space–time continuum may be considered as contrary to nature in view of the molecular structure of everything which happens on a small scale. It is maintained that perhaps the success of the Heisenberg method points to a purely algebraical method of description of nature, that is, to the elimination of continuous functions from physics. Then, however, we must also give up, on principle, the space–time continuum. It is conceivable that human ingenuity will some day find methods which will make it possible to proceed along such a path. At the present time, however, such a program looks like an attempt to breathe in empty space.'

In the Schilpp volume, in the context of a discussion of the geometry to be used in general relativity, Karl Menger[50] suggested dispensing with continuous functions in space-time. Einstein[15] replied that aspects of the quantum theory might suggest the same idea. However, he went on to say that, while there are no new concepts which could be used in a constructive way, his adherence to the continuum did not 'originate in a prejudice, but arises out of the fact that I have been unable to think up anything organic to take its place. How'..., he asked, 'is one to conserve four-dimensionality in essence (or in near approximation) and [at the same time] surrender the continuum?' And a few years later,[5,51] in 1954, he wrote to Bohm that 'My opinion is that if the objective description through the field as an elementary concept is not possible, then one has to find a possibility to avoid the continuum (together with space and time) altogether. But I have not the slightest idea what kind of elementary concepts could be used in such a theory.'

Though Einstein's own attempts at a unified field theory were based solidly on use of space–time and the continuum, it is clear that, at least in principle, he was prepared to look for alternatives to this set of ideas, but he had no idea where to look. Nevertheless, as long as one did retain space-time, he appears to have believed that the idea of locality should not be questioned.

With S_1 and S_2 two partial systems which are spatially separated, though represented by a single wave-function (or in other words the combined system may be entangled), he wrote that: 'Now it appears to me that one may speak of the real factual situation of the partial system S_2. Of this real factual situation, we know ... even less than we know of a system described by the ψ-function. But on one supposition we should, in my opinion, absolutely hold fast: the real factual situation of the system S_2 is independent of what is done with the system S_1, which is spatially separated from the former.' We may describe this statement as Einstein locality. It was somewhat biased to a realistic point of view; it spoke of real factual situations, and it was the real factual situations of one sub-system which should be independent of what may be done to the other sub-system, which is spatially separated from it. Einstein, in fact, effectively regarded ψ_2 as real.

As is well-known, Bell[52] was to give his own definition of locality, which we may call Bell locality. Bell quoted the remark of Einstein just given, but Bell locality differs from Einstein locality in two important ways. First, rather than talking of real factual situations, which of course is only meaningful to those of a broadly realistic persuasion, Bell concentrated on the results of measurement. Rather than Ψ_2 being real, Bell would discuss only the results of a measurement on S_2, given that its wave-function *is* Ψ_2. As was seen in Chapter 6, it is relatively straightforward to translate from one set of ideas to the other; it is just in terminology that Bell was careful to address those of all different philosophical persuasions. Both Einstein and Bell, incidentally, also included choice of measurement by the experimenter as providing indisputably real data, but Bell is more careful to spell details out in terms of directions of magnetic fields or polarisers, while Einstein would content himself by saying what quantity is being measured.

The second point on which Bell[11] went beyond Einstein is that he was keen to take into account the possible sending of signals between the two sub-systems. In his general scheme, the actual timings of decisions on experimental set-ups, and of the appearance of measurement results, are important. Suppose an experimental setting is decided, or a measurement result is obtained at time t_1 in S_1, and a measurement result obtained at later time t_2 in S_2. If the distance between S_1 and S_2 is Δx, then Bell locality requires that $|\Delta x| > c(t_2 - t_1)$, so that no signal may pass from S_1 to S_2 at luminal or sub-luminal speed.

While the problems in connection with determinism and realism were clear from the outset of quantum theory, it was Einstein in the EPR paper who drew attention to the problem with locality. There seems no evidence that he was willing to compromise with locality, at least as long as space-time was to be maintained. On the contrary, he called the violation of locality involved in an EPR situation 'telepathic'[3] or, in a 1947 letter to Born, 'spooky action at a distance'.[30] Bell locality, with its emphasis on signals, draws attention to the demands of special relativity, and obviously one would expect that Einstein would have been willing to endorse this aspect of the argument.

Nevertheless, Einstein, in his own definition of locality, appears to have left aside the question of signals, and concentrated on the more basic notion of 'local realism'. Local realism speaks of real entities existing in space and behaving independently of other real entities at different positions. As Stapp[53] said, local realism was the accepted basis of science for several centuries—right up to the development of quantum theory. It was Einstein's achievement in EPR, practically unrecognised at the time of course, to draw attention to the fact that both aspects of local realism were under threat in quantum theory.

Bell's definition of locality managed to separate locality from realism more convincingly than that of Einstein. Nevertheless, as will be seen in Chapter 9, Bell's main result did refer explicitly to local realism. In fact, Stapp[53] was to call Bell's proof that, if quantum theory was correct, local realism had to be abandoned 'the most profound discovery of science'. There has been a great deal of study of locality and local realism since Einstein's death, in particular the distinction between those aspects of non-locality which allow instananeous transmission of information, and those that do not; this is all described in Chapter 9. In Chapter 14, we discuss how this later work relates to the earlier work of Einstein, and how Einstein's own general views might have been affected by these later developments.

Einstein's Vision for Physics—the Unified Field Theory

In this chapter, we have discussed Einstein's criticisms of the Copenhagen interpretation of quantum theory, and described and assessed his demands for determinism, realism and locality in any satisfactory theory. We must now describe how he did envisage the future for physics, the kinds of theory he hoped for and searched for. Here again we apppear to have a dilemma. Again it is possible to obtain from his writings two rather contrary positions. One, which has been stressed by Leslie Ballentine[54] in particular, is that Einstein believed that a fairly straightforward ensemble approach would remove all the difficulties. The other position could scarcely be more different. Not only was a straightforward addition to quantum theory not possible, there could be no type of addition or amendement to quantum theory that would solve its obvious problems and would enable genuine progress to be made in the understanding of nature and the kinds of theories necessary to describe it.

In his 1936 article 'Physics and Reality',[36] Einstein, as we saw earlier in this chapter, acknowledged the very great successes of quantum theory. It must, he said, be a 'touchstone' for any future theoretical basis, in the sense that it must be deducible from that basis in some appropriate limit, just as electrostatics is deducible from Maxwell's equations, or thermodynamics is deducible from classical mechanics. However, Einstein proceeded to say starkly: 'I do not believe that quantum mechanics can serve as a *starting point* in the search for this basis, just as, vice versa, one could not find from thermodynamics (resp. statistical mechanics) the foundations of mechanics.' Later[15] he was to write: 'I believe ... that this theory offers no useful point of departure for future development.'

In terms of his own past work, Einstein might have said that Newtonian gravitation had been an astoundingly successful theory, with admittedly a few conceptual problems at its basis, and, in Kuhnian terms, an anomaly in terms of the precession of Mercury. General relativity removed the anomaly, and its results were in excellent agreement with those of its predecessor theory in all places where that theory itself was known to agree with experiment. Nevertheless, as Einstein clearly saw, it would naturally have been completely impossible to construct the theory of general relativity from that of Newtonian gravitation, by making use of the anomaly. Neither could one even have attempted to remove the conceptual problems, such as the use of absolute space or time, from Newtons's work, in a piecemeal way.

Rather Einstein had to work from scratch. A totally new theory had to produced, a field theory in fact, based on a few rather general conceptual principles and mathematical operations. The method used in the construction of general relativity was scarcely positivistic, though, as we saw in Chapter 1, Einstein refused to admit this, even to himself, at the time. It would be truer to say that the theory was a free creation of the human mind, and might be best described as Platonic, or in other words based on the concept of mathematical beauty. Indeed the work was heavily mathematical. Einstein[55] later said that: 'The problem of gravitation was thus reduced to a mathematical problem: it was required to find the simplest fundamental equations which are covariant with respect to arbitrary coordinate transformation. This was a well-defined problem that could at least be solved.' It was also triumphantly successful; though the new theory was wholly different from Newtonian gravitation in outlook and methodology, it reproduced the results of the latter theory where it was known to work, it gave the value of the precession of Mercury in agreement with observation, and it made a small number of other new predictions which could be tested.

Einstein thought that the means of producing a replacement for quantum theory would be exactly the same. A completely new theory would have to be produced. While one would not necessarily have to retreat absolutely to first principles, the principles that were involved in the production of the new theory would be totally different from those at the basis of the current quantum theory. In fact, and not surprisingly, Einstein[15] regarded the theory of gravitation, as provided by general relativity, to be the theory which would provide the basis for future development. He aimed to produce a unified field theory, which would bring together the subjects of gravitation and electromagnetism. As he[55] put it: 'Gravitation had indeed been deduced from the structure of space, but besides the gravitational field there is also the electromagnetic field. This had, to begin with, to be introduced into the theory as an entity independent of gravitation .... But the idea that there exist two structures of space independent of each other, the metric-gravitational and the electromagnetic, was intolerable to the theoretical spirit. We are prompted to the belief that both sorts of field must correspond to a unified structure of space.'

It was his expectation that the theory produced would be non-linear, so as to remove the problem with superposition which, while being at the heart of the beauty of quantum theory, was also responsible, as we saw in earlier chapters, for its conceptual problems. These equations themselves would, he hoped, be immune from the difficulties with determinism, realism and locality that troubled many other physicists a little, but distressed Einstein to such a degree that he was determined to look for an alternative. However, in a suitable approximation, he hoped that quantum theory would re-appear, as a series of rules giving correct statistical predictions, at least for the phenomena studied at present, so as to give Einstein's theory all the successes of the present theory. In Fine's pithy phrase,[5] Einstein aimed at replacing quantum theory 'from without', by use of an enirely new theory, rather than 'from within', by tinkering with quantum theory itself.

It was clear that, even compared with general relativity, the new theory would require working with very extensive banks of equations. Wheeler described

Einstein to Bernstein[1] as a 'retail dealer in equations'. And indeed Einstein[3] made his approach crystal clear: 'I have learned something else from the theory of gravitation', he wrote: 'No ever so inclusive collection of empirical facts can ever lead to the setting up of such complicated equations. A theory can be tested by experience, but there is no way from experience to the setting up of a theory. Equations of such complexity as are the equations of the gravitational field can be found only through the discovery of a logically simple mathematical condition which determines the equations completely or (at least) almost completely. Once one has those sufficiently strong formal conditions, one requires only little knowledge of facts for the setting up of a theory.'

It may be remarked that many would believe this change in Einstein's mode of working was highly regrettable. In a meeting to celebrate the centenary of Einstein's birth, Isidor Rabi[56] commented that: 'When you think of Einstein's career from 1903 or 1902 on to 1917, it was an extraordinarily rich career, very inventive, very close to physics, very tremendous insights; and then, during the period in which he had to learn mathematics, particularly differential geometry in various forms, he changed. ... The great originality for physics was altered, well I won't say gone; but it was a different Einstein after 1916 than the 15 or 16 years before that. Whether, perhaps, he had lost interest in physics and had an unfortunate love of mathematics or something else happened ... this original creativity that one gets when one is first introduced to the subject appeared to be absent.'

Einstein made clear his changed view of mathematics in an address of 1933 intended to illustrate his views on the method of theoretical physics.[35] He asks: 'If, then, it is true that the axiomatic basis of theoretical physics cannot be extracted from experience but must be freely invented, can we ever hope to find the right way?' He answers to his own question by saying that 'Our experience hitherto justifies us in believing that nature is the realization of the simplest conceivable mathematical ideas. I am convinced that we can discover by means of purely mathematical constructions the concepts and laws connecting them with each other, which furnish the key to the understanding of natural phenomena ... Experience remains, of course, the sole criterion of the physical utility of a mathematical construction. But the creative princple resides in mathematics. In a certain sense, therefore, I hold it true that our thought can grasp reality, as the ancients dreamed.' It was on this basis that he hoped for success in his self-appointed labour.

But it is a fact that Einstein was not successful. He worked at unified field theories for 35 years, without any real success in relating his mathematical results to physics. In the centennial meeting, Wigner[57] remarked that it would be a mistake to consider the work of Einstein's last 30 years totally ineffective. He described the work Einstein carried out with Hoffmann and Infeld mentioned earlier in this chapter, though he also said that he was not totally convinced of the validity of their argument. Wigner also talked of several other very interesting articles, and also of Einstein's influence on his collaborators and other physicists. Even judged at its most positive, this is a poor return for over half the working life of perhaps the most productive physicist of all time.

Why did he fail? Peter Bergmann, who worked with Einstein from 1936 to 1941 has argued[58] that the effort was premature. Actually in 1920 when Einstein first became interested in the problem, unifying gravitation and electromagnetism did not appear at all an unlikely programme to tackle and even to tackle successfully. It did indeed seem a natural follow-up to the creation of general relativity, and at that time the important point was that gravitation and electromagnetism were the only two known fields. In 1920 Einstein made a casual remark about his work in a letter to Born,[30] and indeed Hermann Weyl had already produced an interesting theory attempting to unite these fields, albeit one which Einstein was able to show was flawed.[33] The kind of work Einstein was starting did not seem totally out of line with the thoughts and ideas of other physicists.

Yet comparatively soon there came the gradual realisation that there were two other fundamental fields or forces in physics, both connected with the burgeoning field of nuclear physics.[59] The strong nuclear field is responsible for binding the protons and neutrons together to form nuclei. (Actually a more modern view is that it binds quarks together to form the protons and neutrons, and that it is only a remnant of this more fundamental force that binds the protons and neutrons.) The weak nuclear field is responsible for radioactive decay. Einstein did not attempt to involve these fields in his schemes. It seems that he was opting to consider, in a way which, in retrospect appears arbitrary, only two of the four known fields. In fact, by including gravitation, he was choosing the field that, 50 years after his death, has been found to be the most difficult to relate to the others.

For there has now been unification of fields,[59] some fully successful, some only partly so. In 1967, Abdus Salam and Steven Weinberg produced a theory successfully unifying the electromagnetic and weak nuclear fields by a technique called gauge theory, totally different from any method used by Einstein. It was actually only gradually realised that the theory of Salam and Weinberg was successful, and some time later still when new elementary particles they had predicted were detected at CERN. Since then a vast amount of work has been done on Grand Unified Theories (GUTs) which also include the strong nuclear interaction, and other theories which involve the gravitational interaction, such as supergravity. However experimental predictions from these later theories, which include the decay of the proton, though with an extremely long lifetime, have not, at least as yet, been confirmed. In particular, gravitation has always been the most difficult of the four fields to include in any theoretical scheme, and there has been no successful quantisation of the gravitation field.

Yet Einstein worked away for several decades on his self-chosen problem. One of his most interesting strategies[16,33] was the use of five-dimensional theory, based on the so-called Kaluza–Klein theory of Theodor Kaluza and Oskar Klein. While many strategies were used by Kaluza, Klein, Einstein and others, the general idea was that the five dimensional formalism was successful in bringing together much of general relativity and Maxwell's equations. The fifth dimension could be rendered unobservable by various mathematical techniques, but its presence might explain some of the features of quantum theory in four dimensional spacetime that Einstein found unpalatable, such as indeterminacy. Einstein considered such theories very seriously in the early 1920s, and returned to them at various times up to the late 1930s, when he finally accepted that, mathematically interesting as all these developments might be, it was not possible to relate them to physics in a meaningful way.

Much the same fate awaited all the other schemes that Einstein tried, some for months, some perhaps for years. In the context of general relativity he[60] had acknowledged 'the years of anxious searching in the dark, with their intense long-ing, their alternations of confidence and exhaustion and the final emergence into the light'. He obviously hoped that the same would result for the search for the unified field theory; unfortunately he was to experience the same struggle but not the same triumphant result. In 1949 he sketched in the Schilpp volume[3] the solution that he considered logically most satisfactory at that time. He presented equations which he believed were the most natural generalisation of the equations of gravitation, but noted that the task of demonstrating their physical usefulness was tremendously difficult. One requires the solutions of the equations that are regular everywhere, and they must be exact—approximations will not suffice. In a footnote, Einstein added that: 'The theory here proposed, according to my view, represents a fair probability of being found valid, if the way to an exhaustive description of physical reality on the basis of the continuum turns out to be possible at all.'

Einstein and Ensembles

So Einstein's desire for the future of quantum theory—to replace it entirely by a new theory, did not come to fruition, but, right through his writings, he appears also to have advocated another position. This is that quantum theory is essentially a theory of ensembles. For instance in 1936, he[36] wrote that: 'The ψ function [terminology Einstein always used in preference to the term 'wave-function'] does not in any way describe a state which could be that of a single system; it relates rather to many systems, to an "ensemble of systems" in the sense of statistical mechanics. If, except for certain special cases, the ψ function fur-nishes only *statistical* data concerning measurable magnitudes, the reason lies not only in the fact that the *operation of measuring* introduces unknown ele-ments, which can only be described statistically, but because of the very fact that the ψ function does not, in any sense, describe the state of *one* single system.' Later in the same paper he discussed the EPR problem, and again concluded that 'coordination of the ψ function to an ensemble of systems eliminates every difficulty'.

Another instructive passage occurs in the Autobiographical Notes[3] in the Schilpp volume. This passage does not speak so directly of ensembles, but does help us to understand what kind of ensemble Einstein had in mind. In this passage he was discussing the views of two physicists, A, one of his own persuasion, and B, one who maintains the orthodox position. They both try to explain the facts of quantum measurement; the initial value of ψ is not mentioned, but the measurement may be of either q, position, or p, momentum. A and B agree, of course, that ψ will give the probability of obtaining any particular value of q or of p in an appropriate measurement, but the all-important question is about the single measured value of q or p. Did the particular system have this value before the measurement was performed? Theory gives no definite answer, Einstein said, since measurement is a process involving a disturbance of the system from outside, so it is conceivable that the system gains the measured value of q or p only through the measurement itself.

This is indeed the view of physicist B, who may thus conclude that ψ is a complete or exhaustive description of the real state of the system. However physicist A believes that the individual system had a definite value of q and of p before the measurement. Indeed, the value of q it possesses before the measurement is the value that will be obtained in a measurement of q, and the same goes for the value of p in a measurement of p. Physicist A therefore must conclude that ψ is not a complete description of the real situation of the system; it merely expresses what we may know about the system on the basis of the results of previous measurements. It is obvious how these remarks are closely related to Einstein's debate with Bohr described in Chapters 5 and 6, and in this article, Einstein indeed then moved on to describe the EPR argument, which, he claimed, would be sure to demonstrate to physicist B the error of his ways.

At the end of the Schilpp volume, in his response to criticisms,[15] Einstein reiterated the ensemble idea. '[D]ifficulties of theoretical interpretation disappear,' he said, 'if one views the quantum-mechanical description as the description of ensembles of systems .... The attempt to conceive the quantum-theoretical description as the complete description of the individual systems leads to unnatural theoretical interpretation, which become immediately unnecessary if one accepts the interpretation that the description refers to ensembles of systems and not to individual systems In that case the whole "egg-walking" performed in order to avoid the "physically real" becomes superfluous.' (We will come back to Einstein's next remarks shortly.)

Clearly there are several questions to be asked. First, what did Einstein mean by his ensemble approach? Second, did the approach work? And third, if the muddles and problems of quantum theory could be cleared up so easily, why did Einstein spend half his working life working on unified field theories to try to by-pass quantum theory altogether? And indeed, why did not the orthodox physicists of the day accept this ensemble idea, which would have avoided a lot of trouble for them as well?

It is important to discuss what kind of ensemble Einstein had in mind, as several different types have been discussed in the quantum mechanical literature. It should be mentioned that the use of ensembles has been particularly popular with those claiming to solve the measurement problem of quantum theory. Its strongest advocate has been Leslie Ballentine, who wrote a much-quoted article on ensembles;[61] in addition, as already stated, he argued that Einstein was himself a supporter of ensemble interpretations, and Ballentine also made ensembles the conceptual framework around which he contructed his well-known textbook[62] on quantum theory. John Taylor[63] has been another important advocate, while Home and Whitaker[64] have presented a more critical review.

Tony Leggett[65] has also criticised the view that ensembles, used in in a rather more sophisticated way and backed up by manipulation of the density-matrix, help to avoid the measurement problem; he wrote that it is 'probably the most favoured by the majority of working physicists', and even described it as 'orthodox', but actually considers it to be 'no solution at all'. Bell,[66] too, has criticised this kind of argument, concentrating on the discussion of Kurt Gottfried;[67,68] the response to Bell's criticism has been described in some detail by Whitaker,[69] and Gottfried[70] has recently published a reformulation of his own position. But, as we have said, ensembles are used in many different ways, so it is important to try to ascertain what was in Einstein's mind when he talked of ensembles.

The most natural type to be considered is what may be called a Gibbs ensemble. This is borrowed from classical physics, where it was introduced by the pioneers of statistical mechanics, in particular Willard Gibbs and Boltzmann. Translated to quantum theory, in a Gibbs ensemble each particle, would have, for example, precise values of both q and p. If ψ corresponds to an eigenvalue of p, corresponding, let us say, to a value p_0, it relates not to a single system, but to an ensemble. In this ensemble, each system has the same momentum, p_0. In addition, though, each system also has a value of q, but these values differ across the ensemble. The distribution of values of q over the ensemble corresponds to the probability distribution of measured values of q if a measurement is performed on the ensemble when its state is ψ. All these values of both q and p are available to become the results of measurements, and measurement, of course, plays no part in creating them.

Use of the Gibbs ensemble certainly seems exceptionally attractive. It appears to remove all the conceptual problems which others have struggled to understand in a natural and straightforward way. It might be said that if it worked, it would make Bohr, Heisenberg and the others look rather foolish for all their conceptual acrobatics! Many of those writing on quantum theory have discussed it, unfortunately giving it many different names. Guy and Deltete[71] have used the term 'Gibbs ensemble', Dugald Murdoch[26] an 'intrinsic-values theory', Fine a 'complete-values thesis'[72] and a 'random-values representation',[5] and Home and Whitaker[64] a 'PIV-ensemble'—an ensemble with pre-arranged initial values before any measurement. Popper,[73] a great supporter of the Gibbs ensemble, called it just '*the* ensemble' and Fine[5] also does use this terminology, preferring to keep the term 'statistical interpretation' for the general ensemble concept. Ballentine,[61] in an article supporting quite strongly a Gibbs ensemble, calls *this* the 'statistical interpretation'.

We reiterate that we consider it is best to use the words 'statistical' and 'probabilistic' unambiguously. Statistics implies counting over fully-characterised alternatives; thus in quantum theory a post-measurement ensemble is undoubtedly statistical; in that sense it is quantum theory, rather than any particular interpretation of it, that is statistical. However the situation is very different when we consider a pre-measurement ensemble. Orthodox interpretations will deny emphatically that, in general, each system has a particular value of the quantity being measured, values which could then be counted statistically; rather each system has an equal probability of the full range of measurement values; we would call the situation probabilistic rather than statistical. However in a Gibbs ensemble approach the pre-measurement ensemble is identical to the post-measurement one, so it is, of course, statistical.

It is interesting that Einstein invariably (in[15] for example) referred to and criticised the orthodox approach as the 'statistical quantum theory'; in our terms his complaint was actually that it was *not* statistical but probabilistic. In a 1954 letter[30] to Born congratulating the latter on his belated Nobel prize, Einstein spoke of the 'statistical intepretation', which, he said, 'decisively clarified our thinking'. In his commentary on this letter, Born remarks that the long delay in his award was unsurprising because his 'statistical interpretation' was opposed by all the great names of the initial period of the quantum theory—he mentions Planck, de Broglie, Schrödinger and, in particular, Einstein. Born expresses gratitude to Bohr and his Copenhagen school for 'lend[ing] its name everywhere to the line of thinking [he] originated'. Again the word 'statistical' seems capable of causing confusion. For Born's great step, the discovery of the statistical nature of the post-measurement ensemble (for a given initial ψ and a particular measurement), was essentially accepted by everybody, including the opponents of Copenhagen. It was the fact that the premeasurement ensemble was, in our terms, probabilistic that was the sticking-point between the two groups.

Having introduced the idea of the Gibbs ensemble, the quotations we have given so far would seem to demonstrate that this is exactly what Einstein required. He wrote of his ensemble being 'in the sense of statistical mechanics', and the quotation from the Schilpp volume suggests that, for each system, all variables should have a value at all times. Also he wrote[36] of a measurement involving 'a transition to a narrower ensemble'; this may well be interpreted as the argument that, before the measurement the ensemble contains systems with several values of the observable O that is being measured, and it is broken down into several ensembles, each narrower than the initial one, by the measurement. We shall for the moment assume that Einstein was thinking specifically of a Gibbs ensemble, though, as so often with Einstein, we may later have to reconsider, at least to an extent.

Our second question, then, is, does the Gibbs ensemble idea work? Does it remove the apparent contradictions of orthodox interpretations of quantum theory? Unfortunately the answer has to be a fairly definite 'no'. Let us start to explain this fact by pointing out that a Gibbs ensemble theory is certainly a hidden-variables theory. If, let us say, the wave-function of a system is an eigenfunction of $\hat{q}$, and yet we say that the system has a value of p as well as a value of q, then that value of p is a hidden variable since it is not given by the wave-function. This fact has been pointed out by several authors.[64,73]

This is important because Gibbs ensemble theorists have often been reluctant to admit that they are dealing with hiden variables. As is well-known, hidden-variable theories have, throughout the period of quantum theory, been treated with, at the very least, great suspicion. Von Neumann's 'proof' that hidden variable theories could not reproduce quantum mechanical results[19] was almost universally accepted up to the time of Einstein's death. Of course it was part of Bell's great achievement[75] to show that von Neumann's proof was not conclusive, and that hidden variable theories were not ruled out. Unfortunately, the other main part of his achievement[52] showed that hidden-variable theories had to have certain properties; in particular they had to be non-local. Certainly for Bell, just as much as for von Neumann, Gibbs ensembles were definite non-starters.

Einstein apparently showed no interest in von Neumann's work, perhaps thinking, as it was to emerge correctly, that such an abstract theorem could well have hidden assumptions, and of course he died before Bell's work. Nevertheless Einstein had not been unaware of many difficulties with Gibbs ensembles. Fine[5] has written that 'the so-called statistical [ensemble] interpretation has been faced with difficulties since 1935, difficulties known by Einstein and simply ignored by him'. One who had taken the responsibility of making Einstein aware of the difficulties with ensembles was Schrödinger. Following the publication of the EPR paper in 1935, Einstein and Schrödinger had an period of intense correspondence,[5] as discussed in Chapter 6. Einstein and Schrödinger were at one in their disdain for Copenhagen, but differed greatly in what they desired to put in its place. Schrödinger[40] aimed at going back to something like his early interpretation, with quantum jumps eliminated; Einstein pushed ensembles. In Schrödinger's paper[76,77] that followed, a section titled 'Can one base the theory on ideal ensembles?' is a critique of Einstein's approach.

In this section, Schrödinger described the Gibbs ensemble, and suggests that: 'At first thought one might well attempt to refer back the always uncertain statements of Q.M. to an ideal ensemble of states, of which a quite specific one applies in any concrete instance—but one does not know which one.' But he concludes that 'this won't work'. His reasons are many, all coming down to the fact that, when each system has to have a value for all observables, it is relatively easy to point out contradictions whatever sets of values are assumed.

Schrödinger gives the example of angular momentum, which one sets up with reference to a specific axis of rotation; with this axis assumed, it is straightforward to assume that only values allowed by quantum theory occur. However, if one refers the motion to a different axis, which is perfectly allowable since choice of axis is arbitrary, values occur which are not allowed by quantum theory. Schrödinger says that 'appeal to the ensemble is no use at all'.

Another example he considers is the simple harmonic oscillator. Since the wave-function extends beyond the classical limit, and the values of x which a Gibbs ensemble attributes to each system must form a distribution related to $\psi\psi^*$, some systems must have potential energy greater than total energy, even though the kinetic energy cannot be lesss than zero. Similarly Jammer[14] has shown that, with the Gibbs assumptions, almost a quarter of the electrons in a hydrogen atom would have potential energy greater than total energy. He states that: '[T]he assumption that a particle has simultaneously well-defined values of position and momentum, even though these may be unknown and unobservable, can be rejected.' Similarly in quantum tunnelling, in, for example, radioactive decay, the essence of understanding of the situation according to quantum theory is that the wave-function must be non-zero in the barrier region, where classically the tunnelling particle is not allowed. A Gibbs ensemble approach insists that all particles have definite values of position, and some of these values will lie in the barrier.

Gibbs ensembles also struggle in interpreting the experiments most characteristic of quantum theory—interference experiments. In a two-slit interference experiment, the Gibbs ensemble would insist that each particle has a precise position at all times, so it must travel in a particular beam and go through a particular slit. Yet the point of interference experiments is that each particle, in at least a sense, samples both paths and goes through both slits, for how else can an interference pattern emerge? Gibbs ideas seem unable to explain any form of interference, and yet interference is at the heart of very many quantum mechanical arguments and calculations. One may develop the ideas further and say that Gibbs ensemble approaches are actually particle-like theories. Each particle has to obey conservation laws independently, and so the linear superpositions at the wave-function level are effectively mixtures.

It might be said that the Gibbs idea is particularly simple, perhaps simplistic, and that we might remove at least some of the problems by making our assumptions more flexible. For example, for the cases mentioned of the simple harmonic oscillator and the hydrogen atom, one might reject the rule that the value the system has for total energy must be the sum of its values of kinetic and potential energy. While some formal success might be achieved along these lines, it would be obtained at the expense of what was the most attractive feature of the Gibbs ensemble; premeasurement values become post-measurement values for each system. Our general conclusion must be that Gibbs ensembles are unable to achieve what their advocates claim.

Since Einstein would have been familiar, at least in general terms, with the problems of Gibbs ensembles in quantum theory, we might check whether it is quite clear that the kind of ensembles he was advocating were indeed of this type. Another type of ensemble that is discussed may be called a minimal ensemble.[64] Here there are no hidden variables. This kind of ensemble may be what Taylor[63] is describing when he writes: '[W]hen we're making a measurement of any observable in a system ... we're making a measurement on an aggregate or *ensemble* of identically prepared systems ... Hence our results take the form of a probability distribution of particular values for that measurement.' In other words, this is just saying that it is the ensemble that is the unit we must work with in quantum theory rather than the individual system.'

There are two possible responses to the concept of the minimal ensemble. The first[64] is to assume that, at some level, it is trying to explain the strange things that happen in quantum theory, to make them seem a little less strange. But how is a minimal ensemble to achieve this? Without hidden variables, all of the systems in the ensemble must be in the same state—otherwise whatever made their states different would be a hidden variable. Then what is the difference between saying that the wave-function relates to a single system and saying that it relates to an ensemble of identical systems? It can be argued that we may certainly think of a post-measurement ensemble, which just means we think of different measurement outcomes, and the number of times each occurs for the ensemble of systems on which the measurement is performed. But the post-measurement ensemble is a part of quantum theory, not of any particular interpretation of quantum theory. What is special about ensemble theories must be that they refer to a premeasurement ensemble as well.

One perhaps sees Abner Shimony[78] struggling with this question when he writes (referring not to Taylor, but to Ballentine, who included both Gibbs and minimal ensembles in his general 'statistical interpretation'): 'There is, for example, Ballentine...He says "I am not a hidden variable theorist, I am only saying that quantum mechanics applies not to individual systems but to ensembles ..." I simply do not understand that position. Once you say that the quantum state applies to ensembles and the ensembles are not necessarily homogeneous you cannot help asking what differentiates the members of the ensemble from each other. And whatever are the differentiating characteristics these are the hidden variables. So I fail to see how one can have Ballentine's interpretation consistently. That is, one can always stop talking, and not answer questions, but that is not the way to have a coherent formulation of a point of view.'

The second response to the advocacy of a minimal ensemble is exactly what Shimony dismisses at the end of his remarks. This is just to restrict attention, before and after measurement to ensembles, and refuse to entertain any mention of individual systems, even to the extent of ignoring what appear to be genuine problems of quantum theory. Taylor's explanation of EPR–Bohm, for example, is that, in each wing of the apparatus, half the spins in the ensemble are 'up', half 'down'. A minimal ensemble theorist will refuse to address the fact that, in each pair, we have one spin 'up' and one 'down' because that would relate to an individual system (a pair of spins being an individual system in this instance). When used in this way, the ensemble interpretation may be called the ensemble *non-interpretation*.

Is there any evidence that Einstein was attracted to the idea of the minimal ensemble? It might be said that, incoherent as it may be, the minimal ensemble could do much that Einstein required—it can refuse to consider what happens at the system level, and draw attention to the statistical nature of the results, without committing itself to any direct physical assumption.

In fact, it is possible to find statements of Einstein that seem in some ways closer to a minimal ensemble than a Gibbs ensemble. After the remarks from the response to criticisms quoted earlier in this chapter, where Einstein introduced the idea of the ensemble, he continued: 'There exists, however, a simple psychological reason for the fact that this most nearly obvious interpretation is being shunned. For if the statistical quantum theory does not pretend to describe the individual system (and its development in time) completely, it appears unavoidable to look for a complete description of the individual system; in doing so it would be clear from the very beginning that the elements of such a description are not contained within the conceptual scheme of the statistical quantum theory. With this, one would admit that, in principle the scheme could not serve as the basis of theoretical physics. Assuming the successes of efforts to accomplish a complete physical description, the statistical quantum theory would, within the framework of future physics, take an approximately analogous position to the statistical mechanics within the framework of classical mechanics. I am rather fairly convinced that the development of theoretical physics will be of this type; but the path will be lengthy and difficult.'

In this long and carefuly prepared quotation, we see several strands of Einstein's thought. At the beginning, the main point is that quantum theory does not relate to the individual system. It may be suggested that we are in the region of the minimal ensemble, and even that this is as far as Einstein wished to take the argument related to the quantum theory of today. The best way of describing quantum theory, he felt, is as (something like) a minimal ensemble. However he did not feel that this is the ultimate answer. He believed in completion by a new theory, as described in the previous section of this chapter. Only having explained this did he mention the Gibbs ensemble idea. This was carefully described only as an 'approximately analogous' position to what he actually espoused, mentioned only to give a general idea of the relationship between the new theory he wished for and the present quantum theory. This exegesis would appear to be a consistent account of the various strands of Einstein's belief and aims, and it is interesting that it occurred in an article where he was making a special attempt to describe his views, it might be said, for posterity. It would be freely admitted that on other occasions he presented his ideas more sketchily and the results were not as clear.

Others have used broadly similar arguments to explain why Einstein did not bother to develop the Gibbs ensemble concept despite its obvious failings. Fine[5] has written that the ensembles 'provided no more than a setting rhetorically apt for calling attention to the incompleteness of quantum theory'. He adds that 'Einstein chose his rhetoric cunningly. For who, learning that a theory is incomplete, could resist the idea that one ought to complete it?' He says that 'Einstein *wanted* his purely *interpretative* scheme to appear heuristically sterile.' Guy and Deltete[71] say that, while Einstein believed that ensembles were the only way to solve the internal problems of quantum theory, they could not describe physical states, so developing the ensemble interpretation would have been a waste of time. They add that 'Einstein is thus far less concerned to show what ensembles can do than he is to insist what they cannot do.'

Einstein and the Bohm Theory

David Bohm's 1952 hidden variable theory is described in Chapter 10, and we have already made brief mention of it earlier in this chapter. Here we concentrate on Einstein's part in the development of the theory, and his response to it.

In 1951, David Bohm wrote an important textbook[79] on quantum theory. It was clearly and thoughtfully written, and is still a useful resource over half a century later. The book was written along orthodox lines; it supported the Copenhagen interpretation, and included an account of EPR which is authoritative but certainly very much along general Bohrian lines. Bohm had been working in Princeton since 1946, and sent a copy of his book to Einstein, who enjoyed the book and asked Bohm to come and talk to him about it. It seems that Einstein convinced Bohm that the Copenhagten interpretation was not able to answer the problems of quantum theory, and for the rest of his life[80,81] Bohm was to be a critic of quantum orthodoxy. From this basis, he was to develop broad and deep analyses of the nature and scope of knowledge.

His 1952 work[82] essentially replaced the wave *or* particle approach by wave *and* particle. By the introduction of hidden variables, Bohm was able to obtain a realistic and deterministic theory. The theory reproduced the results of quantum theory, as it was, at least from a mathematical point of view, an addition to standard quantum theory rather than a replacement of it. It claimed to eliminate wave-function collapse, and, as a realistic theory, made no reference to the observer. In fact in the theory the particle has a precise position and momentum at all times. The method was simple in principle, but quite sophisticated in practice—it linked the Schrödinger equation to the Hamilton–Jacobi equation of classical mechanics.

Bohm's theory was closely related to de Broglie's ideas of 1926–1927, and Bell was to christen the theory 'de Broglie–Bohm'. In his major account of the theory, Peter Holland[83] said that, while de Broglie's ideas remained largely an unfulfilled programme, 'Bohm's model is essentially de Broglie's pilot-wave theory carried to its logical conclusion'. In particular, Bohm was able to answer criticisms of de Broglie's work at the 1927 Solvay conference made by Pauli and others. De Broglie himself had not been able to respond effectively to Pauli, and had abandoned his ideas, being a supporter of Copenhagen for several decades until Bohm's work revived his former ideas.

It was, of course, no surprise that the advocates of Copenhagen, Heisenberg, Pauli and Rosenfeld in particular, criticised Bohm as scathingly as they had criticised de Broglie. Born[30] actually wrote to Einstein that: 'Pauli has come up with an idea ... which slays Bohm not only philosophically but physically as well', though it is difficult to understand Born's point of view, as Pauli's remarks are surprisingly feeble. In general terms, the advocates of Copenhagen conveniently forgot that until recently they had made a great deal of von Neumann's 'proof' which claimed that theories with hidden variables were impossible; now, having had such a theory presented to them, their response was that it was 'superfluous "ideological superstructure"' since it led to the same results as the established 'orthodox' one.

However one would have expected anyone with reservations about the *status quo* to have been exceptionally interested, and indeed delighted, with Bohm's theory. This was certainly the position of Bell. Dramatically he[84] wrote: '[I]n 1952 I saw the impossible done.' Bell was delighted that Bohm showed clearly that the so-called Copenhagen virtues—no hidden variables, no realism, no determinism—had been undermined. For the rest of his life he promoted the de Broglie–Bohm theory from this point of view, and used it as an important illustration of the consequences of his own work. 'Why is the pilot wave picture ignored in text books?' he asked. 'Should it not be taught, not as the only way, but as an antidote to the prevailing complacency? To show us that vagueness, subjectivity, and indeterminism, are not forced on us by experimental facts, but by deliberate theoretical choice?'

As implied in the above quotation, Bell certainly did not regard de Broglie–Bohm as a complete answer. He must have been concerned with its non-locality. As pointed out by Bohm himself, the theory was massively non-local; essentially, in a multi-particle system, the behaviour of each particle depended on the position of all the other particles at that time. This was to lead in the next decade, of course, to Bell's own celebrated work[52] in which he showed that *any* hidden variable theory must have this property. But even apart from this, he recognised that there were good reasons for not especially liking the de Broglie–Bohm picture; he remarks that neither de Broglie or Bohm liked it very much. But as a working hidden variable theory it was of great interest, and as a logical weapon against Copenhagen it was indispensable.

Now we turn to Einstein. In Bohm's paper he referred to Einstein's criticisms of the current interpretation of quantum theory in the very first paragraph, and at the end he thanked Einstein for 'several interesting and stimulating discussions'. He must certainly have hoped that Einstein would help him to promote his ideas as an example of Einstein's wider perspective. Yet this was not to be; Einstein was unimpressed, and never gave any backing to Bohm's work.

There may be two reasons for his failure to support Bohm. First is the non-locality of the theory, which would clearly not have appealed to Einstein. Bell's later work, which was performed, of course, after his death, would have allowed him to see this in a much wider context. However the second possible reason is that Bohm's work was very much the kind of approach to quantum theory that Einstein had rejected, probably as early as 1927, as discussed in Chapter 5. Indeed the failure, as he saw it, of the de Broglie theory, together with that of his own abortive attempt to create a hidden variables model for quantum theory, was a large part of the reason for this rejection. It was hardly likely that he would be enthralled by Bohm turning to the same set of ideas! By 1952, convinced that the solution to the riddles of quantum theory lay in his own unified field theory, still very much under construction, he would have been extremely unlikely to endorse any relatively simple solution.

Bohm was obviously extremely diappointed at Einstein's response, which he blamed on the first of our reasons. He wrote[85]: '[I]t was important that the whole idea did not appeal to Einstein, probably mainly because it involved the new feature of non-locality ... I felt this response of Einstein was particularly unfortunate ... as it certainly "put off" some of those who might otherwise have been interested in the approach.'

In a letter[30] to Born, though, Einstein mentioned the second reason. He wrote: 'Have you heard that Bohm believes (as de Broglie did, by the way, 25 years ago) that he is able to interpret the quantum theory in deterrministic terms? That way seems too cheap to me.' One can well see that, in comparison with the massive scale of Einstein's own efforts, Bohm's would appear too simple, too 'cheap'. (For those who think of Einstein more as a supporter of hidden variables, his response to Bohm must seem much more strange. In his commentary on this letter,[30] Born remarks: 'although this theory was quite in line with his own ideas ....' Bell[11] quotes both Einstein and Born, and agrees with Born.)

Whatever Einstein's thoughts about the de Broglie–Bohm theory, his attitude does appear to have a been major tactical mistake. It would have been quite possible for him to have stressed that it was a clear counter-example to Copenhagen, and it could have acted very much as the kind of debating-point that Einstein would have used to great effect. At the same time, Einstein could have pointed out that it was very far from being a complete and satisfactory solution to the riddles of quantum theory. For that, he might have said, one would have had to wait for the coming to fruition of Einstein's own work. While feeling disappointed that Einstein did not feel able to take this approach, one may agree with Holland that it is unlikely that Einstein's advocacy of Bohm's work would actually have made much difference, because Einstein's own favoured solutions were universally rejected.

Realism, Determinism, Locality

In this last section of the chapter, we return to the discussion of realism, determinism and locality, and ask the question—given that Einstein valued all of these highly, which were the most crucial to him? We will remember the common view, based on 'God does not play dice' that it was determinism, contrasted with Pauli's opinion, as stated to Born, that Einstein's approach was based on realism and not determinism. Fine,[5] incidentally, regards Pauli's remarks as somewhat misleading. While he admits that Pauli was correct in reporting that realism was the central issue for Einstein in the matter discussed with Born, he suggests that Einstein's realism was always of a type that involves determinism as well.

Before proceeding to answer the question, we may note that the question was not necessarily an important one for Einstein. In the days before Bell, the clash between realism and locality was not known. Indeed the point of EPR was to retain locality *and* regain realism. Einstein had every hope that his unified field theory would respect all three of determinism, realism and locality. Perhaps the only occasions where he had to make some sort of choice was when he temporarily lost faith in his ultimate type of solution, which was known to happen,[16] and so was forced to face up to choices about the current quantum theory, or when he was cross-questioned by someone like Pauli. But of course we can attempt to find out where his deepest allegiance lay by analysing his writings.

We will start with locality, and assert that this was an exceptionally important criterion for Einstein. As we have said, occasionally he may have considered, in a very general way, the possibility of being forced to give up working in space-time altogether; this would presumably be on an occasion just mentioned when he lost a little faith in the ultimate success of his own programme. However if; space-time was to be retained, it would seem that an abandonment of locality was not considered. One might cite the demands of special relativity for the strong maintenance of this position, but that would perhaps be more related to Bell locality as defined above. For Einstein locality, the feeling seems to have been that science, as it had been practiced for centuries, would simply be impossible if separated objects did not behave independently.

To get even the slightest hint of Einstein being prepared to question locality, we may note his comment[15] that, of the orthodox attempts to answer the EPR paper, he prefers that of Bohr. Einstein attributes to Bohr the suggestion that, for a system consisting of two systems, there is no requirement to attribute independent existence to either sub-system viewed separately, even if the sub-systems are separated in space. This is a view known as loss of independent existence, which is incidentally discussed in some detail by Murdoch[26] in his study of Bohr's beliefs. It might be described, somewhat loosely, as an attempt to deal a modicum of pain to each of realism and locality, rather than denying one or other altogether. Realism is damaged because localisation in space–time is a usual requirement for a real object; locality is damaged because a cause at one point may have an immediate effect elsewhere, but the damage is minimised because what happens may be blamed on the object itself essentially being non-local.

To the (extremely limited) extent to which any conclusion may be drawn from Einstein's words, it might be that be realism and locality are of equal signifi-cance for him. We would mention, incidentally, that the no-signalling theorem [Ref. 79, pp. 618–9] was known in Einstein's lifetime; this tells us that the kind of locality discussed in EPR cannot be used for sending signals. This criterion will be discussed further in Chapter 9 in terms of the distinction between parameter independence and outcome independence. Whether this restricted form of non-locality would be more palatable to Einstein than one that allowed signalling will be discussed in Chapter 14.

We now move to the comparison of Einstein's support of realism and determin-ism. While post-Bell the main argument would be between realism and locality, before Bell's work the interesting question was this one—which was more central to Einstein, realism or determinism? It is quite clear that at least from the 1930s, Einstein was totally convinced by the realist position. He wrote major papers devoted to realism, and demonstrated total commitment towards it. His accounts of what he meant by realism were well thought-out, deeply felt and articulate. All this has been described earlier in this chapter. The same could not be said of determinism; we do not find any substantial accounts of what Einstein meant by determinsm.

Yet Fine is able to build a good case that for Einstein realism went with deter-minsm, that, for Einstein, realism was to be experienced with non-probabilistic laws. One of his nicest quotes occurs in a letter written to Michele Besso in 1950.[86] Einstein starts a sentence extolling realism over determinism: 'The ques-tion of "causality" is not actually central, rather the question of real existents ....' Yet he then finishes the sentence by coming back firmly to determinism: '... and the question of whether there are some sort of strictly valid laws (not statistical) for a theoretically represented reality.'

Part of the difficulty of addressing the issue is that much of Einstein's rhetoric is devoted to confronting Copenhagen, and for Copenhagen lack of realism and lack of determinism go very much hand in hand. Let us consider, for example, a spin-$1/2$

system initially in an eigenstate of $\hat{S}_z$, say $|S_z = 1/2\rangle$. Restricting ourselves to S_z, we may say that the situation is one of realism; S_z has a distinct value. (We will for the moment ignore the fact that no other component of spin has a distinct value.) We now imagine the system evolving under a Hamiltonian in such a way that, after a period of time, its state-vector is a linear combination of both eigenvectors of S_z, say $(1/\sqrt{2})(|S_z = 1/2\rangle + |S_z = -1/2\rangle)$. At this point, we have lost realism, at least for S_z; it cannot be said that S_z has a particular value. (Of course it may be suggested that S_x now does have one, just to complicate things.) At this stage one may say that we have not experienced indeterminism, because the change in state-vector is deterministic. We might be inclined, though, to argue that there is latent indeterminism, because a measurement of S_z may yield either of two answers, and this latent determinism, of course, becomes manifest when a measurement is performed.

Lack of realism and lack of determinism are thus scrambled together for the Copenhagen interpretation. Einstein must have hated both aspects of what was purportedly happening, and his criticisms were probably generally directed at both. Thus when he criticised the 'statistical' nature of contemporary quantum theory, which at first sight might be seen to be an attack solely on determinism, the renunciation of both determinism *and* realism may have been the intended targets. Even his celebrated 'God does not play dice' may have been related to the unsatisfactory nature of the entire conceptual position, at least according to his own perspective.

A related argument has been made by Mara Beller,[87] who has written that: 'Quantum theory is essentially a probabilistic theory, not because quantum laws, as opposed to classical ones, are statistical ... but because one needs probabilities to describe fully the *state* of the system. The future cannot be known in all its details not because quantum laws are statistical, but because the present cannot be known in all its details, and therefore has to be described probabilistically.'

It is quite easy to construct a theory which is realist but indeterministic if change takes place at a series of discrete values of time. At each of these times, several routes for change from one real situation to another are open, and the probabilities of each are given by parameters in the theory. However it is much more difficult to have a fairly natural realist indeterministic theory if the laws are based on differential equations as in the current quantum theory. One would seem to require an infinity of choices at each instant of time. The alternative that we have for which the usual evolution is deterministic, and the indeterministic elements arise only at specific times of measurement, appears then quite neat, but is non-realist. Again we see that how indeterminism and lack of realism may be linked together in theory construction, at least for common types of theory, and likewise determinism and realism. So when Einstein criticises Copenhagen for its statistical nature, and when he talks of his desired theory being 'free of statistics', it may be realism as much as determinism, or even more than determism, that is being referred to.

Of course this is an argument of plausibility only, and some will doubtless be sceptical of it. It will be readily admitted that Einstein refers also to preferring 'non-statistical laws' where the reference to determinism is clearer and more specific. Indeed it is certainly not in question that Einstein favoured determinism. His ensemble theories were realist *and* deterministic, and he certainly hoped that his unified field theory would be also. However the argument above may make Pauli's position, that Einstein's metaphysics was realist rather than deterministic, seem more plausible than might appear to be the case from some of Einstein's other writings and more famous sayings.

This chapter has contained a survey of Einstein's views on quantum theory. As we have said, events since his death, in particular the coming of Bell's theorem, would necessitate some change in these views, and in Chapter 14 we sketch what these changes might be.

References

1. Bernstein, J. (1991). Quantum Profiles. Princeton: Princeton University Press.
2. Heisenberg, W. (2005). Introduction, In: The Born-Einstein Letters. 2nd edn. (Born, M., ed.) Basingstoke: Macmillan, pp. xxxiv–xxxvii.
3. Einstein, A. (1949). Autobiographical notes, In: Albert Einstein: Philosopher-Scientist. (Schilpp, P.A., ed.), New York: Tudor, pp. 1–95.
4. Hermann, A. (ed.) (1968). Albert Einstein/Arnold Sommerfeld Briefwechsel. Stuttgart: Schwabe; translation by Stuewer, R. and Stuewer, R. included in Ref. [5].
5. Fine, A. (1986). The Shaky Game: Einstein, Realism and the Quantum Theory. Chicago: University of Chicago Press.
6. Kuhn, T.S. (1957). The Structure of Scientific Revolutions. Chicago: University of Chicago Press.
7. Hawking, S.W. (1980). Is the end in sight for theoretical physics? Inaugural lecture for Lucasian Chair, University of Cambridge.
8. Boslough, J. (1985). Stephen Hawking's Universe. New York: Quill.
9. Barrow, J.D. (1990). Theories of Everything. Oxford: Oxford University Press.
10. Durrani, M. and Rodgers, P. (1999). Physics: past, present, future, Physics World **12**(12), 7–14.
11. Bell, J.S. (1976). Einstein-Podolsky-Rosen experiments, Proceedings of the Symposium on Frontier Problems in High Energy Physics. Annali Della Schola Normale di Pisa, pp. 33–45; also in Ref. [12], pp. 81–92, and in Ref. [13], pp. 768–77.
12. Bell, J.S. (2004). Speakable and Unspeakable in Quantum Mechanics. (1st edn. 1987, 2nd edn. 2004) Cambridge: Cambridge University Press.
13. Bell, J.S. (1995). Quantum Mechanics, High Energy Physics and Accelerators. (Bell, M., Gottfried, K. and Veltmann, M., eds.) Singapore: World Scientific; the papers on quantum mechanics were republished on their own in 2001 with the same editors and publisher in John S. Bell on the Foundations of Quantum Mechanics.
14. Jammer, M. (1974). The Philosophy of Quantum Mechanics. New York: Wiley.
15. Einstein, A. (1949). Remarks to the essays appearing in this collective volume, In: Albert Einstein: Philosopher-Scientist. (Schilpp, P.A., ed.), New York: Tudor, pp. 665–93.
16. Pais, A. (1982). 'Subtle is the Lord …': The Science and Life of Albert Einstein. Oxford: Clarendon.
17. Mehra, J. (1999). Einstein, Physics and Reality. Singapore: World Scientific.
18. Penrose, R. (1998). Foreword, In: Einstein's Miraculous Year: Five Papers that Changed the Face of Physics. (Stachel, J., ed.) Princeton: Princeton University Press. pp. 177–98.

19. Bell, J.S. (1981). Bertlmann's socks and the nature of reality, Journal de Physique, Colloque C2, **42**, C2 41–61; also in Ref. [12], pp. 139–59, and in Ref. [13], pp. 420–41.

20. von Neumann, J. (1955). Mathematical Foundations of Quantum Mechanics. Princeton: Princeton University Press.

21. Wigner, E.P. (1961). Remarks on the mind-body question. In: The Scientist Speculates—An Anthology of Partly-Baked Ideas. (Good, L.J., ed.) London: Heinemann, pp. 284–302.

22. Peierls, R. (1991). In defence of 'measurement', Physics World **4**(1), 19–20.

23. Mermin, N.D. (2002). Whose knowledge?, In: Quantum Unspeakables: From Bell to Quantum Information. (Bertlmann, R. and Zeilinger, A., eds.) Berlin: Springer, pp. 271–80.

24. Brukner, C. and Zeilinger, A. (2005). Quantum physics as a science of information, In: Quo Vadis Quantum Mechanics. (Elitzur, A., Dolev, S. and Kolenda, N., eds.) Berlin: Springer, pp. 47–61.

25. Folse, H.J. (1985). The Philosophy of Niels Bohr. Amsterdam: North-Holland.

26. Murdoch, D. (1987). Niels Bohr's Philosophy of Physics. Cambridge: Cambridge University Press.

27. Bohr, N. (1949). Discussion with Einstein on epistemological problems in atomic physics, In: Albert Einstein: Philosopher-Scientist. (Schilpp, P.A., ed.), New York: Tudor, pp. 199–242.

28. Leggett, A.J. (2002). Testing the limits of quantum mechanics: motivation, state of play, prospects, Journal of Physics C – Condensed Matter **14**, R415–51.

29. Arndt, M., Nairz, O., Vos-Andreae, J., Keller, C., van der Zouw, G., and Zeilinger, A. (1999). Wave-particle duality of C_{60} molecules, Nature **401**, 680–2.

30. Born, M. (ed.) (2005). The Born-Einstein Letters. 2nd edn., Basingstoke: Macmillan.

31. Bohm, D. (1952). A suggested interpretation of the quantum theory in terms of 'hidden variables' I and II, Physical Review **85**, 166–93.

32. Einstein, A. (1954). The mechanics of Newton and their influence on the development of theoretical physics, In: Ideas and Opinions. New York: Crown, (1954), pp. 253–61; translated by Sonja Bargmann from der Naturwissenschaften **15**, 273–6 (1927).

33. Fölsing, A. (1997). Albert Einstein. Harmondsworth: Viking.

34. Einstein, A. (1927). Letter to the Royal Society on the occasion of the Newton bicentenary, Nature **119**, 467; Science **65**, 347–8.

35. Einstein, A. (1954). On the method of theoretical physics; collection cited in Ref. [32], pp. 270–6.

36. Einstein, A, (1936). Physics and reality, Journal of the Franklin Institute **221**, 349–82; reprinted in collection cited in Ref. [32], pp. 290–323.

37. Prizbram, K. (ed.) (1967). Letters on Wave Mechanics: Schrödinger, Planck, Einstein, Lorentz. New York: Philosophical Library.

38. Einstein, A. (1948). Quantum mechanics and reality, Dialectica **2**, 320–4; also included in Ref. [30].

39. Einstein, A. (1931). Maxwell's influence on the evolution of the idea of physical reality, In: James Clerk Maxwell: A Commemoration Volume. Cambridge: Cambridge University Press, pp. 66–73; reprinted in collection cited in Ref. [32], pp. 266–70.

40. Schrödinger, E. (1952). Are there quantum jumps? II, British Journal for the Philosophy of Science **3**, 233–42.

41. Heisenberg, W. (1983). Encounters with Einstein: And Other Essays on People, Places and Particles. Princeton: Princeton University Press.

42. Horvath, G.Z.K., Thompson, R.C. and Knight, P.L. (1997). Fundamental physics with trapped ions, Contemporary Physics **38**, 25–48.
43. Leibfried, D., Blatt, R., Monroe, C., and Wineland, D. (2003). Quantum dynamics of single trapped ions, Reviews of Modern Physics **75**, 281–324.
44. Buchwald, D. and Thorne, K.S. (2005). Preface to Ref. [30], pp. vii-xxxiii.
45. Einstein, A. (1940). Considerations concerning the fundamentals of theoretical physics, Science **91**, 487–92, Nature **145**, 920–4; reprinted in collection cited in Ref. [32], pp. 323–35.
46. Einstein, A. and Infeld, L. (1938). The Evolution of Physics. Cambridge: Cambridge University Press.
47. Infeld, L. (1949). On the structure of our universe, In: Albert Einstein: Philosopher-Scientist. (Schilpp, P.A., ed.), New York: Tudor, pp. 475–99.
48. Solovine, M. (ed.) (1956). Albert Einstein: Lettres à Maurice Solovine. Paris: Gauthier-Villars.
49. Heilbron, J.L. (1996). The Dilemmas of an Upright Man: Max Planck and the Fortunes of German Science. Cambridge: Harvard University Press.
50. Menger, K. (1949). The theory of relativity and geometry, In: Albert Einstein: Philosopher-Scientist (Schilpp, P.A., ed.), New York: Tudor, pp. 457–74.
51. Stachel, J. (1983). Einstein and the quantum, In: From Quarks to Quasars. (Colodny, R., ed.) Pittsburgh: University of Pittsburgh Press.
52. Bell, J.S. (1964). On the Einstein–Podolsky–Rosen paradox, Physics **1**, 195–200; also in Ref. [12], pp. 14–21, and in Ref. [13], pp. 701–6.
53. Stapp, H.P. (1977). Are superluminal connections necessary? Nuovo Cimento **40B**, 191–205.
54. Ballentine, L.E. (1972). Einstein's interpretation of quantum mechanics, American Journal of Physics **40**, 1763–71.
55. Einstein, A. (1954). The problem of space, ether and the field in physics, in collection cited in Ref. [32], pp. 276–85.
56. Rabi, I.I. (1980). Untitled comment, In: Some Strangeness in the Proportion: A Centennial Symposium to Celebrate the Achievements of Albert Einstein. (Woolf, H., ed.) Reading, Massachusetts: Addison-Wesley, pp. 485–6.
57. Wigner, E.P. (1980). Thirty years of knowing Einstein, in: Ref. [56], pp. 461–8.
58. Bergmann, R.G. (Peter) (1967). In: Einstein, the Man and his Achievement. (Whitrow, G.J., ed.) London: BBC; New York: Dover, pp. 72–3.
59. Pais, A. (1986). Inward Bound: Of Matter and Forces in the Physical World. Oxford: Clarendon.
60. Einstein, A. (1954). Notes on the origin of the general theory of relativity, in collection cited in Ref. [32], pp. 285–90.
61. Ballentine, L.E. (1970). The statistical interpretation of quantum mechanics, Reviews of Modern Physics **42**, 358–81.
62. Ballentine, L.E. (1998). Quantum Mechanics: A Modern Development. Singapore: World Scientific.
63. Taylor, J. (1986). In: The Ghost in the Atom. (Davies, P.C.W. and Brown, J.R., eds.) Cambridge: Cambridge University Press, pp. 106–17.
64. Home, D. and Whitaker, M.A.B. (1992). Ensemble interpretations of quantum mechanics: A modern perspective, Physics Reports **210**, 223–317.
65. Leggett, A.J. (1987). Reflections on the quantum measurement paradox, In: Quantum Implications: Essays in Honour of David Bohm. (Hiley, B.J. and Peat, F.D., eds.) London: Routledge and Kegan Paul, pp. 85–104.

66. Bell, J.S. (1990). Against 'measurement', Physics World **3**(8), 33–40; also in Ref. [12] (2nd edn.), pp. 213–31, and in Ref. [13], pp. 902–9.

67. Gottfried, K. (1966). Quantum Mechanics. 1st edn., New York: Benjamin.

68. Gottfried, K. and Yan, T.-M. (2003). Quantum Mechanics. 2nd edn., New York: Springer.

69. Whitaker, A. (2006). Einstein, Bohr and the Quantum Dilemma. (1st edn. 1996, 2nd edn. 2006) Cambridge: Cambridge University Press.

70. Gottfried, K. (2000). Inferring the statistical interpretation of quantum mechanics from the classical limit, Nature **405**, 533–6.

71. Guy, R. and Deltete, R. (1990). Fine, Einstein and ensembles, Foundations of Physics **20**, 943–65.

72. Fine, A. (1990). Einstein and ensembles: A response, Foundations of Physics **20**, 967–89.

73. Popper, K. (1967). Quantum mechanics without the observer, In: Quantum Theory and Reality (Bunge, M., ed.), New York: Springer, pp. 7–44.

74. d'Espagnat, B. (1995). Veiled Reality: An Analysis of Present-Day Quantum Mechanical Concepts. Reading, Massachusetts: Addison-Wesley.

75. Bell, J.S. (1966). On the problem of hidden variables in quantum mechanics, Reviews of Modern Physics **38**, 447–52; also in Ref. [12], pp. 1–13, and in Ref. [13], pp. 695–700.

76. Schrödinger, E. (1935). Naturwissenschaften **23**, 807–12, 823–8, 844–9; translation by Trimmer, J.D. published as: The present situation in quantum mechanics, in: Ref. [77], pp. 152–67.

77. Wheeler, J.A. and Zurek, W.H. (eds.) (1955). Quantum Theory and Measurement. Princeton: Princeton University Press.

78. Shimony, A. (1993). Physical and philosophical issues in the Bohr-Einstein debate, In: Proceedings of the Symposia on the Foundations of Modern Physics 1992. (Laurikainen, K.V. and Montonen, C., eds.), Singapore: World Scientific, pp. 79–96.

79. Bohm, D. (1951). Quantum Theory. Englewood Cliffs: Prentice-Hall.

80. Peat, F.D. (1996). Infinite Potential: The Life and Times of David Bohm. Reading, Massachusetts: Addison-Wesley.

81. Pines, D. (1993). David Bohm, 1917–92, Physics World **6**(3), 67.

82. Bohm, D. (1952). A suggested interpretation of the quantum theory in terms of 'hidden variables' I and II, Physical Review. **85**, 166–93.

83. Holland, P. (1993). The Quantum Theory of Motion. Cambridge: Cambridge University Press.

84. Bell, J.S. (1982). On the impossible pilot wave, Foundations of Physics **12**, 989–99; also in Ref. [12], pp. 159–68, and in Ref. [13], pp. 842–52.

85. Bohm, D. (1987). Hidden variables and the implicate order, in: volume cited in Ref. [65] pp. 33–45.

86. Speziali, P. (ed.) (1972). Albert Einstein-Michele Besso, Correspondence 1903–1955. Paris: Hermann.

87. Beller, M. (1990). Born's probabilistic interpretation: a case study of 'concepts in flux', Studies in the History and Philosophy of Science **21**, 563–88.

Interlude

At the time of Einstein's death, it must have seemed that whatever influence he had had on the way in which scientists thought about quantum theory, and that had been minimal for nearly three decades, had gone forever. The kinds of question that concerned him—whether all the properties of a system actually have values at all times, even if they cannot all be measured; whether systems separated in space-time have independent properties; how one might discuss the act of measurement without requiring the observer to be a full participant in the process, maybe even a creator of the observed measurement results; whether determinism could be retained in physical theories; whether classical physics could or indeed should be obtained by straightforward limiting procedures on quantum theory— had all effectively been by-passed. Bohr's complementarity or the Copenhagen interpretation essentially rendered the questions unanswerable by making illegitimate by *fiat* discussion of the kinds of experimental situations that might throw light on them. EPR, effectively Einstein's last attempt at putting over his views, was dismissed practically out of court.

Einstein of course did not accept complementarity, but he had been a member of an exceptionally small minority. Bohr had convinced the overwhelming majority of physicists not only that his ideas were correct, but that it was illegitimate even to question them. Quite simply, to question complementarity was taken to mark one as not being a serious physicist. Most young physicists probably did not take the risk of coming into this category; they may have contented themselves with reading the most superficial account of complementarity, claimed to agree with it though not necessarily to understand it, and got on with their research.

Einstein had not done his position any favours by his dismissal in 1952 of the work of David Bohm, whose hidden variable theory did demonstrate a counterexample to Bohr's position. Without support from Einstein, Bohm's work was pilloried by Bohr' supporters, and, as a warning perhaps to others, Bohm himself, despite several major achievements, was in effect dismissed from the ranks of serious physicists. (Admittedly, American politics of the 1950s played an important role in the latter process—because he had been under investigation by the McCarthy commission, he was not able to obtain a post in the United States—but not, in the end, a crucial role.)

By 1955 it seemed virtually impossible that Einstein's views would ever have any influence on the way in which quantum theory was considered. Our book would have ended at this point, in something of a damp squib!

Such is not the way things worked out. Much of the credit for this must go to John Bell from Ireland, advocate of Bohm's theory and self-confessed follower of Einstein. He was able to exhibit not only great ingenuity as a theoretical physicist, but immense skill and courage in getting his views across despite the sanction explained above. He bided his time, and chose to take Bohr's views on only when it was clear he had unbeatable arguments or clear experimental suggestions to tempt the experimentalists.

Very very slowly – it has taken the full half-century! – it became acceptable to question Bohr's views, even to criticise or dismiss them. Alternative interpretations of quantum theory to Bohr's Copenhagen interpretation have been suggested. Experiments based on Bell's work have been performed with greater and greater sophistication, and have sewn the seed for the new subject of quantum information theory. Einstein's ideas have not necessarily been adopted or even found correct, but, and this is more important, there is now a far more open spirit of tolerance and discussion of these topics. That is the true legacy of Einstein's struggles with quantum theory over fifty years, and we explain how it was achieved in Part C of this book.

Part C
Denouement

I think you must entertain a notion for a long time before you can do something new. If you say, 'I will only think of something the minute you propose an experiment—otherwise I won't think of it,' how will you ever propose anything new? It often takes many, many years to be able to see what sort of experiments could be done.

David Bohm, in *Ghost in the Atom* (P.C.W. Davies and J.R. Brown, eds.) (Cambridge University Press, Cambridge, 1986), p. 126.

[On holding the view in the 1960s that 'quantum mechanics is, at the best, incomplete'] *We emphasise not only that our view is that of a minority, but also that current interest in such questions is small. The typical physicist feels that they have long been answered, and that he will fully understand just how if he can spare twenty minutes to think about it.*

J.S. Bell and M. Nauenberg, The moral aspect of quantum mechanics, in: *Preludes in Theoretical Physics* (A. De Shalit, H. Feshbach and L. Van Hove, eds.) (North Holland, Amsterdam, 1966), pp. 279–286; also in J.S. Bell, *Speakable and Unspeakable in Quantum Mechanics* (Cambridge University Press, Cambridge, 1st edn. 1987, 2nd edn. 2004), pp. 22–8.

9
Bell's Contributions and Quantum Non-locality

Introduction

The question of whether it is possible to supplement the wave-function of a system with extra parameters, known as hidden variables, has been discussed at several points earlier in this book. In Chapters 3 and 4, we saw that, though hidden variables might seem to have the potential to solve many of the apparent problems of quantum theory, orthodox approaches made a major point of excluding them from consideration. In Chapter 5 we discussed why, fairly soon after discovery of (modern) quantum theory, Einstein rejected the idea of producing a simple hidden variable theory. Nevertheless, his project of producing a *complete* quantum theory was likely to lead others in the direction of adding hidden variables, even though his own ideas were considerably more grandiose.

The physicist who did most to promote the ideas and values of Einstein was John Bell, originally from Ireland, who spent his working life at first in Harwell in England, and then at CERN in Geneva. Bell was entirely unconvinced by complementarity, and from a very early point in his career, he was attracted to the idea of hidden variables, which would provide the realism he believed essential in physics. Thus Bell described himself as a follower of Einstein. This was certainly the case as far as realism was concerned, but, as we have said earlier, Bell was probably somewhat misled in believing that Einstein supported hidden variables in any simple form. In the early 1960s, Bell made two extremely important contributions to the theory of hidden variables.

In a hidden variable theory, each individual system, apart from being specified by the wave-function, ψ, is characterized by an additional variable, the so-called hidden variable, designated by, say, λ. As we know, ψ itself is not a directly observable quantity, but λ would be directly observable, at least in the simpler hidden variable theories. If, though, λ is specified together with ψ, the result of any individual measurement is determined. (To be precise, we should say that this requirement is actually a definition of a particular class of hidden variables— deterministic hidden variables. There is another class, stochastic hidden variables, which provide the results of measurements only probabilistically.) In the context of a deterministic hidden variables programme, the statistical predictions of quantum mechanics are obtained by averaging over these additional hypothetical variables λ. This means that conceptually every ensemble of identical quantum systems corresponding to the same ψ can be viewed as consisting of subensembles corresponding to *dispersion-free states*, each with a different value of λ.

However, in Chapter 4 we discussed a theorem formulated by von Neumann in 1932 which apparently showed the logical impossibility of dispersion-free states accounting for quantum mechanical expectation values. On the other hand, in the preceding chapter we mentioned the model developed by David Bohm in 1952 which showed that it was possible to produce an internally consistent interpretation of the quantum mechanical formalism by introducing an additional ontological element in the description of the state of an individual system. (Bohm's work is discussed in some detail in the following chapter on non-standard interpretations.) Before we proceed to discuss how this issue of an apparent incompatibility between Bohm's model and von Neumann's theorem was finally resolved, we briefly discuss from a historical perspective the overall impact of this theorem.

It is interesting that in a 1939 conference report, Bohr referred to von Neumann's proof as the 'most clear and elegant demonstration' of the fact 'that the fundamental superposition principle of quantum mechanics logically excludes the possibility of avoiding the non-causal feature of the formalism by any conceivable introduction of additional variables'.[1] In Born's Waynflete lectures at Oxford in 1948,[2] von Neumann's theorem was cited as one of the important elements in the argument ruling out the possibility of any alternative model of quantum mechanics; according to Born, von Neumann's result showed 'that the formalism of quantum mechanics is uniquely determined by a few plausible axioms; in particular, no concealed parameters can be introduced with the help of which the indeterministic description could be transformed into a deterministic one ... [I]f a future theory should be deterministic, it cannot be a modification of the present one but must be essentially different.'

In 1948, Pauli also talked about 'von Neumann's well known proof that the consequences of quantum mechanics cannot be amended by additional statements on the distribution of values of observables, based on the fixing of values of some hidden parameters, without changing some consequences of the present quantum mechanics'.[3] On the other hand, as mentioned by Cushing,[4] it is a curious fact that in a letter of 2nd July 1935 written by Heisenberg to Pauli, although there is a long addendum titled 'Is a deterministic completion of quantum mechanics possible?', in which von Neumann's book was mentioned, he did not single out von Neumann's 'impossibility' theorem as a key element in his detailed discussion of why a completion of quantum mechanics was not possible.

The possibility of any limitation of von Neumann's theorem was not much discussed before the formulation of Bohm's model in 1952. In his book, Jammer[5] reviewed a few critical discussions of von Neumann's theorem prior to 1952, but none of those were able to bring out sharply any logical fallacy in von Neumann's argument. Then came Bohm's model in 1952 which succeded in providing an internally consistent counter-example to von Neumann's theorem. The cogency of von Neumann's proof became seriously questionable, but opinions on its validity remained greatly divided. Characteristic of this situation was a paper by Nabl,[6] in which, after reviewing how different authors varied in their evaluation of von Neumann's proof, Nabl emphasised the need for a clear-cut answer on the question of whether this proof had any loopholes. We now proceed to discuss how this situation was fully clarified.

Limitation to von Neumann's Theorem and Prelude to Bell's Theorem

It is interesting that in the very paper[7] where Bohm put forward his counter-example to von Neumann's theorem, he also specifically spoke of the 'irrelevance' of von Neumann's theorem with respect to his scheme. Bohm's point was that von Neumann implicitly assumed an *a priori* specified distribution of hidden variables associated with the observed system, that fixes measured values of *all* observables, including those that are mutually non-commuting. However this so-called no-go theorem did *not* prohibit those hidden variables theories in which an individual outcome of a measurement depended 'as much on the observing apparatus as on the observed system' (Ref. [7], p. 187). Since the measurements of non-commuting observables require mutually exclusive experimental arrangements, Bohm argued that hidden variable theories *not* covered by von Neumann's theorem are characterised by the fact that no single *a priori* distribution of hidden variables uniquely fixes the measured values of non-commuting observables.

It was Bell[8,9], though, who provided a more general and decisive criticism of von Neumann's theorem. It might first be mentioned that though, as has been said, Bell was highly appreciative of Bohm's actual model, he was actually rather scathing about Bohm's explanation of how it circumvented von Neumann's theorem. Bell himself argued that von Neumann's theorem was, in fact, more restrictive than Bohm's discussion had made it appear. He pointed out that von Neumann's demonstration depended crucially on the validity of a postulate which von Neumann had referred to as an *ansatz* in the German original of his book.[10] This postulate implies the following relation:

$$\langle aR + bS \rangle = a\langle R \rangle + b\langle S \rangle \tag{9.1}$$

where the Hermitian operators R and S correspond to observable quantities. Eq. (9.1) is true quantum mechanically, irrespective of whether R and S commute or not. But, in von Neumann's proof, this relation is required to be valid even for the hypothetical dispersion-free states relating to the hidden variables themselves, although, as Bell emphasised, there is neither a logical nor any empirical compulsion for this to be the case.

If R and S commute, we can simultaneously measure them for each individual member of any ensemble. Then, for each case, the measured value of $(R + S)$ must equal the sum of the measured value of R and the measured value of S, thereby ensuring the validity of the relation (9.1), even for the dispersion-free ensembles.

However, if observables R and S do not commute, the experimental arrangement to measure $(R + S)$ is, in general, entirely different from individual arrangements required to measure the individual observables R and S separately. Thus from *a priori* considerations, no relation such as Eq. (9.1) should be expected for the corresponding results for the dispersion-free ensembles.

By observing that the applicability of Eq. (9.1) for quantum states is essentially a characteristic feature of the quantum mechanical formalism, and is 'by no means a law of thought', Bell's essential point was that there was no reason to demand the validity of the relation (9.1) 'individually for the hypothetical dispersion-free states whose function is to reproduce the measurable peculiarities of quantum mechanics when averaged over'.[8,9] Hence Bell inferred that, contrary to von Neumann's claim, it is not the 'objective verifiable predictions of quantum mechanics', but rather the generally unjustified *ad hoc* imposition of the relation (9.1) for dispersion-free states, that excludes the possibility of hidden variables within the framework of von Neumann's theorem.

In order to reinforce this point, Bell[10] constructed a simple and instructive hidden variable model which consistently reproduced the quantum mechanically predicted statistical results of any spin measurement on a free spin-$1/2$ particle, but did *not* obey the relation (9.1) at the level of hidden variables. Bell also showed that Bohm's model too did not satisfy relation (9.1) at the level of dispersion-free states. See Jammer[11] for a particularly lucid discussion of Bell's original model; its different versions have been discussed by various authors. (See, for example, Selleri[12] and Cushing [13]).

For completeness of discussion on this point, it should be noted that, interestingly, von Neumann himself had observed that the relation (9.1) called for special attention in the case of non-commuting observables. In footnote 164 of his treatise,[10] he had explicitly referred to the energy measurement corresponding to the Hamiltonian operator H for an electron subjected to, for example, an electric field. Such an energy measurement requires an entirely different procedure from that employed for measuring position and momentum separately. But surprisingly, without any further consideration, von Neumann felt justified in assuming the truth of relation (9.1), even for the case of non-commutation in his 'proof' of the non-existence of dispersion-free states. It was Bell's insight which brought into focus the logical weakness involved in the use of this relation for the purpose of proving von Neumann's theorem.

Having become convinced that a completion of standard quantum mechanics was indeed possible, Bell now concentrated on generalizing Bohm's causal model for the two-particle examples involving correlated states, of the type used in the Einstein–Podolosky–Rosen (EPR) argument, which were discussed in detail in Chapter 6. Bell found that no matter how hard he tried to formulate a two-particle generalization of Bohm-type causal models, while simultaneously maintaining agreement with the quantum mechanical predictions, the model inevitably ended up being *non-local* in the sense that, for a given EPR correlated pair, the individual behaviour of either member of the pair depends upon the setting of the measurement performed on its partner. This applied no matter how large the separation was between the two members. But it is precisely this type of non-locality that the EPR argument found to be unacceptable. Hence Bell started wondering whether such a difficulty was generic to any possible hidden variable model for the EPR problem. Following this line of thinking, Bell ended up discovering a striking result in 1964[14], which later came to be known as Bell's theorem. In the next section we will discuss this theorem.

Here before proceeding further, a little historical digression may be relevant. It is noted that the first of Bell's two major papers on this topic, the one[8,9] which showed the weakness of von Neumann's theorem was written in the summer of 1964, but was not published until 1966, due first to delay in the refereeing process, and then because the manuscript was misfiled at the office of the Journal (*Reviews of Modern Physics*). It is interesting that towards the end of this paper, while drawing attention to the feature of non-locality present in Bohm's model (which Bohm[7] had also alluded to), Bell mentioned his suspicion that *all* causal or hidden variable models might have this property. Subsequently, his second major paper[14] confirming this conjecture was submitted for publication in November 1964, and it was published 2 years ahead of his earlier paper.[8,9]

Bell's Theorem

Bell began the paper[14] which proved his famous theorem by considering Bohm's version[15] of the EPR example, where two spin-$1/2$ particles in a singlet spin state fly apart from each other conserving the total spin. He then proceeded essentially by following the EPR line of argument. He assumed the formalism of quantum mechanics and the locality condition at the level of individual results of measurement. In connection with the latter, Bell quoted Einstein's statement of the locality condition in the form already given in the previous chapter: 'But on one supposition we should, in my opinion, absolutely hold fast: the real factual situation of the system S_2 is independent of what is done with the system S_1, which is spatially separated from the former'.[16] Then since, due to the perfect anti-correlation present in such an example, one can predict in advance the result of measuring *any* chosen component of the spin (say $\vec{\sigma}_2$) of any one member of the pair, by previously measuring the *same* component of the spin (say $\vec{\sigma}_1$) of the other member, Bell inferred, again following EPR, that the result of any such individual measurement must actually be *predetermined*. Now, since the quantum mechanical wave-function does not determine the result of an individual measurement, such a predetermination requires a more complete specification of the quantum mechanical state.

We now imagine that this more complete specification is done by means of the parameter λ. (In general, λ may be discrete or continuous, and it may denote a single variable or a set of variables, but here, for definiteness, λ is taken to be a single continuous parameter) Then the previous consideration suggests the existence of functions $A(\vec{a}, \lambda)$, $B(\vec{b}, \lambda)$ where the result A of measuring $\vec{\sigma}_1 \cdot \vec{a}$ is determined by $\vec{a}$ and λ, and the result B of measuring $\vec{\sigma}_2 \cdot \vec{b}$ for the *same* pair is determined by $\vec{b}$ and λ. Here we have

$$A(\vec{a}, \lambda) = \pm 1, \quad B(\vec{b}, \lambda) = \pm 1 \tag{9.2}$$

The crucial assumption used by Bell in his demonstration is that $A(\vec{a}, \lambda)$ does *not* depend on $\vec{b}$, nor $B(\vec{b}, \lambda)$ on $\vec{a}$; in other words, the result of one measurement does *not* depend on the setting of the apparatus used for the other measurement—this expresses the *locality character* of the class of hidden variable theories considered in this argument.

Now, let $\rho(\lambda)$ be the probability distribution of λ where

$$\int \rho(\lambda) \, d\lambda = 1 \tag{9.3}$$

Then for the type of hidden variable theory under consideration, which satisfies the locality condition, the expectation value of the product of spin variables $\vec{\sigma}_1 \cdot \vec{a}$ and $\vec{\sigma}_2 \cdot \vec{b}$ is given by

$$P_{hv}(\vec{a}, \vec{b}) = \int A(\vec{a}, \lambda) B(\vec{b}, \lambda) \, \rho(\lambda) d\lambda \tag{9.4}$$

This expression is required to equal the corresponding quantum mechanical expectation value which for the spin singlet state used in Bohm's version of the EPR example is given by

$$P_{qm}(\vec{a}, \vec{b}) = \langle (\vec{\sigma}_1 \cdot \vec{a})(\vec{\sigma}_2 \cdot \vec{b}) \rangle = -\vec{a} \cdot \vec{b} \tag{9.5}$$

But Bell was able to show that $P_{hv}(\vec{a}, \vec{b})$ *cannot* equal $P_{qm}(\vec{a}, \vec{b})$ for *all* $\vec{a}$ and $\vec{b}$, that is to say, for all possible orientations of spin measurements performed on the spatially separated particles in the spin singlet state. It therefore follows that *any* local hidden variable theory is inherently incompatible with quantum mechanics. This is Bell's theorem. Moreover, this incompatibility is experimentally testable. Thus the question of whether a completion of quantum mechanics is possible using hidden variables and also satisfying Einstein's locality condition is not merely a metaphysical query, but is an empirically decidable proposition. Therein lies the remarkable feature of Bell's theorem.

This is a rare instance where, by appealing to experiments on the statistical properties of an ensemble, one can draw conclusions about assumptions made at the level of individual measurements. Shimony[17] has thus appropriately given the name *experimental metaphysics* to this type of enterprise which seeks to achieve empirical resolution of what are apparently metaphysical issues.

It may be mentioned that our analysis has used the Pauli spin operator $\vec{\sigma}$, which is equal to the operator for the actual spin $\vec{S}$ divided by $\hbar/2$. So if a spin state is an eigenstate of, say, S_z with eigenvalue $\hbar/2$, it is an eigenstate of σ_z with eigenvalue 1.

After Bell's original proof of his theorem, different versions of this proof have appeared in the literature. Proofs of Bell's theorem have been comprehensively reviewed by various authors.[18-20] Here we discuss two particular demonstrations. We first consider a simple version of the proof for the deterministic case.[21-24] An advantage of this proof over Bell's original proof[14] and its other variants[18-20] is that no explicit use is made of the mathematical machinery of hidden variables. In the following section, we describe a proof using stochastic hidden variables.

For the deterministic case, we use Bohm's version of the EPR example mentioned earlier, and we consider measurements of the spins $\vec{\sigma}_1$ and $\vec{\sigma}_2$ along different directions performed on the correlated particles 1 and 2. For particle 1, we measure either A or A' where $A = \vec{a} \cdot \vec{\sigma}_1$ and $A' = a' \cdot \vec{\sigma}_1$. Measured values of A, A' are ± 1. Similarly, for its partner particle 2, the quantity B or B' is measured, where $B = \vec{b} \cdot \vec{\sigma}_2$ and $B' = \vec{b}' \cdot \vec{\sigma}_2$. Measured values of B and B' must also be ± 1.

Now, let us consider the combination $AB + A'B + AB' - A'B'$. For any given pair, one can measure *only one* of the product quantities $AB, A'B, AB', A'B'$. In each case, the measured value of the particular product is $+1$ or -1. The experiment consists in performing measurements of a large number of pairs, with the setting on one side (particle 1) alternating between $\vec{a}$ and $\vec{a}'$, and that on the other side (particle 2) between $\vec{b}$ and $\vec{b}'$, so that we have a large number of measurements of each of the quantities $AB, A'B, AB'$ and $A'B'$. Basic experimental data are the average values of these quantities which are denoted by $\langle AB \rangle$, $\langle A'B \rangle$, and so on.

We now proceed by discussing the following set of assumptions, which underpin the derivation of Bell's theorem:

(a) Each individual outcome of a measurement is causally determined by a supplementary variable (the so-called hidden variable) that, together with ψ, completely specifies the state of an individual quantum entity—this constitutes the assumption of the *deterministic* form of realism. Note that the notion of hidden variable used in the formulation of Bell's theorem is extremely general, so that, for the purpose of Bell's theorem, hidden variables can be regarded simply as parameters determining the outcomes of measurements on an event-by-event basis. In particular, it is *not* necessary to assume that a premeasurement value of the relevant observable is the same as the value obtained by a measurement.

(b) The value of a hidden variable pertaining to any given entity is *not* affected by measurements or dynamical interventions in any localised region sufficiently separated from the entity in question so that no known physical interaction or influence can causally connect occurrence in that space–time region to the system in question. (This is the locality condition at the level of individual events.) As Bell[14] put it: 'It is the requirement of locality, or more precisely that the result of a measurement on one system be unaffected by operations on a distant system with which it has interacted in the past, that creates the essential difficulty.' In the particular case of spacelike separation, a violation of this condition would entail superluminal causal influence.

(c) The average value of any relevant product observable, such as AB, obtained for the pairs on which measurements are actually performed, is *identical* to the average value pertaining to *all* sets of pairs (with the usual provisos regarding statistical fluctuations). This means that the randomly chosen sample of pairs on which one actually considers the measurement of a quantity such as AB, is typical of the entire ensemble. (This may be called *the principle of induction*).

Postulate (c) is usually considered to be a natural assumption. The conjunction of the assumptions (a) and (b) implies that for each particle, there is a *definite result predetermined* for any observable which is obtained if one measures the observable under consideration. Moreover, this result is not influenced in any way by measurements or other operations in the distant region. Let us now see how these suppositions lead to a testable constraint on the correlation functions. (Note that no input from quantum mechanics is used in the following argument.)

Considering the example of two spin-$1/2$ particles (1 and 2) in a spin singlet state, it follows from the assumptions (a) and (b) that the hidden variable associated with each particle 1 fixes definite values of both $A(+1$ or $-1)$ and $A'(+1$ or $-1)$, that are independent of whether B *or* B' is measured on the particle 2. Similarly, for each particle 2, the predetermined definite values of B and B' are independent of whether A *or* A' is measured on particle 1. Consequently, each particle pair has a value either $+1$ or -1 associated with each of the quantities $AB, A'B, AB'$ and $A'B'$, independent of what measurements are performed on them. Then it may be checked that, for each of the 16 different cases corresponding to the possible choices ± 1 for each of A, A', B, B' separately, we have

$$AB + AB' + A'B - A'B' = \pm 2 \qquad (9.6)$$

Note that Eq. (9.6) refers to a single pair (or a hypothetical group of pairs corresponding to the *same* hidden variable specifying their same initial 'complete state'). For the validity of Eq. (9.6), both the occurrences of, say, A in Eq. (9.6) must have the *same* value; similarly, this holds good for A', B and B'. (This is the input from the locality condition).

Summing Eq. (9.6) over the entire ensemble of pairs and taking the average, we obtain

$$|\langle AB \rangle + \langle AB' \rangle + \langle A'B \rangle - \langle A'B' \rangle| \leq 2 \qquad (9.7)$$

Now, by virtue of the assumption (c), we can identify the averages $\langle AB \rangle, \langle AB' \rangle, \langle A'B \rangle$, and $\langle A'B' \rangle$ with the experimentally measured values of these quantities. Thus we have a clear-cut prediction for the actual measured quantities given by the relation (9.7), which is one of the form of *Bell's inequality*.

However this inequality is violated by the quantum mechanical results for, say, the singlet state where $\langle AB \rangle = -\vec{a} \cdot \vec{b} = -\cos\theta$. The maximum violation occurs when all the directions are coplanar, with the angle between $\vec{a}'$ and $\vec{b}'$ equal to $135°$ and each of the three other angles in (9.7) to $45°$. Then Bell's inequality, as given by the relation (9.7), is violated since the left-hand side of (9.7) is equal to $2\sqrt{2}$. Another way to show this violation is to take $\bar{a} = \bar{b}$, with $\bar{a}'$ rotated from $\bar{a}$ by an angle θ, and $\bar{b}'$ from $\bar{b}$ by an angle $-\theta$, so that

$$\vec{a}' \cdot \vec{b} = \vec{a} \cdot \vec{b}' = \cos\theta \quad \vec{a}' \cdot \vec{b}' = \cos 2\theta$$

Then the left-hand side of Bell's inequality becomes $|-2 - 2\cos\theta\,(1 - \cos\theta)|$, which is greater than 2 for any $\theta < 90°$.

If the quantum mechanical predictions are assumed to be true for all situations, then one may respond to Bell's theorem by abandoning either assumption (a) or (b), or both. If assumption (a) is abandoned, then a realist causal description on an event-by-event basis describing how a particle behaves when subjected to a measuring apparatus is illegitimate. Abandoning assumption (a) would be a straightforward solution of the dilemma arising from Bell's theorem, and, of course, would be welcomed by advocates of Copenhagen. However abandoning assumption (a) would not be satisfactory for Bell and his supporters, who would point to the existence of the quantum measurement paradox. As described in earlier chapters, within the framework of the standard formalism of quantum mechanics it seems difficult to escape the measurement paradox, unless we give up the assumption that ψ provides a complete description of the state of an individual entity. This, in turn, Bell would say, implies the necessity of using supplementary or hidden variables; in other words, unless we modify the quantum mechanical formalism itself, his argument would be that assumption (a) is required to tackle the quantum measurement paradox.

If one grants the necessity of a realist interpretation of quantum mechanics, the only option, if one believes that the theory itself is correct in all details, is to interpret Bell's theorem as implying that *any* realist model of quantum mechanics must necessarily be non-local.

The other alternative, though, is to suspect that if careful experiments plug all possible loopholes, then Bell's inequality will always be satisfied, thereby upholding the validity of the locality condition in the context of realist descriptions at the level of individual events. But this would mean falsifying the standard quantum predictions under certain conditions. Thus it is necessary to appeal to experiments to settle this issue, and the latter part of this chapter will be concerned with experimental work on Bell's inequality.

At this stage, for interest, we discuss an analogous example[22] of correlation between properties of spatially separated objects in classical physics, for which Bell's inequality is always satisfied. The example is of a classical object, initially at rest, disintegrating into two spatially separated fragments 1 and 2 carrying angular momenta $\vec{J}_1$ and $\vec{J}_2$ where $\vec{J}_2 = -\vec{J}_1$. The directions of $\vec{J}_1$, $\vec{J}_2$ vary randomly over an ensemble of such fragments 1 and 2. For each such individual fragment, we consider measurement of the projection sign of its angular momentum along any given direction, corresponding to $r_a = \text{sign}(\vec{J}_1 . \vec{a})$ for fragment 1, and $r_b = \text{sign}(\vec{J}_2 . \vec{b})$ for fragment 2, where $\vec{a}$ and $\vec{b}$ are unit vectors. Possible values for r_a and r_b are ± 1.

Obviously, if $\vec{a} = \vec{b}$, we always have $r_a = -r_b$, so that $\langle r_a r_b \rangle = -1$. Assuming that the directions of $\vec{J}_1$ are uniformly distributed, the probability that $r_a r_b = +1$ is θ/π, and the probability that $r_a r_b = -1$ is $(1 - \theta/\pi)$, where θ is the angle between $\vec{a}$ and $\vec{b}$ $(0 \le \theta \le \pi)$. It therefore follows that the left hand side of relation (9.7) is exactly equal to 2 for $\vec{a} = \vec{b}$, $\vec{a} . \vec{b}' = \vec{a}' . \vec{b} = \cos\theta$, and $\vec{a}' . \vec{b}' = \cos 2\theta$. Note that in the corresponding quantum mechanical case, for this relative orientation of measurement directions, the quantum value of the left hand side of Eq. (9.7) is larger than 2 for any $\theta < 90°$. Thus in the classical example considered here, the correlation between the angular momenta of the two fragments is as strong as possible for a system that obeys the locality condition at the level of individual measurements. In general, the quantum correlation is stronger than or equal to the classical correlation. The form of the angular dependence of the quantum mechanical correlation function is what determines the magnitude of violation of Bell's inequality.

Bell's Theorem Using Stochastic Hidden Variables

In this section we discuss the proof of Bell's theorem based on a general class of stochastic hidden variable models which satisfy the locality assumption. For simplicity, and because of its direct relevance to experimental studies, we follow the Clauser–Horne formulation.[25] The basic idea of such hidden variables theories is that the hidden variable description does not uniquely determine the measured values of local observables for correlated particles, but only the probabilities that particular values occur.

The following argument refers to a source emitting two correlated entities 1 and 2 in opposite directions, where two analyzers can either transmit or absorb them. The dichotomic choice forced in this way on each quantum entity can be used to define corresponding dichotomic observables by prescribing that $A(a) = \pm 1$ and $B(b) = \pm 1$, depending on the choice of transmission $(+1)$ or absorption (-1), pertaining to particle 1 and 2, respectively. This scheme is relevant to actual experiments testing Bell's inequality with pairs of correlated optical photons emitted in atomic cascades. For such photons, the binary choice is between transmission and absorption in a polarizer.

Now, let us invoke a variable to represent the complete physical state of an individual pair of correlated quantum entities (1 and 2) within a general probabilistic scheme in which $p_1(a, \lambda)$ is the probability that an individual particle 1 in the state λ crosses the analyzer with parameter a and then is subsequently detected; and $p_2(b, \lambda)$ is the similar probability for particle 2 crossing the analyser with parameter b. $P(a, b, \lambda)$ represents the joint probability that both 1 and 2 cross their respective analysers having the parameters a and b, and are both detected. The locality condition is expressed by the following factorisability condition:

$$P(a, b, \lambda) = p_1(a, \lambda)p_2(b, \lambda) \tag{9.8}$$

with the obvious extra requirement that the hidden variable distribution function $\rho(\lambda)$ corresponding to the initial joint state of 1 and 2 does not depend on the parameter choice for the analyzers. Note that the quantities $P(a, b, \lambda)$, $p_1(a, \lambda)$, and $p_2(b, \lambda)$ are defined at the individual level. Observable probabilities at the statistical level, in other words quantum mechanically computable quantities, are expressible as weighted averages of individual probabilities:

$$p_1(a) = \int p_1(a, \lambda) \rho(\lambda) \, d\lambda \tag{9.9}$$

$$p_2(b) = \int p_2(b, \lambda) \rho(\lambda) \, d\lambda \tag{9.10}$$

$$P(a, b) = \int P(a, b, \lambda) \rho(\lambda) \, d\lambda \tag{9.11}$$

In order to deduce testable constraints on $p_1(a), p_2(b), P(a, b)$ from the locality condition (9.8) using Eqs. (9.9) and (9.10), following Clauser and Horne,[25] we use this algebraic theorem:

Given real numbers x_1, x_2, X, y_1, y_2, Y such that

$$0 \le x_1, \ x_2 \le X, \quad 0 \le y_1, \ y_2 \le Y \tag{9.12}$$

it follows that:

$$-XY \le x_1 y_1 - x_1 y_2 + x_2 y_1 + x_2 y_2 - x_2 Y - X y_1 \le 0 \tag{9.13}$$

Let us take $x_1 = p_1(a_1, \lambda)$, $x_2 = p_1(a_2, \lambda)$, $y_1 = p_2(b_1, \lambda)$, $y_2 = p_2(b_2, \lambda)$, and assume that the ontological probabilities $p_{1,2}$ at the individual level (which are not directly observable) lie between 0 and 1, so that $X = Y = 1$. Then Eq. (9.13) reduces to

$$-1 \le p_1(a_1, \lambda) p_2(b_1, \lambda) - p_1(a_1, \lambda) p_2(b_2, \lambda) + p_1(a_2, \lambda) p_2(b_1, \lambda)$$
$$+ p_1(a_2, \lambda) p_2(b_2, \lambda) - p_1(a_2, \lambda) - p_2(b_1, \lambda) \le 0 \tag{9.14}$$

By invoking the locality condition (9.8) one can then write

$$-1 \le P(a_1, b_1, \lambda) - P(a_1, b_2, \lambda) + P(a_2, b_1, \lambda)$$
$$+ P(a_2, b_2, \lambda) - p_1(a_2, \lambda) - p_2(b_1, \lambda) \le 0 \tag{9.15}$$

Multiplying each term of the relation (9.15) by the probability density $\rho(\lambda)$, integrating over λ, and using Eqs. (9.9)–(9.11), one obtains

$$-1 \le P(a_1, b_1) - P(a_1, b_2) + P(a_2, b_1)$$
$$+ P(a_2, b_2) - p_1(a_2) - p_2(b_1) \le 0 \tag{9.16}$$

which is a form of Bell's inequality, sometimes referred to as the *inhomogeneous Bell's inequality*, since it is based on both double and single detection probabilities.

An important ingredient in the preceding proof is the condition that the probabilities pertaining to the distribution of hidden variables are positive, and are not larger than 1. One may argue that since such probabilities are not directly observable, it should not be objectionable to invoke negative probabilities for hidden variables. Then, of course, Bell's inequality is no longer derivable for stochastic hidden variable models. In fact, there are explicit examples[26,27] of stochastic hidden variable models which reproduce quantum mechanical violations of Bell's inequality, but at the expense of allowing negative probabilities.

General arguments[28,29] show that one can always reproduce the quantum mechanical results for non-factorisable state vectors of correlated systems by stochastic hidden variable models using negative probabilities. However, for formulating the hidden variable models, it seems conceptually illegitimate to use non-physical negative probabilities, even if the computed testable probabilities are ensured to be positive. This is because, within the framework of a hidden variable model, the quantities interpreted as probabilities at the hidden variable or individual level are considered to have the same reality as the probabilities actually measured at the level of an ensemble.

Contextuality

So far we have described Bell's demonstration[8] that, contrary to von Neumann's stipulation, hidden variable models compatible with the quantum mechanical formalism were possible, but we have also seen that his theorem very much limits such theories—they must be non-local. Bohm's model, of course, fufilled this criterion, as did Bell's own model that he came up with in this same paper.

Also in the same paper, he deduced another criterion for hidden variable theories. In order to ensure *compatibility* with standard quantum mechanics, hidden variable models have to be necessarily *contextual* if the dimensionality of the Hillbert space is greater than 2. To explain what this means, let us imagine a system for which O_1, O_2 and O_3 are three observables, and the operators associated with these observables commute, so that the three observables may be measured simultaneously. Now, consider that O_1 and two other observables, O_2' and O_3' also have mutually commuting operators, so these three observables may also be measured simultaneously. To say that the theory is contextual is to say that the result obtained in the measurement of O_1 depends on whether we measure it simultaneously with O_2 and O_3, *or* with O_2' and O_3'. The requirement for this is analogous to Bell's argument against the unjustified von Neumann assumption. Because neither O_2 nor O_3 commute with either O_2' or O_3', a *different* experimental arrangement is required to measure O_1, O_2 and O_3 from that to measure O_1, O_2' and O_3'. Therefore it is not logically compelling to expect the values of O_1 obtained in the two different joint measurements to be the same.

The need for such contextuality in hidden variable models means that one cannot maintain that in such theories measurements merely record pre-measurements values. In the Bohm model, for example, measurement does not merely record a pre-measurement value, but is a process that changes the pre-measurement value to the value actually obtained in the experiment. It thus has the flexibility to be contextual.

Apart from Bell's argument, there have been a variety of different versions of this no-go theorem showing that if the dimensionality of the Hillbert space is greater than 2, the quantum mechanical formalism is necessarily incompatible with *any* hidden variable model which does *not* satisfy the contextuality condition. Just to mention a few, the argument by Gleason,[30] the formulation of this theorem for the specific case of a spin-1 particle by Kochen and Specker,[31] its extension for a spin-$3/2$ particle shown by Penrose,[32] and an argument[33] showing the incompatibility between quantum mechanics and *any* stochastic non-contextual hidden variable model that requires ascribing probability distributions to hidden variables.

However, an *empirical test* of this theorem was lacking. In this context, a recent development has been the neutron interferometric experimental demonstration[34] of this theorem using a scheme[35] suggested for this purpose by entangling the translational and spin degrees of freedom of a spin-$1/2$ particle which shows quantum mechanical violation of a single particle Bell-like inequality deduced from the condition of non-contextuality.

Signal Locality, and Parameter and Outcome Independence

The presence of non-locality of any type in quantum theory would almost certainly be viewed as problematic by the majority of physicists, and, as we saw in the previous chapter, this would certainly have been Einstein's position. Nevertheless, in quantum theory there are two aspects to non-locality, and the first would have caused a far bigger conceptual upheaval than the second. The first relates to a *measurement setting* at one location being correlated with a measurement result at another, which is separated from the first by a space-like interval ($|\Delta x| > c|\Delta t|$). In this case, an experimenter may choose the meaurement setting, thus influencing the result obtained by a second experimenter at the distant location. This would correspond to the sending of a signal at speed greater than that of light, or *superluminal communication*. In the terms in which Bell's theorem is discussed, it would correspond to loss of *parameter independence*, or loss of *signal locality*. As is known by any student of relativity, it is expected that superluminal communication could result in great conceptual difficulties, such as an effect occurring, at least in some frames, before its cause .

The second aspect of non-locality relates to correlation between two *measurement results* at locations separated by a space-like interval. For this type of non-locality, neither result is in the control of any experimenter, so this situation cannot be used to send signals at speeds greater than that of light. This type of non-locality is referred to as loss of *outcome independence*. While many physicists would have conceptual concerns about loss of outcome independence, it would not lead to the major conceptual problems created by loss of parameter indepependence. Signal locality is maintained. At this point, we note that if one considers an interpretation of quantum theory that includes hidden variables, Bell's theorem tells us that it must be non-local. However, the way in which it is non-local is that it is in violation of outcome independence, but not of parameter independence. One may say that it is a non-locality of the less virulent form, and so signal locality is maintained; messages cannot be sent at a speed greater than that of light. Thus there is no *direct* clash with relativity, and Shimony has indeed remarked that there may be a 'peaceful coexistence' between the two theories. In the remainder of this section, we explore all these ideas in greater depth.

Following Jon Jarrett[36] and Abner Shimony,[37] we may note that the locality condition formulated in Eq. (9.8), can be viewed as a logical conjunction of the following two conditions. First we study parameter independence. We may mention that the terms 'parameter independence' and 'outcome independence' are due to Shimony.[37] Jarrett uses the terms 'locality' and 'completeness' for the same ideas, terminology which is rather confusing, and which we shall not mention again.

We consider the correlated state of two spatially separated particles A and B, and we assume that, given the complete state specification λ for an individual pair, the probability of an outcome x for one of the particles, say A, with the experimental setting i, is independent of the choice j of the experimental setting in the other wing of the experiment, B. This condition can be written as:

$$p_\lambda^A(x|i,j) = p_\lambda^A(x|i) \tag{9.17}$$

A similar condition follows for $p_\lambda^B(y|i,j)$. The condition given by Eq. (9.17) is known as parameter independence. Note that observable statistical distributions are the quantities $p^A(x|i,j)$ and $p^A(x|i)$ integrated over λ, and not the actual $p_\lambda^A(x|i,j)$, and $p_\lambda^A(x|i)$ which occur in Eq. (9.17).

A different condition stipulates that the occurrence of an outcome y in one of the wings, say B, does not affect the probability of the occurrence of any outcome x in the other wing A, with the same λ characterising a given pair. Following the same notation as in Eq. (9.17), this condition can be written in the following form

$$p_\lambda^A(x|\,i,j,y) = p_\lambda^A(x|i,j) \tag{9.18}$$

and similarly for $p_\lambda^B(y|i,j,x)$. The condition given by Eq. (9.18) is called outcome independence.

Now, based on the standard definitions of joint and conditional probabilities, the joint probability $p_\lambda^{AB}(x,y|i,j)$ can be written as:

$$p_\lambda^{AB}(x,y|i,j) = p_\lambda^A(x|i,j,y)p_\lambda^B(y|i,j) \tag{9.19}$$

Following the same notation as above, the locality condition (9.8) can then be written as

$$p_\lambda^{AB}(x,y|i,j) = p_\lambda^A(x|i)p_\lambda^B(y|i) \tag{9.20}$$

Using Eqs. (9.17) and (9.18), we see that Eq. (9.19) reduces to Eq. (9.20). Therefore if *either* parameter independence, Eq. (9.17), *or* outcome independence, Eq. (9.18), is violated, the derivation of Bell's inequality (9.16) is no longer valid. Therefore, the violation of Bell's inequality implies that either parameter independence, or outcome independence, or both are violated at the level of hidden variables.

We now turn to a discussion of signal locality. Bell's theorem implies that, if quantum mechanical non-local correlations in experimental results are explained in terms of hidden connections by using supplementary or so-called hidden variables, superluminal and distance-independent influence is inevitable at the level of individual events. However, since hidden variables are not controllable, as we cannot reproduce an individual quantum system with specified hidden variables, this type of action at a distance is usually considered to be innocuous, because it cannot be employed to transmit information in a usable form. This relates to violation of outcome independence.

We henceforth use the term *signalling* to denote a *controlled way* of sending information that can be used to generate causal paradoxes. For a detailed discussion clarifying the distinction between superluminal causal influence that does not necessarily give rise to causal paradoxes, and actual superluminal signalling, see Maudlin,[38] and Helliwell and Konkowski.[39]

If we define signal locality to mean the absence of faster-than-light signalling, this necessarily implies that any statistically observable property determined by the expectation value of a physical quantity measured in one space–time region should not depend on what happens in another spacelike separated region. Here we note that this assumption of signal locality, formulated independently of quantum theory, implies certain testable inequalities between the observable correlation functions, which were discovered by Roy and Singh.[40] Such inequalities are different from Bell-type inequalities. Non-local hidden variable theories violate the latter, but satisfy signal locality.

In relativistic quantum field theory, it is clear that signal locality will be ensured by the commutativity of observables for spacelike separation. What is interesting is that even non-relativistic quantum theory does not permit superluminal signalling to take place via correlated particles. As Bell[41] states, 'It is as if there is some kind of conspiracy, that something is going on behind the scenes which is not allowed to appear on the scenes.' This relates, as explained above, to parameter independence being upheld.

To explain how signal locality is satisfied in quantum mechanics for two particle correlated states, let us consider a typical two-particle correlated quantum system described by the total wave-function

$$\Psi = a\psi_+(L)\phi_-(R) + b\psi_-(L)\phi_+(R) \tag{9.21}$$

where $\psi_\pm(L)$ and $\phi_\pm(R)$ are wave-functions of particles on the left and right respectively; we assume orthogonality so that $\langle\psi_+(L)|\psi_-(L)\rangle = \langle\phi_+(R)|\phi_-(R)\rangle = 0$. These particles are well separated and become non-interacting. The time evolution operator for this joint system factors into:

$$U = U_L \otimes U_R \tag{9.22}$$

where $U_{L,R}$ operate on particle states on the left (L) and on the right (R), respectively.

Note that since there is no interaction Hamiltonian for these two particles, their individual free evolution Hamiltonians commute with each other. This is because the wave-function domains on which their respective operators act are disjoint. Now, operating U given by Eq. (9.22) on Ψ of Eq. (9.21), it is easily seen that the expectation values of observables and their time evolution in either wing, say, L, are *independent* of U_R in the other wing. In particular, any local change in, say, U_R, through a modification of the Hamiltonian on the right, has no observable statistical effect on the left. This means that although *kinematically* the particle properties in the entangled state (9.21) remain correlated even when the particles are sufficiently separated, they are *dynamically independent* in the sense that the behaviour of one does not depend on what happens to the dynamical evolution of the other.

Let us now show more explicitly that measuring an observable on either member of an entangled pair, say, R, does not affect observable properties of the other member L. For the standard measurement of an observable represented by an operator whose mutually orthogonal states are ϕ_+ and ϕ_-, the final wave-function after measurement on R is given by:

$$\Psi = a\psi_+(L)\phi_-(R)A_1 + b\psi_-(L)\phi_+(R)A_2 \tag{9.23}$$

where A_1 and A_2 are *macroscopically distinguishable* and *mutually orthogonal* states of the measuring apparatus. One can then easily verify that the expectation value of any dynamical variable, say α, pertaining to particle L, evaluated by using Eq. (9.23), is the *same* as calculated from Eq. (9.21) before measurement on R. We also note that the notion of explicit collapse to a mixed state due to a measurement does not make any difference to this conclusion. The earliest version of this demonstration was given by Bohm,[42] and later in different forms by many other authors.[37,43–46]

General Remarks on Local Realism and Entanglement

Entanglement is of great importance in quantum information theory. In fact it would be better to say that entanglement is a great *resource* for quantum information theory. As such, a great deal of very detailed experimental and theoretical work has been performed in the area over the last 30 years or so, and also in the areas of local realism and Bell's inequalities, which are, of course, intimately linked with entanglement. In this section, we give a very brief account of a few points of interest.

(a) A general form of the correlated two-particle wave-function violating Bell's inequality is given by

$$\Psi = \sum_i c_i\phi_i\eta_i \tag{9.24}$$

where $\phi_i\eta_i$ are mutually orthogonal states of the particles 1 and 2, respectively. It has been shown[47] that for *any* correlated state of this form with at least two of the c_is non-zero, Bell's inequality is violated by the quantum mechanical predictions, if one chooses the relevant observables appropriately. This result reinforces the notion that the incompatibility between quantum mechanics and local realism is deep-rooted, and, though Bell's inequality may not fully contain all the restrictions implied by local realism, it is nevertheless sufficient for displaying, with some generality, the inherent inconsistency between multiparticle non-factorisable wave-functions and local realism.

(b) Bell's inequality is one of an infinite set of inequalities that can be deduced from local realism. The first examples of inequalities providing physical restrictions *not* contained in Bell's inequality were given by Roy and Singh,[48,49] followed by other authors.[50–52] Two types of such inequalities have been discussed in the literature, one type valid for arbitrary linear combinations of correlation functions, the other type involving arbitrary linear combinations of joint probabilities. The two sets are not equivalent, and the latter is more general than the former, as shown by Garg and Mermin.[53]

Bell-type Arguments Without Inequalities: Greenberger–Horne–Zeilinger

While the standard way of demonstrating the inconsistency between quantum theory and local realism made use of inequalities, around 1990 two techniques were invented which could perform this task without inequalities.

The first method was produced by Greenberger, Horne and Zeilinger[54–56] (GHZ). In its simplest version, their argument uses a system of three spin-$1/2$ particles, mutually correlated and spatially separated; on each of these particles, measurements of x or y spin components may be made. The so-called GHZ state that they use may be written as

$$|\Psi\rangle = \left(\frac{1}{2}\right)^{1/2} (|1,1,1\rangle - |-1,-1,-1\rangle) \qquad (9.25)$$

Here $|1,1,1\rangle$ and $|-1,-1,-1\rangle$, denote states for which σ_z eigenvalues of the three particles are all $= +1$ or -1 respectively. It is easily verifiable by direct calculation that $|\Psi\rangle$ given by Eq. (9.25) satisfies the following eigenvalue equations:

$$\sigma_x^1 \sigma_x^2 \sigma_x^3 |\Psi\rangle = -|\Psi\rangle \qquad (9.26a)$$

$$\sigma_x^1 \sigma_y^2 \sigma_y^3 |\Psi\rangle = |\Psi\rangle \qquad (9.26b)$$

$$\sigma_y^1 \sigma_x^2 \sigma_y^3 |\Psi\rangle = |\Psi\rangle \qquad (9.26c)$$

$$\sigma_y^1 \sigma_y^2 \sigma_x^3 |\Psi\rangle = |\Psi\rangle \qquad (9.26d)$$

where superscripts 1–3 designate the particles, respectively. In this example it is possible to determine any of these observables $\sigma_x^1, \sigma_y^1, \sigma_x^2 \ldots$ by distant measurement on the other two particles. For example, to know σ_x^1, we need only measure σ_y^2 and σ_y^3 at distinct locations. Applying the notion of local realism it may thus seem legitimate to assume that the observed individual values of σ-operators are *predetermined* (by, say, hidden variables), and that any such individual value is independent of whichever sets of three single particle spin measurements we choose to make on these spatially separated particles. This assumption is precisely along the same lines as the locality assumptions we have made in our various proofs of Bell's theorem. We call these values $m_x^1, m_y^1, m_x^2 \ldots$

Consistent with Eq. (9.26 a–d), we then have the following relations for any set of three such correlated particles whose m values are determined by fixed hidden variables:

$$m_x^1 m_x^2 m_x^3 = -1 \tag{9.27a}$$

$$m_x^1 m_y^2 m_y^3 = +1 \tag{9.27b}$$

$$m_y^1 m_x^2 m_y^3 = +1 \tag{9.27c}$$

$$m_y^1 m_y^2 m_x^3 = +1 \tag{9.27d}$$

Note that in the preceding relations the notion of local realism implies that the individual value of any one of the quantities $m_{x,y}^{1,2,3}$ is the *same* irrespective of the equation in which it occurs [e.g., the value of m_x^1 is the same in Eqs. (9.27a–b)]. But then Eqs. (9.27a–d) are not mutually consistent, since, recalling that $m_{x,y}^{1,2,3} = \pm 1$, we can obtain from Eqs. (9.27b–d):

$$m_x^1 m_x^2 m_x^3 = +1 \tag{9.28}$$

which contradicts Eq. (9.27a).

The GHZ argument demonstrates that, for multiparticle states of the type in Eq. (9.25), quantum mechanical predictions are incompatible with local realism. The chief merit of this form of argument is that it is *non-statistical*: we are not concerned with measurement statistics involving questions about the size of the relevant ensemble, statistical fluctuations, and so on. The GHZ argument reveals that the notion of local realism is *inconsistent* with quantum mechanics even in the case of *perfect correlations*, that is to say even when the relative orientations between the measurement axes are $\theta = 0$ or $\theta = \pi$, cases for which quantum mechanics makes non-probabilistic, definitive predictions regarding the correlated properties.

Another advantage of GHZ is that, for Bell-type arguments with inequalities, the specific form of the correlations as a function of the relative orientation between measurement axes plays a central role, and one requires measurements along various non-orthogonal directions to demonstrate quantum violation of Bell's inequality. In contrast, the GHZ argument is concerned with measured values only along orthogonal axes (say, x or y spin components).

It may be noted, though, that the proof of the GHZ argument is restricted to the deterministic form of local realism, because each individual values $m_{x,y}^{1,2,3}$ is assumed to be uniquely determined by hidden variables, whereas Bell-type arguments can be formulated for stochastic local, hidden-variable theories as well. There is incidentally an equivalence theorem[57,58] which claims to show that one could duplicate predictions of any stochastic local realist theory by using an appropriate deterministic local realist theory.

As a last point, it should be noted that, since a Bell-type inequality is a direct consequence of local realism without reference to quantum mechanics, it provides a direct way of verifying or falsifying local realism, irrespective of the quantum mechanical framework. Compared with this, the structure of the GHZ argument is quite different, since what is essentially demonstrated is an algebraic inconsistency between the relations (9.27a–d), an argument which is based on the local realist interpretation of quantum mechanical results given by Eqs. (9.26a–d).

Bell-type Arguments Without Inequalities:
Hardy's Argument

The GHZ-type demonstration applies to entangled quantum states of three or more particles. Except in the limiting case where the number of posible settings of each local observable is allowed to tend to infinity,[59] it is not possible to use the GHZ method to demonstrate incompatibility between quantum mechanics and local realism for two-particle states. Hardy[60,61] took a different approach and succeeded in formulating a scheme to show that any entangled state of a pair of particles, except the maximally entangled state such as the singlet state used in a Bell-type argument, can be used to demonstrate quantum non-locality without using Bell-type inequalities. Interesting variants of Hardy's argument have been discussed by various authors.[62–64] Here we shall outline the version given by Goldstein.[65]

We note that by so-called Schmidt decomposition, any two-particle entangled state can be written in the form:

$$|\Psi\rangle = \sum_i c_i |e_1^i\rangle |f_2^i\rangle \tag{9.29}$$

where the e states from a basis for particle 1, and the f states a basis for particle 2. If an entangled state is *not* maximally entangled, then there are terms i and j in Eq. (9.29) for which $c_i \neq c_j$. For any such non-maximally entangled state, with a suitable choice of basis states $|u_1\rangle$, $|v_1\rangle$ for particle 1, and $|u_2\rangle$, $|v_2\rangle$ for particle 2, we can write the general state $|\Psi\rangle$ in the following form:

$$|\Psi\rangle = a|v_1\rangle|v_2\rangle + b|u_1\rangle|v_2\rangle + c|v_1\rangle|u_2\rangle \tag{9.30}$$

where a, b and c are non-zero. The state vector (9.30), in which only one term of the product basis, viz., $|u_1\rangle\,|u_2\rangle$, is missing, has the appropriate form for a Hardy-type argument.

Let us define the following observables

$$U_1 = |u_1\rangle \langle u_1| \tag{9.31a}$$

$$U_2 = |u_2\rangle \langle u_2| \tag{9.31b}$$

$$W_1 = |\beta_1\rangle \langle \beta_1| \tag{9.31c}$$

$$W_2 = |\beta_2\rangle \langle \beta_2| \tag{9.31d}$$

$$|\beta_1\rangle = \frac{(a|v_1\rangle + b|u_1\rangle)}{(|a|^2 + |b|^2)^{1/2}} \tag{9.32a}$$

and

$$|\beta_2\rangle = \frac{(a|v_2\rangle + c|u_2\rangle)}{(|a|^2 + |c|^2)^{1/2}} \tag{9.32b}$$

With respect to the state vector (9.30), it is straightforward to verify that the following properties are valid for measurement of U_1, U_2, W_1, and W_2.

1. $U_1 U_2 = 0$.
2. $U_1 = 0$ implies $W_2 = 1$.
3. $U_2 = 0$ implies $W_1 = 1$.
4. There is a non-vanishing probability of finding $W_1 = W_2 = 0$.

Note that while properties 2–4 can be verified by appropriately rewriting Eq. (9.30) using Eqs. (9.32a and b), the validity of property 1 may be seen directly because these is no $|u_1\rangle \, |u_2\rangle$ term in Eq. (9.30). The argument proceeds as follows. We assume that once the quantum state (9.30) is formed, particles 1 and 2 are spatially separated and subjected to interactions with two distinct apparatuses respectively where measurements of U_i or $W_i (i = 1, 2)$ can be made.

Consider an ensemble of experimental runs for which U_1 and U_2 are measured. Property 1 implies three possibilities: $U_1 = 0$, $U_2 \neq 0$; $U_1 \neq 0, U_2 = 0$; and $U_1 = U_2 = 0$. From property 2 we use the locality condition to deduce that, for the subset of runs characterised by a particular hidden variable λ, where $U_1 = 0$, then if W_2 is measured on particle 2 instead of U_2, one would expect to obtain the result $W_2 = 1$. Similarly, from property 3, we use the locality condition for the subset of runs yielding the result $U_2 = 0$ to deduce that one would expect $W_1 = 1$ if W_1 is measured instead of U_1. It then follows that for runs corresponding to the remaining possibility $U_1 = U_2 = 0$, one should obtain $W_1 = W_2 = 1$ if W_1 and W_2 are measured. This would exhaust all possibilities.

Clearly the preceding argument, based on the deterministic form of local realism, in conjunction with properties 1–3, does not permit $W_1 = W_2 = 0$ in an experiment involving the state vector (9.30). But obviously this is inconsistent with the quantum mechanical prediction of property 4. Incompatibility between local realism and the quantum mechanical formalism is thus demonstrated in an ingenious way that in a sense falls halfway between the Bell and GHZ-type arguments. If local realism is valid compared with quantum mechanical predictions, according to the Bell-type argument there must be a statistical violation of quantum mechanics. For GHZ, the violation is not statistical. According to the Hardy approach, validity of local realism implies that we must *either* have a statistical violation of quantum mechanics in the sense that the quantum result

$W_1 = W_2 = 0$ is never seen, *or* a non-statistical violation in that at least one of the quantum mechanical properties 1–3 must be violated in a single run, *or* both. If a single $W_1 = W_2 = 0$ result is seen, that is sufficient to repudiate local realism—an *all or nothing* situation.

Experimental Tests of Bell-type Inequalities

In comparison with other advocates of hidden variables, such as, in particular, Bohm, Bell was able, at least gradually, to obtain attention for his work, because he had been able to suggest experiments which would be able to test fundamental aspects of physics. Indeed such experiments have been discussed and performed for something like 35 years. While great ingenuity has been showed in carrying out the experiments, and results of much interest have been obtained, as yet no totally firm conclusions have been reached.

This is because, although all the experiments, with an early exception, have reported violation of the Bell inequalities and good agreement with quantum theory, there have been two loopholes preventing the results from being conclusive. The first loophole is the so-called detection loophole. Because detector rates for the experiments performed have been low, one cannot be sure that the events that have been sampled have been a fair selection of the total ensemble of events, and if the sampling has not been fair, the results are not necessarily meaningful. A variety of supplementary assumptions have had to be made in interpretation of the results because of the low detection rates, and these assumptions, though plausible, have been subject to discussion and criticism.

The second loophole is the so-called locality loophole. In the early experiments, there was the possibility of exchange of information between the two wings of the system. In recent decades, increasingly stringent means have been used to attempt to eliminate this loophole.

Thus when assessing the experimental tests of Bell-type inequalities, we must distinguish between the following two types of inequalities:

1. An inequality deduced from local realism *alone*, which is incompatible with quantum mechanical results in the case of ideal experiments with highly efficient detectors. This would obviously be the ideal, but progress towards it has been very difficult.

2. An inequality deduced not from local realism alone, but only with the aid of an *additional* or *supplementary assumption*, which is violated by quantum mechanical results. Over nearly the whole of the period in which such experiments have been performed, because only low-efficiency detectors have been available, actual experiments have been of type (1) rather than of type (2).

Typical experimental arrangements have studied the correlation between polarisations on entangled photons, for which the mathematics is identical to that of the entangled spin-$1/2$ systems introduced by Bohm and studied by Bell. A standard form of Bell's inequality for this purpose is given by Eq. (9.16). This equation follows from local realism without any additional assumption. However, inequalities of the type in Eq. (9.16) are *not* violated by quantum mechanical pre-dictions for available photomultiplier detectors in the visible wavelength range, with low efficiencies around 20–30%. For this reason, one usual procedure, first discussed by Clauser and Horne,[25] is to invoke the following supplementary assumption: for every photon in the state specified by a particular hidden vari-able λ, the probability of detection with a polariser placed before a detector is less than or equal to the detection probability with the polariser removed. This is often referred to as the CH, or the 'non-enhancement' assumption.

Note that this is an apparently plausible but actually non-trivial postulate at the level of individual photons for a particular λ, although this assumption is of course true at the observable level of statistical results when averaged over all λs. (For a critical analysis of the non-trivial nature of this hypothesis in the context of local hidden variable models, see Selleri,[66] Ferrero et al.[67] and Santos.[68–70] This additional hypothesis leads to the following inequalities:

$$p_1(a_1, \lambda) \leq p_1(\infty, \lambda); \quad p_1(a_2, \lambda) \leq p_1(\infty, \lambda)$$
$$p_2(b_1, \lambda) \leq p_2(\infty, \lambda); \quad p_2(b_2, \lambda) \leq p_2(\infty, \lambda) \tag{9.33}$$

where the symbol ∞ corresponds to the case when the polarizer is removed.

It is important to understand the differences between the conclusions that may be drawn from analysis of inequalities (9.16) and (9.33). With η the value of the efficiency of the photomultiplier detectors, and ε the efficiency of the polarisers corresponding to the choice between transmission and absorption of a photon, we may use typical numerical values of these parameters as ε close to unity, but η only as high as 0.2–0.3.

Then the inequalities (9.16) and (9.33) may be expressed in terms of a parameter Γ as

$$-0.8 \leq \Gamma \leq 0.2 \tag{9.34}$$

and

$$0 \leq \Gamma \leq 0.08 \tag{9.35}$$

for the cases of local realism alone in Eq. (9.34), and local realism with the sup-plementary assumption in Eq. (9.35). Thus local realism gives the set of possible values of Γ a spread of about 1.0, while the supplementary CH assumption reduces this figure to about 0.08. If we assume that quantum theory is correct, we find that $(\Gamma)_{min} = -0.00138$. This is consistent with the inequality (9.34), but it violates the inequality (9.35) derived with the supplementary CH assumption. It may this be said that the results are solely dependent on this supplementary assumption.

As we have said, this postulate may be inapplicable at the level of hidden variables λs. In fact, explicit but *ad hoc* local hidden variable models have been formulated which are able to reproduce the quantum mechanical results, but at the expense of violating the supplementary CH assumption.[71,72] Since this assumption at the level of unobservable λs cannot be tested directly, one way of making further progress is to subject its various implications to appropriate scrutiny. For a review of the pertinent studies, see Selleri,[66] and also an article by de Caro and Garuccio,[73] who have discussed a general framework for providing insights into how the CH assumption can be violated within the local realist models.

It should be noted that the point that for a direct demonstration of the quantum mechanical violation of Bell's inequality, derived without using any additional assumption, one would need photomultiplier detectors of visible light of a very high efficiency, was first discussed by Clauser et al.[74] in a paper of 1969. In the same paper, the authors introduced a different supplementary assumption, known as the CHSH postulate, that, if a pair of photons emerges from the polarisers, the probability of their joint detection is independent of the polariser orientations. However, a difficulty with this form of supplementary assumption was later pointed out by Clauser and Shimony.[18]

In 1987, it was argued that, for decisive experimental tests of Bell's inequality using polarised photons without using any auxilary condition, the required efficiency of visible photon detectors is of the order of 84%.[75] Subsequently, Eberhard[76] suggested ways to reduce this lower bound on the required efficiency to around 67%, but even this value of efficiency is well above that of any detectors that have been used in such experiments so far.

However, although these experiments remain far from ideal because of the detector inefficiency loophole, many find it difficult to believe that quantum mechanics would agree with all the experiments so far for inefficient detectors, yet would fail for more efficient set-ups. Nevertheless, there is another fundamental aspect of such experiments that too merits careful scrutiny, the locality loophole.

Bell was concerned about this problem from the very first paper[24] in which he formulated his famous theorem. It is to do with ruling out the possiblity of signalling between the two wings of the apparatus. Bell believed that it was possible that quantum theory would be obeyed if, but only if, the detectors in the two wings could reach some kind of 'mutual rapport' by exchange of a signal at a speed less than or equal to that of light. He therefore felt that experiments of the EPR–Bohm type in which the settings of the detectors were changed during the flight of the particles would be crucial. This idea was actually suggested even before Bell's work in terms of EPR by Bohm and Aharonov,[77] but the first generation of experiments attempting to probe this question was by Aspect and his co-workers in the early 80s.[78–81] An account of these atomic cascade experiments using polarisation measurements on correlated photons, and making full use of laser technology, has been given by Aspect and Grangier.[82] However these experiments could not achieve genuinely random switching of the settings of the polarisers in the two wings while the correlated photons were in flight.

In the late 1980s, a new technique, that of parametric down-conversion was used to provide the entangled photons for use in tests of Bell-type inequalites, initially by Shih and Alley[83] and soon afterwords by Ou and Mandel.[84] Another interesting development was pioneered by Rarity and Tapster,[85] who reported a violation of Bell's inequality based on phase and momentum rather than spin or polarisation. Their work was similar to an arrangement suggested independently by Horne and co-workers.[86,87] In their experiment, the photons in a correlated photon pair were separated and recombined with a variable phase delay, and an apparent non-local fourth-order interference effect was seen, which violated Bell's inequality by 10 standard deviations.

A recent important set of experiments has been performed by Tittel and colleagues[88-90] in Gisin's group at Geneva. They tested the strong quantum correlations between time-entangled photons separated by more than 10 km. In this experiment, a source containing pairs of energy-time entangled photons was situated at a telecommunication station in Geneva. The photons were sent along installed standard telecom fibers to analysers in Bellevue and Bernex. The lengths of the fibers were 8.1 and 9.3 km, while the distances of Geneva from Bellevue and Bernex, and from Bellevue to Bernex were 4.5, 7.3 and 10.9 km, respectively. Absorption in the two fibers was suitably controlled. The analysers consisted of all fiber-optic interferometers with equal differences in pathlength. The photons were detected by photon counters, and classical signals were transmitted back to the source where the coincidence electronics was located.

Tittel et al. performed a wide range of experiments, the relevant details being given in Ref. [90]. They found that Bell's inequality was violated by up to 25 standard deviations, and all the experiments were in good agreement with quantum theory. The fringe visibility was around 86%, reduced comparatively little from the 94% for a similar experiment carried out in the laboratory. The loss of 14% in this experiment could be attributed to detector noise, and to factors related to the rather large bandwidth of the down-converted photons (around 70 nm), effects such as birefringence and chromatic dispersion in the interferometers, and possibly slightly unequal transmission amplitudes for the interferring paths. It was thus demonstrated that increasing the distance of separation at least up to 10 km did not affect the non-locality of quantum theory.

Another very important work to be noted is that of Kwiat et al.[91] performed in 1995; they overcame the problem of most of the previous work using down-conversion, the fact that a the down-conversion actually produced not an entangled state but a product state. An effectively entangled state was obtained by post-selecting only half of the final ensemble. Whereas most previous experiments had used a phase matching in which the correlated photons have the same polarisation, and a pair of photons with equal wavelength emerge on a single cone, Kwiat et al. used an improved method of phase matching in which the photons were emitted into two cones, one photon having ordinary polarisation and the other extraordinary polarisation. In this experiment, the polarisation-entangled state was produced directly out of a single non-linear crystal of beta-barium borate. No extra beam splitters or mirrors were needed, and there was no requirement to discard any part of the final ensemble. Indeed, the source used in this experiment was more than an order of magnitude brighter than sources for polarisation-entangled photons, and so has considerable application in the field of quantum information, as well as the foundations of quantum mechanics.

We now come to the recent work and extremely important of Weihs et al.[92] in the group of Anton Zeilinger, who have claimed to remove at long last and completely the signalling possibility from the Aspect type of experiments. These workers fully enforced the condition of locality by maintaining sufficient distance between the measurement stations, and by achieving ultrafast and random setting of the analysers, and completely independent data registration. As Weihs et al. commented, it had taken more than 15 years for major improvements of technology to allow significant improvement on the work of Aspect and his colleagues.

In order to remove fully any possibility of communication between the two wings of the experiment, Weihs et al. insisted that, with an individual measurement defined to last from the first moment which could influence the choice of the analyser setting, till the actual registration of the photon, such a measurement must be performed quickly enough so that no information can travel to the other observer, before that observer completes his measurement. For this purpose, selection of an analyser direction must be completely unpredictable, and this requires the use of a perfect random number generator, not a pseudo random number generator. Also, any common activity pertaining to the two wings should be avoided; individual events should be registered independently on each side, and only compared after the experiment has terminated. This requires independent and highly accurate time bases on each side.

In the experiment of Weihs and colleagues, the observers were separated by 400 m, which means that the time available for a measurement as discussed above was shorter than 1.3 μs. Polarised entangled photons were sent to the observers through optical fibers, the difference in fiber lengths being less than 1 m, so that the photons were registered simultaneously within 5 ns. The duration of a measurement was kept far below the 1.3 μs limit by using high-speed random number generators and fast electro-optic modulators. Individual data registration was performed by each observer having his own time interval analyser and atomic clock, synchronised only once before each experimental run. Results yet again confirmed the predictions of quantum theory, and violated Bell's inequality by around 30 standard deviations. It seems that at last the locality loophole has been closed.

For the detection loophole, Tittel et al. have suggested various experimental ideas could have the potential to close this loophole. The proposal of Kwiat et al.[93] involves a two-crystal downconversion source, Fry et al.[94] suggested using a pair of Hg_{19} atoms, and Hagley et al.[95] using a pair of atoms in a high Q microwave superconducting cavity. While some of the suggestions involve working with photons with novel techniques, most use heavy particles, for which detector efficiencies can be very much higher than for photons.

Then in 2001 there was announced the first claim of an experiment succeeding in closing the detector loophole. This was performed by Mary Rowe and collaborators[96] in Wineland's group at the National Institute of Standards and Technology in Boulder, Colorado. This experiment uses pairs of beryllium ions initially confined in an ion trap, and Bell's inequality is violated by around 8 standard deviations. What is more significant is that a measurement outcome is recorded for every experiment, so no fair sampling assumption is required and the detector loophole is closed.

It indeed seems possible that, more than 40 years after Bell's celebrated work, we may now be approaching a decisive empirical verdict on the question of locality in quantum mechanics.

Experimental Tests of Quantum Non-locality Without Inequalities

An experimental observation of Greenberger–Horne–Zeilinger (GHZ) entanglement has been reported by Bouwmeester et al.[97] of the Innsbruck group. The technique used is to transform two pairs of polarisation entangled photons into three entangled photons and a fourth independent trigger photon. The GHZ entanglement has been observed in the sense that both the trigger photon and the three entangled photons are actually detected.

Tests of Hardy's proof of non-locality have been made by several groups.[98–100] In the experiment by Togerson et al.[98] photons with orthogonal polarisations are incident on a non-polarising but unsymmetrical beam-splitter, which produces a suitable entangled state with a fixed degree of entanglement.

In the rather more versatile scheme of Di Giuseppe et al.[99] two photons with the same wavelength and the same polarisations were generated by parametric downconversion, and their polarisations were then rotated by angles ϕ_1 and ϕ_2 respectively by two independent rotators. These photons were then injected to the two input modes, orthogonal to two plane sides of a cubic polarising beam splitter. On emergence, pairs of photons giving rise to a coincidence in detection were in an entangled state, and it was possible to tune the degree of entanglement by adjusting ϕ_1 and ϕ_2. The polarisations of each photon was then rotated through angles θ_1, θ_2 by transmission through two identical analysers, and each photon was then passed through a cubic polarising beam splitter with fixed orientation, followed by detectors coupled to each of the output ports.

The flexibility of the arrangement was achieved because ϕ_1, ϕ_2, θ_1, and θ_2 could be varied independently, permitting a fairly complete study of the properties of the system. Hardy's proof was confirmed by detailed analysis of the data, and this demonstration was later confirmed in the work of Boschi et al.,[100] who used more settings of the apparatus at each wing of the experiment, and were able to utilise a greater proportion of the pairs of photons.

Finally, before concluding this discussion on experimental tests of quantum non-locality without using Bell-type inequalities, for completeness it should be noted that interesting variants of the GHZ and Hardy arguments suggested by Cabello[101,102] have not yet been experimentally studied.

Quantum Teleportation

Quantum teleportation (QT) is one of the major branches of quantum information theory, which is the topic of Chapter 11, but it also is a good example of many of the arguments about non-locality, so we describe it briefly in this chapter.

We suppose that a two-level quantum system, which in quantum information is a way of representing the basic unit of information, known as a qubit, is in a state

$$|\psi\rangle = a|0\rangle + b|1\rangle \tag{9.37}$$

This qubit is initially given to Alice, who with her partner, Bob, are the universal participants in any scheme in quantum information theory. Alice may, or may not have any information about the state. Her task is to send all the information encoded in such a state to Bob without physically transporting the system. If she has no *a priori* information about the state, Alice can neither get any information about the state without disturbing it, nor produce copies of this state so that she can determine the state by measurements on an ensemble. (The latter is not possible because of the *no-cloning theorem*[103] which is discussed in Chapter 4, and also mentioned in Chapter 11.) Therefore Alice has no way of telling Bob what the state is.

However, there is a way out which was discovered by Charles Bennett et al.[104] The scheme for this purpose operates by letting Alice and Bob share a singlet EPR state of two particles A and B

$$\psi_{AB} = \frac{1}{\sqrt{2}}(|01\rangle - |10\rangle)_{AB} \tag{9.38}$$

where $|01\rangle_{AB} = |0\rangle_A|1\rangle_B$, $|10\rangle_{AB} = |1\rangle_A|0\rangle_B$. Particle A goes to Alice, and particle B to Bob. Denoting the initial qubit given to Alice by Q, the initial state of the three particles Q, A, B is given by

$$(a|0\rangle + b|1\rangle)_Q \frac{1}{\sqrt{2}}(|01\rangle - |10\rangle)_{AB} \tag{9.39}$$

This state may be rewritten as

$$\frac{1}{2}(|01\rangle - |10\rangle)_{QA}(-a|0\rangle - b|1\rangle)_B + \frac{1}{2}(|01\rangle + |10\rangle)_{QA}(-a|0\rangle + b|1\rangle)_B$$

$$+ \frac{1}{2}(|00\rangle - |11\rangle)_{QA}(b|0\rangle + a|1\rangle)_B + \frac{1}{2}(|00\rangle + |11\rangle)_{QA}(b|0\rangle - a|1\rangle)_B$$

$$\tag{9.40}$$

where the particles available to Alice, A and Q, have been separated out from B, which is in the possession of Bob. The next ingredient in this scheme is that Alice performs a joint measurement on Q and A with respect to the Bell basis of states. The Bell states may be described as a complete set of maximally entangled states of Q and A, and are given by:

$$\psi_{\mp} = \frac{1}{\sqrt{2}}(|01\rangle_{QA} \mp |10\rangle_{QA}) \tag{9.41}$$

$$\phi_{\mp} = \frac{1}{\sqrt{2}}(|00\rangle_{QA} \mp |11\rangle_{QA}) \tag{9.42}$$

We note, for example, that the state from Eq. (9.41) with the minus sign is just the EPR-Bohm state.

When Alice performs this measurement, she collapses the state into one of the four terms in Eq. (9.40). This entails not only particles A and Q going into one of the Bell states, but particle B going into one of four states, each of which is correlated with a particular one of the Bell states for particles A and Q. Subsequently, Alice sends a signal by classical means, for example she telephones Bob, and tells him the outcome of her above measurement. This required two bits of classical information. This information tells Bob which of the four states of particle B he is in possession of, and he then applies a unitary operation on his qubit according to the following table:

Result of Alice's measurement	Bob's unitary operation
ψ_-	I
ψ_+	σ_z
ϕ_-	σ_x
ϕ_+	$i\sigma_y$

In this table, σ_z, σ_x and σ_y refer to rotations about the z-, y- and x-axes respectively. For the first state in Eq. (9.40), it might seem that a minus sign has been omitted, but it should be remembered that this sign corresponds merely to a change of phase, which does not entail any physical change in the actual state. In each of these cases, Bob ends up with particle B in state $|\psi\rangle$ which is the original state (9.37) of particle Q that Alice had. In the above scheme, quantum non-local correlations embodied in the joint wave-function (9.38) of A and B, in conjunction with the Bell-basis measurement on Q and A have led to a *disentanglement* of the originally entangled state (9.38) in such a way that B is ultimately left with the state that Q originally had.

In order actually to implement the above scheme, performing the Bell-basis measurement on two independent particles is a formidable experimental challenge, which has been overcome by a number of experiments using photons.[105–108] Teleporting states of *massive objects* remains the next important goal. For this, entanglement of massive objects needs to be prepared. The *generic scheme* for producing entangled states of *any* system, bosonic or fermionic, suggested by Bose and Home[109] could be useful for this purpose.

Note that the teleported quantum information encoded in the state does not travel materially from Alice to Bob. All that travels materially is the message about Alice's measurement result, which tells Bob how to process his particle but carries no information about Alice's state itself.

In one out of four cases, Alice is lucky with her measurement, and Bob's particle immediately becomes an identical replica of Alice's original. Then it might seem as if information has traveled instantly from Alice to Bob. Yet this strange feature cannot be used to send *usable information intantaneously*, because Bob has *no* way of knowing that his particle is already an identical replica. Only when he *learns* the result of Alice's Bell-state measurement, which is transmitted to him via classical means, can he exploit the information in the teleported quantum state. Suppose he tries to guess in which cases the teleportation was instantly successful. He will be wrong 75% of the time, and moreover he will not know, of course, which guesses were correct. If he uses his particles based on such guesses, the results will be the same as if he had taken a beam of particles with random distribution of $|1\rangle$ and $|0\rangle$ states. Thus, in this scheme, too the 'spooky' instantaneous action at a distance implied by quantum mechanics at the level of individual events fails to send any usable information faster than the speed of light.

However, there remain certain intriguing questions about the conceptual consequences of QT. What, if anything, is transferred in the first stage of the process in which Alice and Bob receive the entangled particles and Alice performs the Bell-state measurement? *How* does the overall transfer of information actually take place? These questions have been much debated and a variety of views have been expressed.[110-115]

One way of discussing the conceptual significance of QT has been suggested by Zeilinger.[116] Before the first stage, Bob has absolutely no knowledge about the original state of particle A. After the first stage, Bob at least knows that one of four operations applied to his particle 3 will definitely duplicate the initial state of particle A. Only at the end of the next stage after receving the classical signal, does Bob know *which* of these operations must be used. But though at the end of the first stage, Bob has no usable information, it is plausible to suggest that Bob is in possession of, what may be called, '*potential information*' or '*latent information*' which becomes accessible in the next stage *after* the classical signal arrives. However, the precise nature of such 'latent information' remains unclear.

In this context, it is of particular significance to critically examine the question as to what extent the QT process can be modelled by *local hidden variable models*. Such a study has been attempted by Zukowski,[114] and also by Clifton and Pope.[115] However, more investigations are needed for a deeper understanding of this question. Another recent direction of study has been to study a variant of the QT scheme in terms of an entangled tripartite system with non-orthogonal states which has interesting implications for quantum non-locality.[117,118] Such schemes using *entanglements* involving *non-orthogonal states* require more illustrations in terms of specific dynamical models for the preparation of this type of entanglement.

Bell and Einstein

In this chapter we have examined at some length the important papers of John Bell, and some of the work performed in the decades since to analyse the significance of his work, to provide experimental checking of its conclusions, and, in the case of quantum teleportation, to make application of the basic ideas. More applications are included in the other areas of quantum information theory discussed in Chapter 11. In the context of this book, the natural question is: how does this work affect one's conclusions of Einstein's own struggles with quantum theory over so much of his life?

An obvious answer, and quite a common one, though, in our opinion, an extremely trite one, is to suggest that Bell's work did considerable damage to Einstein's position. Einstein, it might be said, pushed above all else the idea of local realism. This was the conclusion he hoped to achieve from EPR, and it is the set of ideas that Bell clearly and definitively ruled impossible to attain.

It will be readily admitted that Einstein's desired solution to the problem of quantum interpretation was local realism. However, as stressed in Chapter 6, the impossibility of local realism was by no means contrary to the actual conclusion of the EPR analysis, which was to rule out the combination of locality and *non-realism*. This conclusion, as also pointed out in Chapter 6, was well understood by Einstein. It may be remarked that, if a scientist is to be damned for having strongly held ideas, for suggesting experimental or theoretical ways to check these ideas, and finding that, though the results are of great interest, they are not along the lines they had hoped for, few scientists would escape unscathed, and the most brilliant ones—those capable of coming up with the clearest and most forward-looking hypotheses—would often be subject to the greatest criticism.

In fact, not only is Einstein's contribution to the present greatly enhanced interest, in and understanding of the fundamental ideas of quantum theory and quantum information theory immense, but also the work of the past 40 years has served, in a very broad sense, to justify his approach to quantum theory, and to demonstrate the depth of his arguments.

For a start, Einstein's influence on Bell was enormous. Bell described himself as a follower of Einstein,[119] and he followed Einstein in being extremely anxious to uphold realism and locality. For Bell, determinism was not of very great importance; here there would be a difference with Einstein, though we argued in Chapter 8 that determinism, though admittedly very important for Einstein, may have not been quite as important as often assumed. While Bell's personality was such that he would almost certainly have had the courage of his own convictions, the example of Einstein in firmly rejecting Copenhagen, and seeking for a completion of quantum theory must have been stimulating and encouraging.

And, of course, much of Bell's analysis was developed using Einstein's own work. His most important paper of all,[14] the one developing Bell's theorem, was, after all, called 'On the Einstein–Podolsky–Rosen paradox', and the whole of the work on the theorem has been centred around entanglement, a concept which was brought to the attention of the world in the EPR paper. So Einstein must deserve a large amount of credit for the intellectual fruits which have resulted from Bell's work—the sharper focus on issues such as local realism, the questioning of Copenhagen and the rise of novel interpretations of quantum theory, the development of quantum information theory.

However, just as Einstein deserves credit for these developments, the very existence of this substantial new effort given to understanding quantum theory, and the realisation that the solutions given by Copenhagen were certainly not complete and final, demonstrate clearly that Einstein's approach to quantum theory in the last 30 years of his life, far from showing his lack of appreciation of what was required to make progress in physics, showed that his perceptions and beliefs were more acute than those of the mass of scientists who were content to follow where Bohr and von Neumann led. In every way, the work of John Bell greatly facilitated renewed interest in the ideas of Einstein.

Note added in proof

Simon Gröblacher[1] and others in Zeilinger's group have recently published the results of an experimental test which rule out a broad class of *non-local* realistic theories of quantum mechanics. Previously untested correlations between entangled photons violate an inequality based on one proposed by Leggett[2] for such theories. Aspect[3] has commented on this result. Many other non-local realist theories, such as that of Bohm, are *not* included in this result.

1. Gröblacher, S., Paterek, T., Kaltenbaek, R., Brukner, C., Żukowski, M., Aspelmeyer, M. and Zeilinger, A. (2007). An experimental test of non-local realism, Nature **446**, 871–5.
2. Leggett, A.J. (2003). Nonlocal hidden-variable theories and quantum theory: An incompatibility theorem, Foundations of Physics **33**, 1469–93.
3. Aspect, A. (2007). To be or not to be local, Nature **446**, 866–7.

References

1. Bohr, N. (1939). The causality problem in atomic physics, In: New Theories in Physics. Paris: International Institute of Intellectual Cooperation, pp. 11–45.
2. Born, M. (1949). Natural Philosophy of Cause and Chance. Oxford: Clarendon Press, pp. 108–9.
3. Pauli, W. (1948). Editorial, Dialectica **2**, 307–11.
4. Cushing, J. (1994). Quantum Mechanics. Chicago: University of Chicago Press, p. 131.
5. Jammer, M. (1974). The Philosophy of Quantum Mechanics. New York: Wiley, pp. 272–7.
6. Nabl, H. (1959). Die Pyramide **3**, 96.
7. Bohm, D. (1952). A suggested interpretation of the quantum theory in terms of 'hidden' variables, Physical Review **85**, 166–79, 180–93.
8. Bell, J.S. (1966). On the problem of hidden variables in quantum mechanics, Reviews of Modern Physics **38**, 447–52.
9. Bell, J.S. (2004). Speakable and Unspeakable in Quantum Mechanics. (1st edn. 1987, 2nd edn. 2004) Cambridge: Cambridge University Press, pp. 1–13.
10. von Neumann, J. (1932). Die Mathematische Grundlagen der Quantenmechanik. Berlin: Springer-Verlag.
11. Ref. [5], pp. 304–5.
12. Selleri, F. (1990). Quantum Paradoxes and Physical Reality. Dordrecht: Kluwer, pp. 48–51.
13. Cushing, J.T. (1994). Quantum Mechanics—Historical Contingency and the Copenhagen Hegemony. Chicago: University of Chicago Press, pp. 140–3.
14. Bell, J.S. (1964). On the Einstein-Podolsky-Rosen paradox, Physics **1**, 195–200; also in Ref. [9], pp. 14–21.
15. Bohm, D. (1951). Quantum Theory. Englewood Cliffs, New Jersey: Prentice-Hall, p. 614.

16. Einstein, A. (1949). Autobiographical notes, In: Albert Einstein: Philosopher-Scientist. (Schilpp, P.A., ed.) Evanston, Illinois: Library of Living Philosophers, p. 85.

17. Shimony, A. (1989). Searching for a world view which will accommodate our knowledge of microphysics, In: Philosophical Consequences of Quantum Theory: Reflections on Bell's Theorem. (Cushing, J.T. and McMullin, E., eds.) Notre Dame: Notre Dame University Press, pp. 25–37.

18. Clauser, J.F. and Shimony, A. (1978). Bell's theorem: experimental tests and implications, Reports of Progress in Physics **41**, 1881–1927.

19. Selleri, F. (ed) (1988). Quantum Mechanics versus Local Realism: The Einstein-Podolsky-Rosen Paradox. New York: Plenum.

20. Home, D. and Selleri, F. (1991). Bell's theorem and the EPR paradox, Rivista del Nuovo Cimento **14**(9), 1–95.

21. Eberhard, P.H. (1977). Bell's theorem without hidden variables, Nuovo Cimento B **38**, 75–80.

22. Peres, A. (1978). Unperformed experiments have no results, American Journal of Physics **46**, 745–7.

23. Redhead, M. (1987). Incompleteness, Nonlocality, and Realism. Oxford: Clarendon Press, Ch. 4.

24. Leggett, A.J. (1987). Problems of Physics. Oxford: Oxford University Press, Ch. 5.

25. Clauser, J.F. and Horne, M.A. (1974). Experimental consequences of objective local theories, Physical Review D **10**, 526–35.

26. Ivanovic, J.D. (1978). On complex Bell's inequality, Lettere al Nuovo Cimento **22**, 14–6.

27. Muckenheim, W. (1982). A resolution of the Einstein-Podolsky-Rosen paradox, Lettere al Nuovo Cimento **35**, 300–4.

28. Home, D., Lepore, V.L. and Selleri, F. (1991). Local realistic models and non-physical probabilities, Physics Letter A **158**, 357–60.

29. Agarwal, G.S., Home, D., and Schleich, W. (1992). Einstein-Podolsky-Rosen correlation—parallelism between the Wigner function and the local hidden variable approaches, Physics Letters A **170**, 359–62.

30. Gleason, A.M. (1957). Measures on the closed subspaces of a Hilbert space, Journal of Mathematics and Mechanics **6**, 885–93.

31. Kochen, S. and Specker, E.P. (1967). The problem of hidden variables in quantum mechanics, Journal of Mathematics and Mechanics **17**, 59–87.

32. Penrose, R. (1994). On Bell non-locality without probabilities: some curious geometry, In: Quantum Reflections (Ellis, J. and Amati, D., eds.) Cambridge: Cambridge University Press, pp. 1–22.

33. Roy, S.M. and Singh, V. (1993). Quantum violation of stochastic noncontextual hidden-variable theories, Physical Review A **48**, 3379–81.

34. Hasegawa, Y., Liodl, R., Badurek, G., Baron, M., and Rauch, H. (2003). Violation of a Bell-like inequality in single-neutron interferometry, Nature **425**, 45–8.

35. Basu, S., Bandyopadhyay, S., Kar, G., and Home, D. (2001). Bell's inequality for a single spin 1/2 particle and quantum contextuality, Physics Letters A **279**, 281–6.

36. Jarrett, J.P. (1984). On the physical significance of the locality conditions in the Bell arguments, Noûs **18**, 569–89.

37. Shimony, A. (1984). Controllable and uncontrollable non-locality, In: Proceedings of the 1st International Symposium on the Foundations of Quantum Mechanics in the

Light of New Technology (Kamefuchi, S. et al., eds.), Tokyo: Physical Society of Japan, pp. 225–30.

38. Maudlin, T. (1994). Quantum Nonlocality and Relativity. Oxford: Blackwell.

39. Helliwell, T.M. and Konkowski, D.A. (1983). Causality paradoxes and nonpara-doxes: classical superluminal signals and quantum measurements, American Journal of Physics 51, 996–1003.

40. Roy, S.M. and Singh, V. (1991). Tests of signal locality and Einstein-Bell locality for multiparticle systems, Physical Review Letters 67, 2761–4.

41. Bell, J.S. (1986). In: Ghost in the Atom (Davies, P.C.W. and Brown, J.R., eds.) Cambridge, UK: Cambridge University Press, p. 50.

42. Bohm, D. (1951). Quantum Theory. Englewood Cliffs: Prentice-Hall, pp. 618–9.

43. Eberhard, P.H. (1978). Bell's theorem and the different concepts of locality, Nuovo Cimento B 46, 392–419.

44. Eberhard, P.H. and Ross, R.R. (1989). Quantum field theory cannot provide faster-than-light communication, Foundations of Physics Letters 2, 127–49.

45. Ghirardi, G.C., Rimini, A., and Weber, T. (1980). A general argument against super-luminal transmission through the quantum mechanical measurement process, Lettere al Nuovo Cimento 27, 293–8.

46. Page, D.N. (1982). The Einstein-Podolsky-Rosen physical reality is completely described by quantum mechanics, Physics Letters A 91, 57–60.

47. Gisin, N. (1991). Bell's inequality holds for all non-product states, Physics Letters A 154, 201–2.

48. Roy, S.M. and Singh, V. (1978). Experimental tests of quantum mechanics versus local hidden variable theories, Journal of Physics A 11, L167–71.

49. Roy, S.M. and Singh, V. (1979). Completeness of tests of local hidden variable theories, Journal of Physics A 12, 1003–9.

50. Garuccio, A. and Selleri, F. (1980). Systematic derivation of all the inequalities of Einstein locality, Foundations of Physics 10, 209–16.

51. Garuccio, A. (1980). All the inequalities of Einstein locality, In: Quantum Mechanics Versus Local Realism: The Einstein-Podolsky-Rosen Paradox. (Selleri, F., ed.) New York: Plenum, pp. 87–113.

52. Lepore, V.L. (1989). New inequalities from local realism, Foundations of Physics Letters 2, 15–26.

53. Garg, A. and Mermin, N.D. (1982). Correlation inequalities and hidden variables, Physical Review Letters 49, 1220–3.

54. Greenberger, D.M., Horne, M.A., and Zeilinger, A. (1989). Going beyond Bell's theorem, In: Bell's Theorem, Quantum Theory and Conceptions of the Universe. (Kafatos, M., ed.) Dordrecht: Kluwer, pp. 73–6.

55. Greenberger, D.M., Horne, M.A., Shimony, A., and Zeilinger, A. (1990). Bell's theorem without inequalities, American Journal of Physics 58, 1131–43.

56. Mermin, N.D. (1990). Quantum mysteries revisited, American Journal of Physics 58, 731–4.

57. Fine, A. (1982). Hidden variables, joint probability, and the Bell inequalities, Physical Review Letters 48, 291–5.

58. Fine, A. (1982). Joint distributions, quantum correlations, and commuting observables, Journal of Mathematical Physics 23, 1306–10.

59. Hardy, L. (1991). A new way to obtain Bell inequalities, Physics Letters A 161, 21–5.

60. Hardy, L. (1992). Quantum mechanics, local realistic theories, and Lorentz-invariant realistic theories, Physical Review Letters 68, 2981–4.

61. Hardy, L. (1993). Nonlocality for two particles without inequalities for almost all entangled states, Physical Review Letters **71**, 1665–8.
62. Stapp, H.(1993). Mind, Matter, and Quantum Mechanics. New York: Springer-Verlag, pp. 5–8.
63. Mermin, N.D. (1994). What's wrong with this temptation? Physics Today **9**(6), 9–11.
64. Jordan, T.F. (1994). Testing Einstein-Podolsky-Rosen assumptions without inequalities with two photons or particles with spin $\frac{1}{2}$, Physical Review A **50**, 62–6.
65. Goldstein, S. (1994). Nonlocality without inequalities for almost all entangled states for two particles, Physical Review Letters **72**, 1951.
66. Selleri, F. (1990). Quantum Paradoxes and Physical Reality. Dordrecht: Kluwer, Ch. 6.
67. Ferrero, M., Marshall, T.W. and Santos, E. (1990). Bell's theorem: local realism versus quantum mechanics, American Journal of Physics **58**, 683–8.
68. Santos, E. (1989). Relevance of detector efficiency in the optical tests of Bell inequalities, Physics Letters A **139**, 431–6.
69. Santos, E. (1991). Interpretation of the quantum formalism and Bell's theorem, Foundations of Physics **21**, 221–41.
70. Santos, E. (1996). Unreliability of performed tests of Bell's inequality using parametric down-converted photons, Physics Letters A **212**, 10–14.
71. Marshall, T.W., Santos, E. and Selleri, F. (1983). Local realism has not been refuted by atomic cascade experiments, Physics Letters A **98**, 5–9.
72. Home, D. and Marshall, T.W. (1985). A stochastic local realist model for the EPR atomic-cascade experiment which reproduces the quantum-mechanical coincidence rates, Physics Letters A **113**, 183–6.
73. de Caro, L. and Garuccio, A. (1995). Correlation functions and Einstein locality, In: Frontiers of Fundamental Physics—Proceedings of the Olympia Conference 1993. (Barone, M. and Selleri, F., eds.), New York: Plenum, pp. 529–49.
74. Clauser, J.F., Horne, M.A., Shimony, A. and Holt, R.A. (1969). Proposed experiment to test hidden-variable theories, Physical Review Letters **23**, 880–4.
75. Garg, A. and Mermin, N.D. (1987). Detector inefficiencies in the Einstein-Podolsky-Rosen experiment, Physical Review D **35**, 3831–5.
76. Eberhard, P.H. (1993). Background level and counter efficiencies required for a loophole-free Einstein-Podolsky-Rosen experiment, Physical Review A **47**, R747–50.
77. Bohm, D. and Aharonov, Y. (1957). Discussion of experimental proof for the paradox of Einstein, Rosen and Podolsky, Physical Review **108**, 1070–6.
78. Aspect, A. Imbert, C. and Roger, G. (1980). Absolute measurement of an atomic cascade rate using a two photon coincidence technique: application to the $4p^2$ $^1S_0$—$4s4p$ $^1P_1$ - $4s^2$ $^1S_0$ cascade of calcium excited by a two photon absorption, Optics Communications **34**, 46–52.
79. Aspect, A., Grangier, P. and Roger, G. (1981). Experimental tests of realistic local theories via Bell's theorem, Physical Review Letters **47**, 460–3.
80. Aspect, A., Grangier, P. and Roger, G. (1982). Experimental realization of Einstein-Podolsky-Rosen-Bohm gedankenexperiment: a new violation of Bell's inequalities, Physical Review Letters **49**, 91–4.
81. Aspect, A., Dalibard, J. and Roger, G. (1982). Experimental tests of Bell's inequalities using time-varying analysers, Physical Review Letters **49**, 1804–7.
82. Aspect, A. and Grangier, P. (1985). Tests of Bell's inequalities with pairs of low energy correlated photons: An experimental realization of Einstein-Podolsky-Rosen

type correlations, In: Symposium on the Foundations of Modern Physics: 50 Years of the Einstein-Podolsky-Rosen Gedankenexperiment. (Lahti, P. and Mittelstaedt, P., eds.), Singapore: World Scientific pp. 51–71.

83. Shih, Y.H. and Alley, C.O. (1988). New type of Einstein-Podolsky-Rosen-Bohm experiment using pairs of light quanta produced by optical parametric down conversion, Physical Review Letters **61**, 2921–4.

84. Ou, Z.Y. and Mandel, L. (1988). Violation of Bell's inequality and classical probability in a two-photon correlation experiment, Physical Review Letters **61**, 50–3.

85. Rarity, J.G. and Tapster, P.R. (1990). Experimental violation of Bell's inequality based on phase and momentum, Physical Review Letters **64**, 2495–8.

86. Horne, M.A. and Zeilinger, A. (1985). A Bell-type EPR experiment using linear momenta, In: Symposium on the Foundations of Modern Physics: 50 Years of the Einstein-Podolsky-Rosen Gedankenexperiment. (Lahti, P. and Mittelstaedt, P., eds.) Singapore: World Scientific, p. 435–9.

87. Horne, M.A., Shimony, A., and Zeilinger, A. (1989). Two-particle interferometry, Physical Review Letters **62**, 2209–12.

88. Tittel, W., Brendel, J., Gisin, B., Herzog, T., Zbinden, H., and Gisin, N. (1998). Experimental demonstration of quantum correlations over more than 10 km, Physical Review A **57**, 3229–32.

89. Tittel, W., Brendel, J., Zbinden, H., and Gisin, N. (1998). Violation of Bell inequalities by photons more than 10km apart, Physical Review Letters **81**, 3563–6.

90. Tittel, W., Brendel, J., Gisin, N., and Zbinden, H. (1999). Long-distance Bell-tests using energy-time entangled photons, Physical Review A **59**, 4150–63.

91. Kwiat, P.G., Mattle, K., Weinfurter, H., Zeilinger, A., Sergienko, A.V. and Shih, Y.H. (1995). New high-intensity source of polarisation-entangled photon states, Physics Review Letters **75**, 4337–40.

92. Weihs, G., Jennewein, T., Simon, C., Weinfurter, H., and Zeilinger, A. (1998). Violation of Bell's inequality under strict Einstein locality conditions, Physical Review Letters **81**, 5039–43.

93. Kwiat, P.G., Eberhard, P.H., Steinberg, A.M., and Chiao, R.Y. (1994). Proposal for a loophole-free Bell inequality test, Physical Review A **49**, 3209–20.

94. Fry, E.S., Walther, T., and Li, S. (1995). Proposal for a loophole-free test of the Bell inequalities, Physical Review A **52**, 4381–95.

95. Hagley, E., Maitre, X., Nogues, G., Wunderlich, C., Brune, M., Raimond, J.M., and Haroche, S. (1997). Generation of Einstein-Podolsky-Rosen pairs of atoms, Physical Review Letters **79**, 1–5.

96. Rowe, M.A., Klepinski, D., Meyer, V., Sackett, C.A., Itano, W.M., Monroe, C., and Wineland, D.J. (2001). Experimental violation of a Bell's inequality with efficient detection, Nature **409**, 791–4.

97. Bouwmeester, D., Pan, J.-W., Daniell, M., Weinfurter, H., and Zeilinger, A. (1999). Observation of three-photon Greenberger-Horne-Zeilinger entanglement, Physical Review Letters **82**, 1345–9.

98. Togerson, J.R., Branning, D., Monken, C.H., and Mandel, L. (1995). Experimental demonstration of the violation of local realism without Bell inequalities, Physics Letters A **204**, 323–8.

99. Di Giuseppe, G., de Martini, F., and Boschi, D. (1997). Experimental test of the violation of local realism in quantum mechanics without Bell inequalities, Physical Review A **56**, 176–81.

100. Boschi, D., Branca, S., de Martini, F., and Hardy, L. (1997). Ladder proof of nonlocality without inequalities: theoretical and experimental results, Physical Review Letters **79**, 2755–8.

101. Cabello, A. (2001). Bell's theorem without inequalities and without probabilities for two observers, Physical Review Letters **86**, 1911–4.

102. Cabello, A. (2001). 'All versus nothing' inseparability for two observers, Physical Review Letters **87**, 010403.

103. Wootters, W.K. and Zurek, W.H. (1982). A single quantum cannot be cloned, Nature **299**, 802–3.

104. Bennett, C.H., Brassard, G., Crepeau, C., Jozsa, R., Peres, A., and Wootters, W.K. (1993). Teleporting an unknown quantum state via dual classical and Einstein-Podolsky-Rosen channels, Physical Review Letters **70**, 1895–9.

105. Bouwmeester, D., Pan, J.-W., Mattle, K., Eibl, M., Weinfurter, H., and Zeilinger, A. (1997). Experimental quantum teleportation, Nature **390**, 575.

106. Bouwmeester, D., Pan, J.-W., Daniell, M., Weinfurter, H., Zukowski, M. and Zeilinger, A. (1998). A posteriori teleportation—reply, Nature **394**, 841.

107. Boschi, D., Branca, S., de Martini, F., Hardy, L., and Popescu, S. (1998). Experimental realization of teleporting an unknown pure quantum state via dual classical and Einstein-Podolsky-Rosen channels, Physical Review Letters **80**, 1121–5.

108. Furusawa, A., Sorensen, J.L., Braunstein, S.L., Fuchs, C.A., Kimble, H.J., and Polzik, E.S. (1998). Unconditional quantum teleportation, Science **282**, 706–9.

109. Bose, S. and Home, D. (2002). Generic entanglement generation, quantum statistics, and complementarity, Physical Review Letters **88**, 050401.

110. Braunstein, S.L. (1996). Quantum teleportation without irreversible detection, Physical Review A **53**, 1900–2.

111. Jozsa, R.O. (1998). Entanglement and quantum computation, In: The Geometric Universe: Science, Geometry and the Work of Roger Penrose. (Huggett, S.A., Mason, L.J., Tod, K.P. and Woodhouse, N.M.J., eds.) Oxford: Oxford University Press, pp. 369–79.

112. Penrose, R. (1998). Quantum computation, entanglement and state reduction, Philosophical Transactions of the Royal Society A **356**, 1927–39.

113. Deutsch, D. and Hayden, P. (2000). Information flow in entangled quantum systems, Proceedings of the Royal Society A **456**, 1759–74.

114. Zukowski, M. (2000). Bell's theorem for the nonclassical part of the quantum teleportation process, Physical Review A **62**, 032101.

115. Clifton, R. and Pope, D. (2001). On the non-locality of the quantum channel in the standard teleportation protocol, Physics Letters A **292**, 1–11.

116. Zeilinger, A. (2000). Quantum teleportation, Scientific American **282**(4), 50–9.

117. Corbett, J.V. and Home, D. (2000). Quantum effects involving interplay between unitary dynamics and kinematic entanglement, Physics Review A **62**, 062103.

118. Corbett, J.V. and Home, D. (2004). Information transfer and non-locality for a tripartite entanglement using dynamics, Physics Letters A **333**, 382–8.

119. Bell, J.S. (1981). Bertlmann's socks and the nature of reality, Journal de Physique Colloque C2, **42**, 41–61; also in Speakable and Unspeakable in Quantum Mechanics. (1st edn. 1987, 2nd edn. 2004) Cambridge: Cambridge University Press, pp. 139–58.

10
Non-standard Quantum Interpretations

Introduction

From the time of Bohr's Como paper in 1928, the Copenhagen interpretation reigned supreme for several decades. Indeed supporters of the Copenhagen interpretation, such as Leon Rosenfeld and Rudolf Peierls,[1] disliked even the use of the term 'Copenhagen interpretation', for it suggested that this was one interpretation, possibly among many conceivable ones. For them, the conceptual structures of Bohr and Heisenberg were not an adjunct to quantum theory, but a clear and indispensable component of the theory. Einstein, as we have seen at length, disliked the Copenhagen interpretation. Perhaps as much as or even more then the interpretation itself, he disliked the fact that its supporters considered it immune from criticism, so that no questioning would be taken seriously, and certainly any alternative interpretation would be ruled out of court without being given even fair attention.

Today, not only is the Copenhagen interpretation severely criticised by many, though, of course, many others still support it, but there is also a wide range of alternative or non-standard interpretations. Of these, two particularly important ones pre-date the work of John Bell; these are the Bohm interpretation of 1952, and the many-worlds interpretation or MWI, dating from 1957. We have seen that Einstein had encouraged Bohm to perform his work, but rather unfortunately felt unable to give the final result even partial support. The MWI was a result principally of disagreement with the idea of wave-function collapse at a measurement; this disagreement was expressed by Einstein, but also others such as Margenau.[2]

However it was the work of Bell that stimulated most of the interpretations from the 1960s and later; one does not here mean specifically Bell's theorem, so much as his general criticism of the ideas of Bohr and von Neumann. Overall it would be fair to say, though, that most of the ideas we describe here owe a lot in general terms to the original criticisms of Einstein of the 1930s and afterwards, though it would also be the case that little would probably have emerged had not Bell essentially acted as Einstein's spokesman from the 1960s on. To say that Einstein deserves some credit for the present feast of interpretations does not, of course, mean that he would necessarily have supported *any* of them. We saw this for the

Bohm case, and we must remember that, as stressed in Chapter 8 in particular, for the last three decades of his life Einstein was not concerned with interpreting or improving quantum theory, but with replacing it. In this chapter, we review some of the most significant non-standard interpretations, and then sketch a few others.

Many-Worlds Interpretation

The many-worlds interpretation or MWI attempts to solve the quantum measurement problem by retaining, not only the standard formalism, but also as much as possible of the standard interpretational framework. There are different presentations of MWI, starting from the original version in the 1957 paper by Everett.[3] Here we follow mainly the exposition by Squires,[4,5] which is often called a many-minds approach. A range of other possible approaches are described in the recent book by Whitaker.[6]

As explained in Chapters 3 and 4, the standard theory results in the combined wave-function for system and apparatus being a superposition of different classical outcomes. The von Neumann collapse postulate then reduces the wave-function to a form corresponding to a single outcome. The MWI seeks to eliminate the inconsistency between the universal applicability of quantum mechanics and the definiteness of an individual outcome, so it does not use a reduction of the wave-function, thus retaining the superposition of final states. It then postulates a special relationship between the final wave-function and the conscious observer's state of awareness. Each term in the final combined wave-function for system and apparatus is assumed to correspond to a definite state of awareness registering a particular outcome. Experiences of all the different outcomes are thus considered to be part of the final wave-function. This is often expressed graphically by saying that as a result of any observation, the 'world' branches into different worlds, and awareness of each outcome belongs to one particular world. This is the origin of the name 'many worlds'.

As Squires[4,5] emphasised, here we are really envisaging an interpretation incorporating perceptions of different outcomes, where each perception involves a relationship between a state of awareness and a corresponding state of the measured system entangled with the apparatus. Using Everett's terminology, we summarise this approach by saying that any macroscopically discernible part of the total wave-function space has a perceptible meaning only in relationship to a frame consisting of the observer's *mind*. Just as there are many permissible coordinate frames in relativity, there are many observer frames (in the preceding sense) in quantum theory.

At this stage there is a problem. The system-apparatus combined final wave-function can be expanded in *any basis*; for example individual branches of the wave-function might have a fixed value of S_z for a particular system, or they might have a fixed value of S_x. Therefore the MWI must assume some *preferred basis*. This has been noted by several authors,[7-9] although the problem was not mentioned in Everett's original paper.[3] Lockwood[10] argues that within the framework of MWI, the preference for a particular basis is 'rooted in the nature of consciousness'. This again means assuming a favoured set of eigenstates of a suitable observable in the relevant brain system, designated by what is called *consciousness*; these eigenstates are correlated with macroscopically distinguishable states of macroobjects.

Awareness of a definite outcome is thus interpreted as a conscious subject perceiving the correlated macroobject as having a definite state relative to a designated state of consciousness. This form of MWI is thus sometimes called the *relative states interpretation*. According to this point of view, states preferred in the arena of macroscopic systems are essentially preferred from the point of view of awareness itself. This point of view has phenomenal perspectives, as its windows on reality are associated with the eigenstates of a preferred brain observable;[10] in other words, definite outcomes take place only in individual minds.

Clearly in any of its various versions, the MWI does not require a change in either the formalism or the basic premise of the standard interpretation that a wave-function provides a complete description of the state of a system. What is added to the standard framework is the putative registration of a particular outcome by the conscious awareness of an observer, with weight factors specified by the standard quantum mechanical rule. This selection envisaged in MWI is a one-way process: As an observer becomes aware of one result, this does not change the wave-function. In the words of Squires,[5] 'The conscious mind does not change physics; it selects an experienced world from a range of options provided by the physics. Physics as defined by quantum theory is contained entirely within a wave-function. It is through an interplay of consciousness with a wave-function that definite outcomes are supposed to emerge'.

There is thus a fundamental difference between the MWI and the speculation by some that consciousness somehow collapses the wave-function; Squires[4] illustrates this point by using the analogy of selecting a television channel, which does not affect the physics of the incoming radiation, but enables a picture to appear on the television screen. Similarly, according to the MWI, without conscious awareness there are no definite outcomes, only a total wave-function containing the possibility of various outcomes. Furthermore, for two observers making successive conscious registrations of a particular outcome recorded in a given apparatus to necessarily agree on the result, we must ascribe some sort of universal nature to consciousness. In terms of the television set analogy, this corresponds to the fact that all observers see the same program if a particular channel is selected.

A number of proponents of MWI defend this approach as being economical in concepts and principles; for example Primas[11] claims that MWI is 'superior in logical economy'. However, an obvious criticism is that the problem is simply moved from the domain of physics to the speculative area of theory of mind, for which we do not have as yet a sound formulation. For seriously considering the possibility of such a theory, we must make some assumptions about the functioning of *brains*. We must decide whether present-day physics, classical or quantum, suffices to describe how the mind perceives reality.

Thus pursuing MWI involves some metaphysical baggage. Moreover, there seems to be no strong *a priori* reason for going beyond physics to seek a solution to the quantum measurement problem. A more attractive approach, at least to the majority of the physicists concerned with the measurement problem, may be to obtain a solution from within physics unless and until this is shown to be impossible. The two most widely discussed approaches in this direction, that of Bohm, and that of the spontaneous collapse of the wave-function are discussed in the following sections. After that, there are shorter treatments of a number of other novel interpretations.

Bohm's Model: The Ontological Interpretation

From 1926 to 1927 de Broglie[12-14] advocated the position that a wave-function can be interpreted as having a dual role, in the sense that it not only determines the probability of the actual location of a particle, but also influences the behaviour of the particle by exerting what may be called a quantum force on it. According to de Broglie, the wave-function acts as a 'pilot wave' to guide the particle. For this purpose he visualised a particle as a mathematical singularity in a field which acts nonlinearly on the particle. This approach is often referred to as the double solution interpretation. (For more details see Jammer[15]). De Broglie[16] used this approach to obtain two related wave solutions, one for the singularity or particle, and the other for a continuous wave. The purpose of this scheme was to show that the motion of the singularity was determined by the phase of the continuous solution of the relevant wave equation.

De Broglie presented an elaborate version of this scheme, in fact a many-particle pilot wave theory, at the 1927 Solvay Conference. There was a strong negative reaction. Pauli's[17] specific objection referred to an example of the inelastic scattering of a particle with a given incident energy, represented for simplicity as a plane wave, by a rigid rotator. If the interaction potential between the rotator and the incident particle vanishes except in a small region around the fixed axis of rotation of the rotator, the final wave-function after scattering is a superposition of states corresponding to different values of energy of both the rotator and the scattered particle. However, when measured, both the rotator and the scattered particle must have definite final energies. Pauli claimed that this showed an internal inconsistency in de Broglie's approach. However, as discussed below, Pauli's objection indicated a deficiency of de Broglie's scheme only inasmuch as it did not outline a well-formulated way of addressing the measurement problem. This was, however, a general failing of quantum theory, rather than being a technical flaw of this particular scheme.

After the inelastic scattering of a particle by a rotator, an intervention by a macroscopic measuring device is required to complete the measurement on the rotator and the scattered particle. Thus it is essential to consider the role of such a measurement process, which must lead to a definite result in an individual run, regardless of what interpretation is adopted. The inability to explain the definiteness of a particular outcome, without use of a scheme such as collapse, specifically designed for the purpose, is a generic feature of all variants of the standard interpretation, as discussed in detail in Chapter 4. It is therefore a curious historic fact that Pauli's objection was widely considered to strike a mortal blow against the very idea of an objectively real event-by-event description of quantum phenomena in terms of particle trajectories in ordinary space. De Broglie himself was so disheartened by the criticism he received that he gave up working along this line. For an incisive analysis of the relevant socio-cultural factors that may have prompted the general disinterest, and often the rather dogmatic refusal to entertain even the logical possibility of an interpretation other than the standard one, see Cushing,[18] Selleri[19] and Beller.[20]

In spite of the then prevailing scepticism, in his seminal papers of 1952, Bohm[21] succeeded in showing systematically and quite decisively that an ontological interpretation of quantum mechanics was possible. His approach described the action of quantum mechanical wave-function ψ, treated as a field, on localized particles which themselves were considered to have well-defined values of dynamic variables at all instants. Bohm showed that such an interpretation was fully consistent with the standard formalism of quantum mechanics and all experimentally verified quantum predictions. Some call this approach the de Broglie–Bohm model. However, Bohm's formulation in the form we discuss here has some major technical differences with de Broglie's original approach, although the basic spirit is the same; see Holland.[22] Note that Bohm himself in an appendix to his paper with Hiley[23] mentioned that he had developed his model without being aware of de Broglie's 1926–1927 work; only after sending the preprint of his 1952 papers to de Broglie was he told about the latter's work, which he then acknowledged in the introduction to his paper.[21]

In this book the ontological scheme we discuss will be referred to as Bohm's model. How this scheme seeks to solve the quantum measurement paradox is discussed below. For more comprehensive expositions of Bohm's model, and its specific applications in computing particle trajectories in various cases, such as stationary states of atoms, scattering from barriers, and double-slit interference and tunnelling, see the books by Holland,[22] and Bohm and Hiley;[24] for less detailed treatments, see, for example, Cushing,[25] Albert[26] and Whitaker.[6] For a critical appraisal of the status of ongoing studies on Bohm's model and further avenues of research, see a very useful collection of articles.[27] In what follows we confine our discussions mainly to broader aspects of the Bohmian interpretation. (For detailed treatments on how particles with spin can also be handled, see for example, Ref. 22, Chapters 9 and 10.)

In the Bohmian model, each individual particle has a definite observation independent position, the so-called hidden variable. Over an ensemble these positions will give the probability density $\rho(\vec{r}, t) = |\psi(\vec{r}, t)|^2$, where the wave-function, ψ, evolves with time according to the Schrödinger equation. ψ is treated as a field which choreographs the motion of the particles. The essential difference between the quantum field ψ and Maxwell's electromagnetic field is that the former does not have sources, nor, unlike the electromagnetic field, is it dynamically affected by the particles. Also Maxwell's field is in three dimensional space, in contrast to the ψ field, which is in configuration space, which is of $3N$ dimensions, where N is the number of particles.

Since the Bohmian model is concerned with the motion of actual particles, then the quantum mechanical continuity equation is interpreted as corresponding to an actual flow of particles, similar to a flow in hydrodynamics . This is in contrast to the usual interpretation of the quantum continuity equation, which is of a flow of probability. Thus in the Bohmian model, the probability current density can be written as $\vec{J}(\vec{r}, t) = \rho(\vec{r}, t)\,\vec{v}(\vec{r}, t)$, where $\vec{v}(\vec{r}, t)$ is identified as the velocity of an individual particle corresponding to the spacetime trajectory of the particle concerned, and $\rho(\vec{r}, t)$ is the density of particles.

The central postulates of the Bohmian model can then be summarised as follows:

(i) The so-called hidden variables in this model are the ontological position space coordinates of particles, which are regarded as dynamical variables having definite values, irrespective of observation. Thus the position coordinate of an individual particle (as a function of time) is taken to be an additional variable that must be specified, along with the wave-function $\psi(\vec{r}, t)$, which itself satisfies the time-dependent Schrödinger equation, in order to provide a complete description of the state of a single particle.

(ii) The squared modulus of a wave-function $\rho(\vec{r}, t) = |\psi(\vec{r}, t)|^2$ is interpreted as the probability of particles actually being *present* within a region of space, instead of being interpreted (à *la* Born) as the probability of *finding* particles within a specified region.

(iii) Since in the Bohmian approach, the *particle trajectories* are taken to be ontologically real, they should at all instants satisfy an equation of continuity derived from the time-dependent Schrödinger equation. This equation of continuity is given by

$$\frac{\partial \rho}{\partial t} + \vec{\nabla}.\vec{J} = 0 \tag{10.1}$$

where

$$\vec{J} = \frac{\hbar}{2mi}(\psi^*\vec{\nabla}\psi - \psi\,\vec{\nabla}\psi^*) \tag{10.2}$$

Then since in the Bohmian model, $\vec{J}(\vec{r}, t) = \rho(\vec{r}, t)\vec{v}(\vec{r}, t)$, it follows that the equation of motion of an individual particle is determined by the following guidance equation

$$\vec{v}(\vec{r}, t) = \frac{\vec{J}(\vec{r}, t)}{\rho(\vec{r}, t)} = \frac{\vec{\nabla}S(\vec{r}, t)}{m} \tag{10.3}$$

Here we have used Eq. (10.2), along with $\rho(\vec{r}, t) = |\psi(\vec{r}, t)|^2$. We have written $\psi(\vec{r}, t)$ in the polar form $\psi(\vec{r}, t) = \sqrt{\rho(\vec{r}, t)}\,\exp[iS(\vec{r}, t)\,t/\hbar]$, where $S(\vec{r}, t)$ is a real function of space and time. It then follows that if the distribution of initial positions of particles is taken to be $\rho(\vec{r}, t = 0) = |\psi(\vec{r}, t = 0)|^2$, the particles move in such a way that preserves this agreement for all subsequent times, so that $\rho(\vec{r}, t) = |\psi(\vec{r}, t)|^2$ for any value of t.

However, it is important to note that the quantum current density is inherently nonunique. This is because, if one adds any divergence free term to the expression for $\vec{J}$, as given by Eq. (10.2), then the new current density also satisfies the same equation of continuity. Since the Bohmian trajectories are calculated from the local velocity $\vec{v}(\vec{r}, t)$, which explicitly relies on the expression for $\vec{J}(\vec{r}, t)$, the quantum *non-uniqueness* of current density implies *non-uniqueness* of quantum trajectories in the Bohmian model.

However Holland[28] has demonstrated a way of obtaining a unique result. He pointed out that, if one moves to a relativistic approach, the probability current density derived from the Dirac equation for a spin-1/2 particle is unique, and this uniqueness is preserved in the non-relativistic regime, where it contains a spin-dependent term to be added to $\vec{J}$ as given by Eq. (10.2). One can further argue that this property of non-uniqueness is *not* specific to the Dirac equation; rather it is a consequence of *any* consistent relativistic equation.[29,30] For spin-1/2 particles, the expression for unique current density from Dirac equation in the non-relativistic limit is given by

$$\vec{J} = \frac{\hbar}{2mi}(\psi^* \nabla \psi - \psi \nabla \psi^*) + \frac{1}{m}\nabla\rho \times \vec{s}(t) \qquad (10.4)$$

where $\vec{s}(t) = \chi^\dagger(t)\vec{s}\chi(t)$, $\vec{s} = s_x\hat{x} + s_y\hat{y} + s_z\hat{z}$, and $\chi^\dagger(t)\chi(t) = 1$. Here $\chi(t)$ is the spin eigenstate of the spin-1/2 particle.

The above expression for $\vec{J}$ leads to the *unique* modified Bohmian velocity for a spin-1/2 particle which has a spin-dependent term:

$$\vec{v}(\vec{r}, t) = \frac{\vec{J}(\vec{r}, t)}{\rho(\vec{r}, t)} + \frac{1}{m}(\vec{\nabla}\log\rho(\vec{r}, t) \times \vec{s}(t)) \qquad (10.5)$$

The implications of the spin-dependent contribution in the expression for $\vec{J}$ have been studied for an eigenstate transition of the hydrogen atom,[31,32] in the Bohmian trajectories for the two-slit experiment,[33] and in calculating the arrival time distribution for free particles,[29] as well as for the particles passing through a spin-rotator containing a magnetic field.[30]

Here it may also be mentioned that in recent years an interesting direction of study has been initiated towards exploring the possibilities of *alternative onto-logical models;* for example, in the Roy–Singh model,[34,35] both position and momentum are treated on the same footing; their ontological premeasurement values are the *same* as their measured values. From a different perspective, Holland[36,37] has also suggested an interesting twist to the Bohmian approach by formulating a scheme for computing the quantum mechanical wave function directly from the Bohmian trajectories.

Bohm's Model: A Simple Illustrative Example

In order to illustrate the Bohmian model, it is instructive to consider a specific example, that of the *free evolution* of a Gaussian wave packet for a particle without spin. For simplicity, we restrict our discussion to one dimension.

The one dimensional Gaussian wave-function at $t = 0$ is given by:

$$\psi(x, t = 0) = \frac{1}{(2\pi\sigma_0^2)^{1/4}} \exp\left[\frac{-x^2}{4\sigma_0^2} + ikx\right] \tag{10.6}$$

where σ_0 is the initial width of the associated wave packet, which is peaked at $x = 0$ and is moving freely along the positive x-direction with the initial group velocity $u = (\hbar k/m)$.

The Schrödinger time evolved wave-function and probability-density for free motion at any instant t are given by:

$$\psi(x, t) = \frac{1}{(2\pi S_t^2)^{1/4}} \exp\left[-\frac{(x - ut)^2}{4\sigma_0 S_t} + ik\left(x - \frac{ut}{2}\right)\right] \tag{10.7}$$

$$\rho(x, t) = |\psi(x, t)|^2 = \frac{1}{(2\pi\sigma_t^2)^{1/2}} \exp\left[\frac{(x - ut)^2}{2\sigma_t^2}\right] \tag{10.8}$$

where $S_t = \sigma_0(1 + (i\hbar t/2m\sigma_0^2))$ and $\sigma_t = |S_t| = \sigma_0(1 + (\hbar^2 t^2/4m^2\sigma_0^4))^{1/2}$; σ_t is the width of the wave packet at any instant t.

Using Eqs. (10.3) and (10.8) we obtain the Bohmian velocity from Eq. (10.1) which is given by

$$v(x, t) = u + \frac{(x - ut)bt}{(1 + bt^2)} \tag{10.9}$$

where $b = (\hbar^2/4m^2\sigma_0^4)$

Integrating Eq. (10.9) and further simplifying one can calculate the Bohmian trajectory equation of a freely evolving ith particle of the ensemble with its initial position x_{0i}, which is given by

$$x_i(t) = ut + x_{0i}\sqrt{1 + bt^2} \tag{10.10}$$

where u, the initial group velocity of the wave packet, is the initial velocity of any individual particle of the ensemble, and x_{0i} is the initial position of the ith individual particle corresponding to the wave-function given by Eq. (10.6).

Now the velocity of an individual particle as a function of the initial position $x = x_{0i}$ is given by

$$v_i(x, t) = \frac{dx_i(t)}{dt} = u + \frac{x_{0i}Bt}{\sqrt{1 + Bt^2}} \tag{10.11}$$

From Eq. (10.11) we can then infer the following points:

1. If the initial position x_{0i} for the ith particle is the point $x = 0$ at which the wave packet is peaked, then $v_i(x, t) = u$, so any particle in the ensemble having such an initial position will follow the Newtonian trajectory for free motion.

2. If the initial position x_{0i} lies within the right half of the initial wave packet (i.e., we take $x_{0i} = +\varepsilon$ where $\varepsilon > 0$), then $V_i(x, t) = u + (\varepsilon bt/\sqrt{1 + bt^2})$. Since $b > 0$, then it is seen that the particles in the ensemble whose initial positions are distributed within the *right half* of the initial wave packet are *all accelerated*.

3. If the initial position x_{0i} lies within the left half of the initial wave packet (i.e., we take $x_{0i} = -\varepsilon$ where $\varepsilon > 0$) then $V_i(x, t) = u - (\varepsilon bt/\sqrt{1 + bt^2})$, so the particles in the ensemble whose initial positions are distributed within the left half of the initial wave packet are *all decelerated*.

Thus the Bohmian equation of motion in this case given by Eq. (10.11) implies that any particle in the ensemble starting from any position lying within the left half of the Gaussian wave packet will take a *longer time* to travel a given distance, say d, compared with the time taken by any particle in the initial ensemble whose initial position lies within the right half. Eq. (10.9) implies that the premeasurement value is the same for all particles in the ensemble, but after the measurement the momentum distribution is determined by the Fourier transform of Eq. (10.6), illustrating the contextuality (discussed in Chapter 9) inherent in the Bohmian model.

Bohm's Model: Approach to the Measurement Problem

We first note that, in the Bohmian model, the premeasurement value of a dynamical variable is, in general, different from the value obtained by the measurement of that dynamical variable, except for the *special* status given to the position coordinate, whose premeasurement value is taken to be *the same* as its post-measurement value. How the measurement interaction changes the value of a dynamical variable in the Bohmian model has been discussed elaborately in the books of Holland[22] and Bohm and Hiley.[24]

In the Bohmian model, the quantum measurement problem is addressed by assuming that *all* measurements of microphysical attributes ultimately reduce to observations of the ontological position variables of macro-objects, such as the centre-of-mass coordinate of pointers, serving as indicator variables of an apparatus. In other words, it is assumed that all instrument outputs are in the end readings in position space. This assumption is claimed to cover a large class of standard measurements. Even measurements of such physical variables such as mass and wavelength, which are not directly associated with Hermitian operators, are ultimately measurements of position; for instance, mass is inferred from a position in a mass spectrograph, and wavelength is obtained from the fringe spacing in an interference experiment.

According to this approach, a definite outcome in an individual measurement is determined by the relevant ontological position variable associated with an apparatus, which has an objective value at all instants. Interpreted in this way, one seeks to eliminate the intrinsic inexactness of quantum theory, by ensuring a correspondence between a definite outcome of an individual measurement and the ontological position coordinate introduced in the theory at a fundamental level.

However there are a few critical aspects which we may briefly mention here:

1. In the Bohmian scheme, macroscopically distinct states of an apparatus are taken to mean position localised states, for example Gaussian distributions, with their peaks well separated. Then the problem of mutually overlapping 'nonzero tails' arises. No matter how sharp the apparatus states are, there will always be, *in principle*, an ill-defined domain of the position space which would correspond to an individual measurement yielding no definite outcome, with the pointer having no definite position.

2. It is crucial to the Bohmian argument that although the post-measurement final wave-function is still a superposition involving different outcomes, an individual outcome is characterised by the position coordinate of a measuring device, for example the position of the centre of mass of a pointer. This seems to imply that, in the Bohmian scheme, the ontological position of a particle is ascribed a more fundamental reality than the wave-function. Moreover, in order to make the Bohmian explanation of the measurement problem entirely consistent with an actual measurement situation, it may be necessary to *assume* the unobservability of interference between *macroscopically distinct states* of the apparatus. Such questions require a clearer understanding.

Dynamical Models of Spontaneous Wave-Function Collapse

The two preceding approaches, MWI and Bohm, accept the standard formalism and introduce new ingredients only at the interpretational level. On the other hand if we address the measurement problem by adjusting the formalism, it is evident from earlier discussions in Chapters 3 and 4 that the modified time evolution must be *non-linear*, since a pure state is doomed to remain a pure state under any linear evolution. As already pointed out, the von Neumann idea of a collapse of wave-function induced by measurement lacks precision, because it assigns a special role to measurement interactions, without specifying at what point of complexity an interaction establishing a correlation between the observables of a microsystem and a macrosystem actually becomes a measurement.

Though the transition from micro to macrosystem is gradual in the actual physical world, the difference between Schrödinger evolution the von Neumann postulated collapse dynamics is sharp, so that it is difficult to comprehend how at some point the linear unitary Schrödinger evolution is suddenly suspended, allowing the collapse dynamics to take over. It is thus clear that a logically coherent scheme for accommodating the notion of wave-function collapse must have a seamless mathematical description with no dichotomy between measurement and other interactions; this also means no arbitrary split between microsystems and macrosystems.

It has been stressed that, from the time of Bohm's presentation of his model in 1952, Bell was a keen advocate of its importance, without ever, of course, suggesting that it was necessarily a final answer to the problems of quantum theory. Through the 1980s he also encouraged the production of the kind of theory discussed in this section. For example, in 1981 he wrote:[38,39] 'One line of development towards greater physical precision would be to have the "jump" in the equations and not just the talk—so that it would come about as a dynamical process in dynamically defined conditions. The jump violates the linearity of the Schrödinger equation, so the new equation (or equations) would be non-linear.' And in 1986 he wrote[40] 'Surely the big and small should merge smoothly with one another? And surely in fundamental physical theory this merging should be described not by vague words but by precise mathematics? ...The necessary technical theoretical development involves introducing what is known "non-linearity" and perhaps what is called "stochasticity" into the basic "Schrödinger equation".'

In such developments it follows that the collapse process must be *spontaneous*, in the sense of being present in the fundamental equation, without being induced by an external stimulus, such as the system-apparatus interaction. In recent years, there has been a systematic development of such *spontaneous dynamical collapse* (SDC) models. Before delving into specifics of one particular version of SDC, we discuss their general attributes.

Additional *non-linear* terms which, as we shall show, must be *stochastic*, that is to say probabilistic rather than deterministic, are incorporated into the Schrödinger equation; this entails a modified time evolution of the system. Such new terms in the Schrödinger equation are postulated to satisfy the two divergent requirements of having a practically negligible effect for all microsystems, which is necessary due to the extremely high degree of validity of all tested predictions of the standard quantum formalism in the microdomain, and of being able to induce an appropriately rapid suppression of superpositions of macroscopically distinguishable states in the macrodomain. The latter feature is required in order to eliminate the quantum mechanical manifestations of superpositions of macroscopically distinct states, and also to ensure the definiteness of an individual outcome.

Another important feature of these schemes is that the postulated non-linearity implies a *preferred basis states* in ordinary position space, thereby destroying the inherent equivalence between different basis of states. Incorporating these features, various versions of SDC ensure that a measurement has a definite outcome as follows. During evolution of the microsystem, the Schrödinger term totally dominates, and the non-linear term must be ineffective. However a measurement interaction leads to an entangled state entailing a superposition of macroscopically distinct states of a macroapparatus. It is then that the non-linear SDC process becomes effective for the macroapparatus, resulting in the breaking up of the superposition to give a single measurement outcome, provided the following criterion is satisfied.

For the occurrence of a definite outcome through the SDC process, relevant macroscopically distinct states of an apparatus, comprising a sufficiently large number of particles, must be localized in position space and mutually separated by a large distance, compared with a suitable microscopic length scale, usually taken to be around 10^{-7} m. Then any superposition of such states must be intrinsically unstable, reducing rapidly to *any one* of the superposed states under the action of the SDC-inducing non-linear terms in the Schrödinger equation.

To complete the desired correspondence between the intersubjective reality of a definite outcome and an appropriate theoretical representation, we must ascribe an *objective reality* to the wave-function of an individual system, since SDC approaches consider a wave-function to be a complete description of the state of an individual system. We now analyse some salient technical aspects of the SDC approach.

To modify the standard formalism by introducing non-linearity, the most natural possibility would be to incorporate *non-linear deterministic* corrections into the Schrödinger equation. However, as shown by Gisin[41] and Polchinsky,[42] such a modification turns out to have a serious internal difficulty, since faster-than-light signalling (FLS) is then allowed. These authors showed that if a non-linear deterministic component is introduced into the Schrödinger evolution, the expectation value of an observable in one wing of an EPR experiment can be affected by the type of distant measurement performed in the other wing, thus allowing FLS.

The demonstration that any non-linear deterministic correction to the Schrödinger evolution permits FLS was interpreted by Weinberg[43] as indicating that it was impossible 'to change quantum mechanics by a small amount without wrecking it altogether'. He said that: 'This theoretical failure to find a plausible alternative to quantum mechanics, even more than the precise experimental verification of linearity, suggests to me that quantum mechanics is the way it is because any small change in quantum mechanics would lead to logical absurdities.' This is an extremely interesting argument, as it would strongly discourage anybody from attempting to do exactly what is the subject of this section—to maintain most aspects of the Schrödinger equation, but to make small adjustments to achieve change in the aspects of the theory that are seen as causing conceptual difficulties.

However, Weinberg overlooked the possibility that the non-linear corrections may be not deterministic but stochastic. In fact, *non-linear stochastic* corrections do prohibit FLS, provided the probabilities with which a pure state wave-function evolves stochastically into mixed states are taken to be those specified by the Born–Dirac rule. This will be the type of model we examine for the remainder of this section.

The most celebrated model, that of Ghirardi, Rimini and Weber[44,45] (GRW), gained strong support from Bell.[46,47] It is based on the assumption that, in addition to standard evolution, a quantum mechanically described physical system is subject to spontaneous localisations in position space occurring at random times. In such processes, the wave-function is localised around a value q_0 with a spatial spread denoted by a localisation parameter a_0, which is typically given a value of 10^{-7} m. This is small on a macroscopic scale, but large with respect to typical atomic distances. The probability of the wave-function collapsing to any particular region is related to the corresponding probability density for a von Neumann collapse to the same region. This ensures that the criterion of the previous paragraph is obeyed, but also makes it clear that the results of the von Neumann scheme, which are in agreement with experiment, are retained, though now they are achieved using a model which is set up rigorously from a mathematical point of view.

We have not yet mentioned the rate of localisations, and the way this is defined is one of the very attractive aspects of the theory. For a single particle, the rate is given as λ, which is typically given a value of $10^{-16}\,\text{s}^{-1}$, so localisation for a single particle will take around 3×10^8 years, a satisfactorily long time. However, for a macro-object composed of N constituent atoms, the localisation rate λ_N is given by $N\lambda$, so clearly for macroscopic objects localisation will be very fast. Macroscopic pointers containing say 10^{23} particles will have a localisation time T_N around $10^{-7}\,\text{s}$. This means that any superposition of their states in position space collapses in a time substantially less than human perception times, which are of order of $10^{-2}\,\text{s}$. Furthermore, for the pointer in a superposed state to yield discernible definite outcomes following collapse whenever the superposition involves observably different states, the spatial spread a_0 must be smaller than separations distinguishable by visible light, which are around $10^{-6}\,\text{m}$, a condition clearly satisfied by the chosen value of $a_0 = 10^{-7}\,\text{m}$.

There is very much more to be said about the GRW technique. There are important conceptual problems still remaining. For example, the spontaneous localisations do not conserve energy exactly, so the first law of thermodynamics is violated. Also the localisation is to a region which is sharply peaked but does nevertheless extend with non-vanishing amplitude far from the peak, this giving the problem of 'tails'. There is also the interesting question of whether the departure from the strict Schrödinger equation might be observable. For further discussion of these points, the reader is invited to study other texts.[6,48]

Consistent Histories and Decoherent Histories

Standard interpretations of quantum theory imply the occurrence of some strange wave-functions for an individual particle. Robert Griffiths, who developed the idea of *consistent histories* called them 'grotesque'! For example a particle may reach a beam-splitter, from which there are two possible paths at right angles to each other, each ending up at a particle detector. A standard interpretation will say that, once the particle is past the beam-splitter, the wave-function is straddled between two regions, each quite localised, but separated by a steadily increasing macroscopic distance. Only when either of the detectors registers a particle does the wave-function collapse to the region of that detector. A histories approach aims to build up a more complete and perhaps more 'sensible' account or 'history' of what has taken place. It will decide that the particle was always on one particular path, the one that leads to the detector which actually registers the particle.

It is immediately obvious that this type of approach will only avoid conceptual problems in some types of case. If, for example, the beams were to be recombined so that they would interfere, it would be quite illegitimate to decide that a particle has gone along one path or the other; the possibility of interference requires that the particle must 'sample' both paths. Griffiths[49,50] called his approach 'consistent histories'. The decision that a particle has travelled along one particular path gives a consistent history if each path leads to its own detector, but not if the paths combine and interference is possible. The term 'decoherent history' which was introduced by Roland Omnès,[51,52] is technically different from a 'consistent history' but conveys the same basic idea, that such histories do not interfere and therefore will evolve separately. Only for a case of consistent or decoherent histories may each history occur with a particular probability.

Murray Gell-Mann and James Hartle[53] have extended the concept of consistent or decoherent history much further than Griffiths and Omnès. They aim to provide a complete cosmology for the Universe. Indeed, as Bohm and Hiley[24] say, 'The theory of Gell-Mann has to be understood, from the very outset, as dealing with nothing less than the entire history of the Universe from the beginning to the end.' Gell-Mann and Hartle explain how the observer, whom they refer to as an *IGUS* or *information-gathering- and -utilising-system*, has evolved through history. They say that the IGUS must concentrate on situations of decoherence, because only in that way can useful predictions be made, and these are essential for survival.

While consistent and decoherent histories are undoubtedly interesting, it is not quite clear what their status is relative to other interpretations of quantum theory, or, indeed, precisely what are the claims of those who have produced them. Omnès has stated[51] that 'The logical structure of quantum mechanics [decoherent histories] ... led to a theory of classical phenomenology containing the proof that classical determinism is a consequence of quantum mechanics. Together with decoherence, it generated an apparently complete and consistent interpretation [of quantum theory].' This is a large claim, but it is clear that he regards the approach as merely a logically respectable means of recovering the Copenhagen methodology. He says that: 'Although [the new approach] relied on some new principles, it soon turned out to be in fact a reformulation of the Copenhagen interpretation well suited to the treatment of consistency.' In like manner, Griffiths says that[49] 'nothing can inspire a greater respect for Bohr's insights than rediscovering them with the help of a deductive method'.

If the approach reaches the same ideas as Copenhagen, but by a logical route, it seems that it might be expected to solve the measurement problem. Indeed, Omnès says that, in this approach, wave-function collapse 'is now found to be a convenient calculation recipe with no specific physical content', which appears to suggest that no conceptual discussion is required. However, he admits that the theory gives only probabilities for various outcomes, but says that it would be unreasonable to expect more. Griffiths explicitly refrains from saying whether the approach solves the measurement problem, while Jonathan Halliwell,[54] another proponent of this approach, categorically states that it does not. It would seem that the approach may claim to be a conceptually more satisfactory way of approaching the ideas of Copenhagen, but does not actually solve its conceptual problems.

Another interesting question is the relationship between consistent and decoherent histories, and hidden variable theories. To say that a particle goes along one of two paths, while the wave-function encompasses both, would normally be regarded as a claim of a hidden variable. However this would be the simplest type of hidden variable theory, which, for a general case, would clearly be ruled out by Bell's theorem, as it is local. Actually Griffiths specifically states that the approach does *not* involve hidden variables. Clearly there are still many fundamental questions to be answered about the conceptual status of this approach.

Knowledge and Information Interpretations

It is often felt to be a straightforward solution to the riddles of quantum measurement to say that the wave-function represents, not the state of a system, but knowledge of the state of the system. Then it is not surprising that the wave-function collapses at a measurement. Beforehand, we may know only that the observable takes one of several values, while at the measurement we discover which is the actual value. Clearly an immediately repeated measurement must give the same answer.

While the approach seems very persuasive at first sight, there are a number of problems, or at least questions that need to be answered. First, we may talk of 'knowledge', but whose knowledge is this? Many people may perform a variety of measurements on a system; how are the various elements of knowledge that each may gain from their measurements related to each other? Also the simplest type of statement about such interpretations makes them seem to be hidden variable interpretations. In the previous paragraph, for example, there is a not-so-tacit assumption that the observable does actually have a value even before the measurement, and this value is, of course, a hidden variable. We may describe the approach more carefully, so as to avoid this assumption, but this does render the argument less convincing and appealing. We have to admit that the measurement does not just put into the knowledge of somebody a value that already existed; the value must have been created in the measurement, and this makes the ideas rather similar to those of more standard approaches. We could also ask: What about knowledge of the early Universe when there were no observers?

Indeed until recently the most famous exponent of such ideas was Rudolf Peierls, a very famous traditional theoretical physicist, who would certainly have regarded himself as merely presenting the standard approach of Bohr. Indeed his most famous paper on this topic[55] was explicitly written to defend orthodoxy against a severe attack from Bell.[47] In this paper he attempted to give a clear answer to the question of whose knowledge should be represented in a mathematical description of a state. All observers, he said, may have their own descriptions, but all descriptions must be consistent. We cannot have a situation where one observer 'knows' the value of S_z, another of S_x, because simultaneous knowledge of both is forbidden by the laws of quantum theory. As to knowledge of the early Universe, he said that the observer does not have to be contemporary with the actual event; when we draw conclusions about the early Universe, we are acting as observers.

However, until recently, this approach of Peierls would not have been popular, probably for the reason we mentioned above—it seems to be more natural when coupled with hidden variables. Nevertheless, with the rise of quantum information theory, and the consequent focus of attention on the concept of 'information', there has been interest in the idea that information is the fundamental quantity in the Universe, and that the task of the wave-function should be to convey information.

David Mermin[56] was present at a famous meeting at Amherst, when, very nearly at the end of his life, Bell[47] fulminated against the use of the word 'measurement' as a primary term when discussing quantum foundations. Among the other words he rejected were 'system', 'apparatus', 'environment', 'microscopic', 'macroscopic', 'reversible', 'irreversible' and 'observable'. They did not include 'knowledge' but did include 'information'. On the last term, Bell's sardonic comments were '*Whose* information? Information about *what*?'

Mermin considers that the words 'knowledge' and 'information' should be regarded as synonymous, at least as they are used in quantum theory. While initially in support of Bell, with the coming of quantum information theory he has felt obliged to reconsider the matter. He has paid considerable attention to the question 'Whose information', and has shown that Peierls's answer is not true in detail. He has actually made some progress in search of a more complete answer, but there are remaining issues to be resolved. As to the question: 'Information about what?', he has come to think that it is metaphysical; he suggests that we cannot know whether information is about anything, or just information.

The view that the fundamental concept of quantum physics is information, and that quantum physics is only indirectly a science of reality but basically a science of knowledge, has been particularly argued recently by Brukner and Zeilinger.[57] They claim that their approach leads to a natural understanding of complementarity and entanglement, and a means of deriving an expression for the quantum evolution of a two-state system. They even claim that they may be able to obtain an answer to the most fundamental question—Why the quantum? Since information must be quantised in an integral number of propositions, and if quantum theory is just a way of describing information, then it would seem natural that it must indeed exhibit quantisation.

Stochastic Interpretations

As mentioned in Chapter 5, there has been a very long history,[15] going right back to Schrödinger in 1932, of attempting to demonstrate an analogy between quantum theory and a classical stochastic theory such as that of Brownian motion. If this could be achieved, so that quantum physics is, in a sense, put into a classical format, it would seem that the conceptual problems specifically of quantum theory must be removed. The physicist who has made the largest contributions to this task over several decades is Edward Nelson,[58,59] but many other realist opponents of Copenhagen have also contributed.

Considerable success has been achieved from the mathematical point of view, and some interesting physical pictures have also been developed. For example, the zero-point field may play an important role. While in conventional approaches to quantum theory, this may be looked on as rather formal and mathematical, in stochastic quantum theory it performs a directly physical task as a real physical field. Any atomic system must exchange energy with this field, and so its own energy will appear to fluctuate randomly. Thus the zero-point field causes the quantum fluctuations, but also the fluctuations create the field in a self-consistent way.[60]

Despite this apparent achievement, stochastic interpretations of quantum theory have also been subject to persistent and fundamental questioning and criticism. Schrödinger pointed out some of the difficulties in his original work. In quantum theory the wave-function is complex, while in the classical theories it is (mathematically) real. In classical physics it is the probability density that is subject to a differential equation, while in quantum physics it is the amplitude. Also Wallstrom[61] has claimed that, in order to obtain the Schrödinger equation starting from a classical stochastic equation, a quantisation condition must be applied, and there is nothing in the classical theory to justify such a condition.

More fundamentally a major problem has become clear since the publication of Bell's theorem. It must be stated that, for practically all the proponents of stochastic interpretations, the aim was to show that quantum theory was basically realist, and a component of that aim that was largely unwritten but certainly taken for granted was that the theory should be local. But essentially a stochastic interpretation of quantum theory is a hidden variable theory, and so it cannot escape Bell's theorem—and so must be non-local. Nelson, in fact, reached this result for himself when he applied his method to entangled systems. In 1985 he[59] wrote 'I have loved and nurtured ... stochastic mechanics for 17 years, and it is painful to abandon it. But the whole point was to construct a physically realistic picture of microprocesses, and a theory that violates locality is untenable.'

Others may not feel it necessary to abandon the theories merely because they must be non-local. Indeed Bacciagaluppi[62] has recently produced a substantial account of their properties. However it is now clear that they cannot be regarded as providing a fully classical picture of quantum theory. Rather they must be regarded as being a similar type of theory to that of Bohm, having broadly realist features but being non-local.

The Quantum State Diffusion Model

In recent years a technique for visualising quantum processes and quantum measurement has been developed, principally by Nicholas Gisin and Ian Percival.[63,64] Most descriptions of quantum systems use a technique called the density-matrix, which is discussed very briefly in Chapter 12. This technique relates, not to a single system but to an ensemble of systems. While, with the use of the density-matrix, one may discuss quantum evolution, quantum transitions and quantum measurement very effectively from the mathematical point of view, the physical meaning is often not very clear. In addition, although the mathematics is undoubtedly elegant, the actual justification of the various steps may be obscure; this has been one of the main messages of this book!

Gisin and Percival[63] contrast this with the fact that many physicists, particularly experimentalists, view the experimental results in terms of individual systems. Indeed a single run of a laboratory experiment will have a natural interpretation in terms of individual systems making 'quantum jumps' or transitions from one state to another.

Gisin and Percival have attempted to provide a theoretical structure that brings us closer to this experimentalist's picture. They consider an individual quantum system interacting with an environment. The Schrödinger equation used is for the system itself, but it is modified from the standard equation by the addition of extra terms corresponding to the interaction with the environment. These extra terms are stochastic or random, because the interaction itself is random.

The results they produce[63] are quite startling. A good example is for a two-state system that may have one photon or none. In the appropriate equation, terms are inserted for absorption and stimulated emission processes. The results show the system spending time in each state, with transitions between them. These transitions are not instantaneous but typically take a few per cent of the time the system spends in one of the states. The result is far more dramatic than the one that would be given by the density-matrix method, which would give a fairly dull curve heading to an average photon occupation number of 1/2.

Another interesting example given is for a measurement process. The initial state is pure; it is a superposition of states with one, three, five, seven and nine photons. In the measurement, there is a clear progression to systems with one *or* three *or* five *or* seven *or* nine photons, exactly the result which one hopes a measurement will produce. In contrast, since the density-matrix method follows the unadulterated Schrödinger equation exactly, it can never describe a genuine measurement process, since it can never allow a pure state to evolve into a mixture.

Critics of the method might contend that, interesting as the results are, the method is empirical rather than rigorous, and that the results may relate to no more than the particular additional terms that are added to the basic Schrödinger equation in order to simulate the interaction of the individual system with the environment. Gisin and Percival argue, though, that there is no need to carry out a detailed justification of the precise form of these terms, any more than one would refuse to use the general idea of the electrical resistance of an electric circuit, just because calculation of the resistance of a particular material would be extremely difficult.

Many Hilbert Space Approach

This approach, proposed by Machida and Namiki,[65-67] seeks to derive the wave-function collapse on measurement from what the authors call statistical fluctuations in the measuring macroapparatus, resulting from the effectively infinite number of particles that constitute the apparatus. The basic physical idea is that different incoming particles interact one by one with a given macrodetector in different macroscopic states, and undergo randomly different phase shifts. The randomness results from the fact that the internal motion of the particles constituting the detecting device changes randomly between successive interactions with incoming particles. Machida and Namiki[65-67] argue that the interactions of measured particles with the whole ensemble of rapidly fluctuating microscopic states of a macrodetector cannot be properly treated within the framework of a single Hilbert space, the usual mathematical structure used in quantum theory, so they suggest using what they call a direct sum of many Hilbert spaces for the mathematical representation. The technicalities of this proposed extension of the formalism are comprehensively reviewed by Namiki and Pascazio.[68]

The outcome of a rather lengthy formal exercise is the strictly mathematical result that the combined pure state of a measured system and an apparatus actually becomes a mixed state, or in other words exact wave-function collapse occurs, only in the limit when the apparatus consists of an infinite number of particles. Then for actual measurements performed by a real apparatus with a finite number of particles, a so-called 'partial collapse' occurs. Advocates of this approach[65-68] try to quantify the extent to which the collapse makes a pure state behave like a mixed state by defining what they call a 'decoherence parameter'. Nevertheless, the central difficulty of the measurement paradox remains unalleviated because, as stressed in Chapter 3, even if a pure state is ensured to behave almost like a mixed state, it cannot be interpreted as a mixed state unless it actually becomes one.

The interpretational shift between a pure state or homogeneous ensemble, and a mixed state or heterogeneous ensemble is sharp, not gradual, so that measurements either have definite outcomes or not. Therefore it would seem that we cannot solve the fundamental conceptual problem of explaining the occurrence of a definite outcome by regarding wave-function collapse as a sort of asymptotic phenomenon of the many Hilbert space scheme, unless additional interpretational arguments are introduced that the present approach lacks.

Gravitationally Induced Wave-Function Collapse Approach

This approach, put forward by Penrose,[69-72] attributes the difficulties of reconciling quantum theory with the general relativistic theory of gravity in part to the fact that standard quantum theory requires additional inputs in order to solve the measurement problem. It suggests a solution to the problem that involves the effects of gravitation in a non-trivial way. Note that like other non-standard approaches, this one assumes the primacy of position-localised states in providing a preferred basis with respect to which the measurement outcomes are registered.

Following Penrose we consider a typical measurement process, the initial stage of which produces a superposition of two states that involve quite different positions for the detecting device, which may be, for example, a macroscopic spherical lump. Penrose says that the fact that the lump has its own gravitational field indicates that the final state 'involves a superposition of two different gravitational fields'. This implies that 'two different space-time geometries are superposed,' and the suggestion is that this situation results in the spontaneous collapse of the wave-function as an objectively real process.

The crucial question for this scheme is thus whether there is a suitable criterion for deciding when the two superposed space-time geometries are sufficiently distinct that the standard Schrödinger evolution must be is replaced by wave-function collapse. One possibility is that the scale of the difference between these superposed geometries must be at least of the order of 10^{-35} m, the Planck scale that characterises the distance for which quantum gravity effects become significant. To make this idea more concrete, Penrose sketches the following criterion.

Considering the case of the detector lump in a superposition of two spatially displaced states, Penrose suggests calculating the energy required to displace the lump from one configuration to the other, taking only gravitational effects into account. By calculating the gravitational energy required to bring about this displacement, taking the reciprocal of this energy and using the time-energy Heisenberg principle, one can then obtain a rough estimate of the time required for the spontaneous collapse of the superposed wave-function into one localised state or the other.

If the lump is taken to be spherical with mass M and radius R, we obtain a quantity of the general order of GM^2/R for the relevant gravitational energy, and accordingly the collapse time can be estimated in this scheme. Though the actual value of the relevant gravitational energy depends on the separation between spatial configurations, additional energy required to move away from the contact position, even to a very distant position, is of the same order of magnitude as that involved in moving from coincidence to the contact position.

Following this prescription, Penrose estimates the predicted collapse time in different situations. If a water droplet of radius 10^{-7} m is in a superposed state, its collapse time is predicted to be on the order of a few hours; for a radius of 10^{-6} m, this collapse time is reduced to about 0.05 s. For a neutron or proton, taking R to be the strong interaction scale, 10^{-15} m, the collapse time is estimated to be nearly 10^7 years. Thus this proposed scheme, based on an apparently esoteric notion of linking the quantum measurement problem with gravity, does not immediately lead to an absurd consequence. Given that there is no *a priori* logical objection to the basic idea of Penrose's proposal, and even though it is not yet formally developed with adequate rigor, it should be worthwhile to examine more closely the implications of such a scheme. In this context, we note that the possibility of a setup which may be able to test this type of wave-function collapse model experimentally will be indicated in Chapter 13 (Refs. [32]), in the course of discussing various examples which have been proposed as tests of the quantum superposition principle for macrosystems.

Here for completeness it should be noted that prior to Penrose, Karolyhazy[73] had initiated preliminary discussions about such a model for wave-function collapse, followed by Diósi[74] who had a specific scheme with a dynamical stochastic input, but this scheme has not yet been further developed. On the other hand, Penrose's arguments are independent of any particular kind of dynamics, while seeking to justify the requirement of a gravitationally induced wave-function collapse on the ground of ensuring consistency between the general principles of quantum mechanics and the general theory of relativity.

In this context we note a particularly interesting point recently made by Penrose.[75] The nature of the wave-function transformation from a freely falling frame to the one fixed in a gravitational field implies that these two frames correspond to different vacua. This feature may be used to argue that for a strict consistency between quantum mechanics and the principle of equivalence, the superposition of two gravitational fields should be unstable since it involves the superposition of different vacua. If such an argument is formulated in more detail, it can possibly provide a more rigorous justification for the ideas underlying the gravitationally induced wave-function collapse.

Another relevant question which calls for more study is whether the problem of energy nonconservation inherent in the dynamical models of wave-function collapse also suggests that gravitational effects may be relevant to the dynamics of wave-function collapse. This possibility stems from the counterintuitive feature in general relativity that both the gravitational energy in the mutual attractive field of massive objects and that residing in empty space cannot be computed merely by adding up local energy density contributions, nor can such energy even be localised in any particular space-time region. As Penrose has remarked, it is rather tempting to relate the equally tricky problem regarding the energy of the dynamical collapse process to that of gravity, and to offset one against the other so as to provide a coherent overall picture.

Wave-Function Collapse Models Based on Irreversibility

This approach seeks to resolve the quantum measurement problem by extending the quantum mechanical formalism for describing the interaction between a microsystem and a macroapparatus to incorporate irreversibility at a fundamental level. Detailed work along this line has been pursued by Prigogine and his coworkers.[76–79] The basic idea is that a measuring apparatus is a complicated many-body system having an inherent dynamic instability. Technically this is called non-integrability, which means that it is not possible to express the Hamiltonian with respect to the canonical coordinates in such a way that the derivative of the Hamiltoniam with respect to the canonical coordinates vanishes. This condition is claimed to lead to its irreversible behaviour. The extension of Hilbert space envisaged for incorporating this feature is used to describe the dynamic process of wave-function collapse. For more details and relevant references, see Prigogine and Stengers.[78,79]

Like the many Hilbert space approach, in this approach too, the wave-function collapse occurs only in the mathematical limit of an infinite number of constituent particles of an apparatus. Thus discussions about the many Hilbert space approach apply in this case as well. But while it is crucial to this scheme that a measuring apparatus be a non-integrable dynamic system, the measured system is treated as an integrable system. Thus an awkward dichotomy is introduced. Prigogine attempts to justify the necessity of regarding the apparatus as a non-integrable system from a broader philosophical perspective of understanding irreversibility as, in principle, an exact feature, rather than allowing it to emerge as an approximation. In the absence of a specific justification of the criterion used to characterise a measuring apparatus, it is difficult to accept this approach as a strong candidate for providing a solution to the quantum measurement problem.

Conclusions

In this chapter we have seen almost a perhaps bewildering number of proposed interpretations of very many different types. All have some advantages, but all too have difficulties and problems remaining to be resolved. In particular, whether any of the approaches suggested so far has been able to provide a completely satisfactory resolution of the quantum measurement problem remains a contentious issue. Hence it is important to explore the extent to which the non-standard interpretations can be empirically discriminated, apart from seeking further theoretical refinements of these schemes. But we must stress that the number and variety of approaches is an extremely healthy sign, as it suggests that many physicists with different backgrounds and ideas have recognised the central nature of the problems of quantum theory and quantum measurement.

The situation is particularly pleasing when one contrasts it with the all-but monolithic acceptance of the Copenhagen interpretation not so many years ago, together with the stigma that would have been felt by anybody even daring to question it. Also one must contrast the industrious and detailed work put into the development of many of today's interpretations with the set of rather sententious aphorisms which constituted the Copenhagen interpretation.

Whatever Einstein would think of any one of the various interpretations if he were alive, he should be glad that discussion is proceeding and lively, and that many of the proposed interpretations are based on specifically physical principles and concepts, rather than only on ideas borrowed from philosophy. In his own right, and through his influence on Bohm and Bell, he deserves much of the credit for stimulating active studies in this earlier ignored area.

References

1. Rosenfeld, L. (1962). In: Observation and Interpretation in the Philosophy of Physics. (Körner, S., ed.) New York: Dover; Peierls, R. (1986). In: The Ghost in the Atom. Cambridge: Cambridge University Press.

2. Margenau, H. (1936). Quantum mechanical description, Physical Review **49**, 240–2.

3. Everett, H. (1957). 'Relative state' formulation of quantum mechanics, Reviews of Modern Physics **29**, 454–62.

4. Squires, E.J. (1989). An attempt to understand the many-worlds interpretation of quantum theory. In: Quantum Theory without Reduction. (Cini, M. and Levy–Leblond, J.M., eds.) Bristol, UK: Adam Hilger, pp. 151–61.

5. Squires, E.J. (1993). Quantum theory and the relation between the conscious mind and the physical world, Synthese **97**, 109–23.

6. Whitaker A. (2006). Einstein, Bohr and the Quantum Dilemma: From Quantum Theory to Quantum Information. 2nd edn., Cambridge: Cambridge University Press.

7. De Witt, B.S. and Graham, N. (1973). Many-Worlds Interpretation of Quantum Mechanics. Princeton: Princeton University Press. pp. 155–65.

8. Ballentine, L.E. (1973). Can the statistical postulate of quantum theory be derived?—a critique of the many-universes interpretation, Foundations of Physics **3**, 229–40.

9. Deutsch, D. (1985). Quantum theory as a universal physical theory, International Journal of Theoretical Physics **24**, 1–41.

10. Lockwood, M. (1989). Mind, Brain, and the Quantum. Oxford: Blackwell, Ch. 13.

11. Primas, H. (1981). Chemistry, Quantum Mechanics, and Reductionism. Berlin: Springer-Verlag, Sect. 3.6.

12. de Broglie, L. (1926). Possibility of relating interference and diffraction phenomena to the theory of light quanta, Comptes Rendus **183**, 447–8.

13. de Broglie, L. (1926). Remarques sur ta nouvellle mecanique ondulatoire [Remarks on the new wave mechanics] Comptes Rendus **184**, 272–4.

14. de Broglie, L. (1927). La mecanique ondulatoire at la structure atomique de la matière et du rayonnement [Wave mechanics and the atomic structure of material and waves], Journal de Physique et le Radium **8**, 225–41.

15. Jammer, M. (1974). Philosophy of Quantum Mechanics. New York: Wiley, pp. 44–9.

16. de Broglie, L. (1928). In: Electrons et Photons: Paris: Gauthier-Villars, pp. 105–132.

17. Pauli, W. (1928). In: Electrons et Photons. Paris: Gautier-Villars, pp. 280–282.

18. Cushing, J.T. (1994). Quantum Mechanics — Historical Contingency and the Copenhegen Hegemony. Chicago: University of Chicago Press, Chs. 6, 10 and 11.

19. Selleri, F. (1990). Quantum Paradoxes and Physical Reality. Dordrecht: Kluwer, Chs. 1 and 7.

20. Beller, M. (1996). The rhetoric of antirealism and the Copenhagen spirit, Philosophy of Science **63**, 183–204.

21. Bohm, D. (1952). A suggested interpretation of the quantum theory in terms of 'hidden' variables, Physical Review **85**, 166–79, 180–93.

22. Holland, P. (1993). Quantum Theory of Motion. Cambridge: Cambridge University Press, Cambridge, pp. 15–20.

23. Bohm, D. and Hiley, B.J. (1982). The de Broglie pilot wave theory and the further development of new insights arising out of it, Foundations of Physics **12**, 1001–16.

24. Bohm, D. and Hiley, B.J. (1993). The Undivided Universe. London: Routledge.

25. Cushing, J.T. (1994). Quantum Mechanics — Historical Contigency and the Copenhegen Hegemony. Chicago: University of Chicago Press, Ch. 4.

26. Albert, D.Z. (1992). Quantum Mechanics and Experience. Cambridge: Harvard University Press, Ch. 7; Bohm's alternative to quantum mechanics, Scientific American **270**(5), 32–9 (1994).

27. Cushing, J.T., Fine, A. and Goldstein, S. (eds.) (1996). Bohmian Mechanics and Quantum Theory: An Appraisal. Dordrecht: Kluwer.

28. Holland, P. (1999). Uniqueness of paths in quantum mechanics, Physical Review A **60**, 4326–30.
29. Ali, M.M.,. Majumdar, A.S., Home, D. and Sengupta, S. (2003). Spin-dependent observable effect for free particles using the arrival time distribution, Physical Review A **68**, 042105.
30. Pan, A.K., Ali, M.M. and Home, D. (2006). Observability of the arrival time distribution using spin-rotator as a quantum clock, Physics Letters A **352**, 296–303.
31. Colijn, C. and Vrscay, E.R. (2002). Spin-dependent Bohm trajectories for hydrogen eigenstates, Physics Letters A **300**, 334–40.
32. Colijn, C. and Vrscay, E.R. (2003). Spin-dependent Bohm trajectories associated with an electronic transition in hydrogen, Journal of Physics A **36**, 4689–702.
33. Holland, P.R. and Philippidis, C. (2003). Implications of Lorentz invariance for the guidance equation of two-slit quantum interference, Physical Review A **67**, 062105.
34. Roy, S.M. and Singh, V. (1995). Causal quantum mechanics treating position and momentum symmetrically, Modern Physics Letters A **10**, 709–16.
35. Roy, S.M. and Singh, V. (1999). Maximally realistic causal quantum mechanics, Physics Letters A **255**, 201–8.
36. Holland, P.R. (2005). Computing the wavefunction from trajectories: particle and wave pictures in quantum mechanics and their relation, Annals of Physics (New York) **315**, 505–31.
37. Holland, P.R. (2005). Hydrodynamic construction of the electromagnetic field, Proceedings of the Royal Society A **461**, 3659–79.
38. Bell, J.S. (1981). Quantum mechanics for cosmologists, In: Quantum Gravity 2. (Isham, C., Penrose, R. and Sciama, D., eds.), Oxford: Clarendon, pp. 611–37; also in Ref. [39], pp. 117–38.
39. Bell, J.S. (2004). Speakable and Unspeakable in Quantum Mechanics. (1st edn. 1987, 2nd edn. 2004) Cambridge: Cambridge University Press.
40. Bell, J.S. (1986). Six possible worlds of quantum mechanics, In: Proceedings of the Nobel Symposium 65: Possible Worlds in Arts and Sciences (Allén, S., ed.) Stockholm: Nobel Foundation; also in Ref. [39], pp. 181–95.
41. Gisin, N. (1990). Weinberg non-linear quantum mechanics and supraluminal communications, Physics Letters A **143**, 1–2.
42. Polchinski, J. (1991). Weinberg's nonlinear quantum mechanics and the Einstein-Podolsky-Rosen paradox, Physical Review Letters **66**, 397–400.
43. Weinberg, S. (1993). Dreams of a Final Theory. London: Vintage, pp. 69–70.
44. Ghirardi, G.C., Rimini, A., and Weber, T. (1985). In: Quantum Probability and Applications. (Accardi, L. and Von Waldenfels, W., eds.) Berlin: Springer-Verlag, pp. 223–32.
45. Ghirardi, G.C., Rimini, A., and Weber, T. (1986). Unified dynamics for microscopic and macroscopic systems, Physical Review D **34**, 470–91.
46. Bell, J.S. (1987). Are there quantum jumps? In: Schrödinger: Centenary of a Polymath. Cambridge: Cambridge University Press; also in Ref. [39], pp. 201–12.
47. Bell, J.S. (1990). Against 'measurement', Physics World 3(8), 33–40; also in Ref. [39] (2nd edn.), pp. 213–31.
48. Home, D. (1997). Conceptual Foundations of Quantum Physics: An Overview from Modern Perspectives. New York: Plenum, pp. 97–119.
49. Griffiths, R.B. (1984). Consistent histories and the interpretation of quantum mechanics, Journal of Statistical Physics **36**, 219–72.

50. Griffiths, R.B. (2003). Consistent Quantum Theory. Cambridge: Cambridge University Press.
51. Omnès, R. (1992). Consistent interpretations of quantum mechanics, Reviews of Modern Physics **64**, 339–82.
52. Omnès, R. (1999). Quantum Philosophy: Understanding and Interpreting Contemporary Science. Princeton: Princeton University Press.
53. Gell-Mann, M. and Hartle, J.B. (1993). Classical equations for quantum systems, Physical Review D **47**, 3345–82.
54. Halliwell, J. (1994). Aspects of the decoherent histories approach to quantum mechanics, In: Stochastic Evolution of Quantum States in Open Systems and Measurement Processes. (Diósi, L. and Lukáks, B., eds.) Singapore: World Scientific, pp. 54–68.
55. Peierls, R. (1991). In defence of measurement, Physics World **3**(1), 19–20.
56. Mermin, N.D. (2002). Whose knowledge? In: Quantum [Un]speakables: From Bell to Quantum Information. (Bertlmann, R.A. and Zeilinger, A., eds.) Berlin: Springer, pp. 271–80.
57. Brukner, C. and Zeilinger, A. (1999). Operationally invariant information in quantum measurements, Physical Review Letters **83**, 3354–7.
58. Nelson, E. (1967). Dynamical Theories of Brownian Motion. Princeton: Princeton University Press.
59. Nelson, E. (1985). Quantum Fluctuations. Princeton: Princeton University Press.
60. Santos, E. (1991). Comment on 'Source of vacuum electromagnetic zero-point energy', Physical Review A **44**, 3383–4.
61. Wallstrom, T.C. (1994). Inequivalence between the Schrödinger equation and the Madelung hydrodynamic equation, Physical Review A **49**, 1613–7.
62. Bacciagaluppi, G. (1999). Nelsonian mechanics revisited, Foundations of Physics Letters **12**, 1–16.
63. Gisin, N. and Percival, I.C. (1992). The quantum-state diffusion model applied to open systems, Journal of Physics A **25**, 5677–91.
64. Percival, I.C. (1998). Quantum State Diffusion. Cambridge: Cambridge University Press.
65. Machida, S. and Namiki, M. (1980). Theory of measurement in quantum mechanics: mechanism of reduction of wave packet, Progress of Theoretical Physics **63**, 1457–73, 1833–47.
66. Namiki, M. (1986). Annals of the New York Academy of Sciences **480**, 78.
67. Namiki, M. (1988). Many-Hilbert-spaces theory of quantum measurements, Foundations of Physics **18**, 29–55.
68. Namiki, M. and Pasacazio, S. (1993). Quantum theory of measurement based on the many-Hilbert-spaces approach, Physics Reports **231**, 301–411.
69. Penrose, R. (1994). Shadows of the Mind. Oxford: Oxford University Press, Ch. 6.
70. Penrose, R. (2000). Wavefunction collapse as a real gravitational effect, In: Mathematical Physics 2000. (Fokas, A., Kibble, T.W.B., Grigouriou, A. and Zekarlinski, B., eds.), London: Imperial College Press, pp. 266–82.
71. Penrose, R. (1998). Quantum computation, entanglement and state reduction, Philosophical Transactions of the Royal Society A **356**, 1927–39.
72. Penrose, R. (2004). The Road to Reality. London: Jonathan Cape, Ch. 30.
73. Karolyhazy, F., Frenkel, A., and Lukács, B. (1986). On the possible role of gravity in the reduction of the wave function, In: Quantum Concepts in Space and Time. (Penrose, R. and Isham, C.J., eds.) Oxford: Oxford University Press, pp. 109–28.

74. Diósi, L. (1984). Gravitation and QM localization of macro-objects, Physics Letters A **105**, 199–202.
75. Penrose, R. (2004). The Road to Reality. London: Jonathan Cape, p. 867.
76. Prigogine, I. and George, C. (1983). The second law as a selection principle—the microscopic theory of dissipative processes in quantum systems, Proceedings of the National Academy of Sciences **80**, 4590–4.
77. Petrosky, T. and Prigogine, I. (1997). The Liouville space extension of quantum mechanics, Advances in Chemical Physics **99**, 1–120.
78. Misra, B., Prigogine, I. and Courbage, M. (1979). Lyapounov variable — entropy and measurement in quantum mechanics, Proceedings of the National Academy of Sciences **76**, 4768–72.
79. Prigogine, I. and Stengers, I. (1984). Order Out of Chaos. New York: Bantam Books.

11
Einstein and Quantum Information Theory

The Rise of Quantum Information Theory

In the first decade of the twenty-first century, quantum information theory is indisputably a 'hot topic'. A great deal of theoretical work is being performed in the main branches of the subject—quantum computation, quantum cryptography and quantum teleportation. Many different experimental techniques are being explored with the eventual aim of producing the first useful quantum computer, though it is recognised that this will almost certainly be decades away. In the other two main branches, though, considerable progress has been made; quantum teleportation has been demonstrated in the laboratory, while quantum cryptography has reached the stage where it is capable of being applied to ensure the security, say, of the financial district of a large city; this will probably happen quite soon.

The subject[1-9] came into prominence quite dramatically with two very important discoveries in 1994 and 1995. One[10] was an unexpected proof by Peter Shor that a quantum computer could perform tasks impossible for a classical computer. The second was a theoretical demonstration by Juan Cirac and Peter Zoller[11] of a method to construct a crucial quantum gate using an ion trap. One might say that the first discovery showed that quantum computation could be very useful, which had been far from clear until that point, and the second gave a boost in confidence that building a quantum computer would just be exceptionally difficult, not absolutely impossible.

However there had been much steady if less spectacular development before that date. One might just conceivably go back as far as 1959, when Richard Feynman[12] made a famous speech titled 'There's plenty of room at the bottom' in which he called for the miniaturisation of physics and technology. As well as fore-telling the coming of nanotechnology,[13] half a century later another hot topic, and indeed quite a controversial one, Feynman suggested that computers, which at that time still occupied very large rooms, should come down to an atomic scale, with the obvious corollary that they would have to be analysed using quantum not classical methods. We could go back to 5 years later, when Gordon Moore, co-founder of Intel, announced the famous 'Moore's Law', which suggested that computing power, as measured by the number of transistors on a silicon chip, was doubling, and would continue to double, roughly every 18 months.[2] This prediction, which has broadly come true, implies that, by perhaps 2010, the number of electrons in a transistor would be down to three or four, so certainly quantum methods of analysis would be required. In very different early work of 1970, Stephen Wisener came up with notions analogous to today's quantum cryptography, but unfortunately these remained unpublished.[2]

So it was not till the 1980s that much genuine progress in the general area of quantum information theory was made, and the central person in the first half of this decade was again Feynman, who was, of course, a great physicist,[14–16] but also, according to his close friend, the important computer scientist, Marvin Minsky,[17,18] loved every aspect of computation. In these years Feynman gave an interdisciplinary course[19] at Caltech called 'The Physics of Computation'. As well as covering such topics as computer organisation, the theory of computation, coding and information theory, and the physics of computation, he also discussed some aspects of the quantum computer, and also of reversible computation, which is an essential aspect of quantum computation.

Feynman was supported in this course by guest lecturers including John Hopfield, Carver Mead, Gerald Sussman, Rolf Landauer, Minsky, Charles Bennett, Paul Benioff, Norman Margolus, Ed Fredkin and Tom Toffoli. During this period, he interacted with all these scientists to produce what may be called a physics of computation, including some aspects of quantum computation. Recently, as well as editing a version of Feynman's lectures,[19] Tony Hey has encouraged all these friends of Feynman to contribute to a second book,[18] which gives an account of their work with Feynman in the 1980s, and also a discussion of recent developments.[20] However, as Julian Brown[2] has suggested, for quantum computation this period may best be thought of as analogous to the days of the so-called old quantum theory[21] between Planck's initial discovery and the rigorous developments of Heisenberg and Schrödinger. Important contributions were made, and lasting results were achieved, but there was no underlying theory, and no genuine understanding of the fundamental nature of what was being discovered.

Feynman himself produced two very important papers. The first[22] was a keynote speech he had to be persuaded to give to a 1981 conference at MIT on The Physics of Computation, which Landauer[23] said was an almost invisible field at the time. Feynman's interest helped to change this. His lecture was one of the very first to discuss using quantum ideas in computation, and it introduced the term 'quantum computer'. Benioff[24] had already shown that it was possible to use quantum systems to simulate classical systems; it had previously been suspected that the uncertainty principle might be a barrier to this. Feynman now asked the reverse question: Can a classical computer simulate a quantum system exactly? He built in a requirement that the amount of computer memory required should *not* increase exponentially with the size of the quantum problem, and he found that this requirement would be violated. Feynman suggested, though, that, even with this requirement included, quantum systems could be simulated by other quantum systems, and wondered whether there might be such a system as a universal quantum simulator that might simulate anything, including the physical world.

Feynman's second important paper[25] dated from 1985, and dealt in much more detail with the structure of the proposed quantum computer. In particular it considered the reversibility of computers, which had been studied for a very considerable period by Landauer, Bennett, Margolus, Toffoli, Fredkin and Feynman himself. Reversibility is important for all types of computer, because irreversibility implies heating, which becomes a greater and greater problem as computers become smaller. However it is additionally important for a quantum computer, because quantum theory (the Schrödinger equation) is itself reversible, so a quantum computer should use reversible components (reversible gates). Early work of Szilard in 1929 and von Neumann in 1949 had suggested that such operations as measurement or manipulation of information expend an amount of energy $kT \log_e 2$, which would suggest that reversible computation would be impossible. But as early as 1961, Landauer[26] showed that it was actually erasure or discarding of energy that caused energy dissipation. If this could be avoided, reversibility could be achieved.

However it must seem that, if the computer is not to be choked by garbage, there must be erasure. Bennett[27] showed how to solve this dilemma by performing a computation in three stages. In the first the computation is performed leaving the answer, but also information that is not required further, or just garbage. In the second stage, the required answer is copied onto a blank tape reversibly. In the third, the steps of the first stage are reversed; during this stage all garbage is removed by the program itself, not by erasure, and the final state of the computer memory is exactly the same as the initial state.

Given this clue that computation may, at least in theory, be reversible, Fredkin, Toffoli and Feynman set themselves the task of designing a reversible classical computer. The Fredkin and Toffoli gates, which were both reversible, were designed, and the so-called billiard ball computer[2,28] was designed. This was something like a series of sophisticated pinball machines. Billiard balls could be rolled in to the machine at particular angles to the principal axis, and reflected at mirrors parallel to this axis. The timings of the input of billiard balls and the positions of the various mirrors could be arranged so that each pinball machine acted as a particular reversible gate, and the whole would be a reversible computer. While friction and departures from perfect reflection and perfectly elastic collisions would inevitably prevent the billiard ball computer from working efficiently in practice, this work demonstrated that the laws of physics themselves did, as Bennett had claimed, allow reversible computation.

The work of Feynman and his colleagues anticipated many aspects of quantum computation, but was *restricted* to the quantum computer acting as a *simulator* of quantum systems. It was David Deutsch,[29] in a seminal paper of 1985, who launched quantum computation with a theoretical base of its own, generalising, or actually *replacing*, the 50-year old approach to computing produced by Alan Turing, which, as Deutsch pointed out, was related to classical physics, not quantum physics. Quantum computers would be in no way restricted to quantum simulations, but could be applied to any problem, in competition with classical computers. Unfortunately, though Deutsch was able to describe a problem where a quantum computer could do better than a classical one, the so-called Deutsch algorithm, and this itself was an important result from the theoretical point of view, the actual problem was not really important and the improvement not dramatic. (It was not till 1994 and Shor[10] that this was to change.)

It was about the same time as the work of Deutsch that the first approach to quantum cryptography was developed by Bennett and Giles Brassard,[30] though it was some years before there was any experimental progress.[31] An alternative approach to quantum cryptography due to Artur Ekert[32] was produced in 1991, while quantum teleportation[33] was invented in 1993. From the time of Deutsch's paper, progress in quantum computation itself was steady rather exciting, until the events of the mid-1990s referred to at the beginning of the section.

Einstein and Quantum Information Theory

The preceding section has presented a sketch of the history of quantum information theory, particularly of the earliest part. But the immediate question might be — what has this to do with Einstein? After all, all the events took place after his death, the great majority 30 to 40 years after. And the physical content of quantum information theory relies purely on quantum theory, certainly not on any particular interpretation of it, so, to the extent that the argument between Einstein on the one hand, and Bohr, Heisenberg or anybody else on the other hand, was a matter purely of interpretation, it could have had no direct effect on the rise of quantum information theory.

Yet there is far more to it than that. At one level, one may say that the elements of quantum theory to which Einstein drew attention in EPR, in particular entanglement and the discussion of locality, are central in nearly every aspect of quantum information theory. (When we talk here about locality, we are concerned with the fact that a measurement at one point is related immediately to an effect at a distant point. It is not relevant whether or not such a relationship contradicts Bell locality by being inexplicable by local means even if one includes realism in the form of hidden variables.) Bell[34,35] considered himself a follower of Einstein, and his most important work was based on entanglement, and specifically on the EPR problem; it was actually titled 'On the EPR Paradox'.[36] In turn Bell's ideas also fed through directly to quantum information theory. This is totally explicit in, for example, Ekert's method for quantum cryptography[32] and quantum teleportation,[33] but, as will be shown below for one particular case, they are also central in all the basic techniques of quantum computation. Also the experimental methods that were established to study the Bell inequalities could very often be adapted in a straightforward way to demonstrate quantum cryptography and teleportation as well;[9,37] examples will be given later in this chapter.

Of course it might well be argued that followers of the orthodox approach to quantum theory could well have produced quantum information theory themselves, that quantum information theory might have arisen even if Einstein and Bell had restricted themselves to general relativity and particle physics, respectively. After all, it might be said that Feynman had made good progress with quantum computation, despite the fact that, as John Clauser[38] has made totally clear, at least in the 1970s he was by no means sympathetic to any questioning of the current approach to quantum theory. By the 1980s, actually, Feynman had taken on board Bell's theorem, which played a part in his demonstration[22] that a classical computer cannot simulate a quantum system. (If it were not for Bell's theorem, hidden variables would allow the quantum system to be simulated probabilistically, the classical computer mapping onto the hidden variables.) Strangely Feynman managed to present Bell's theorem with no mention of Bell himself!

And in fact it seems unlikely that the orthodox approach could have been successful in producing quantum information theory. An interesting pointer to this is provided by comments of the great physicist John Wheeler to Jeremy Bernstein[39] in the late 1980s. Wheeler remained a convinced supporter of Bohr and did not question the exact truth of quantum theory; for Wheeler the intensely interesting question was *why* it is as it is, and, as he said, you would not be concerned with that question if you were not convinced that it is exactly as we think it is. For Wheeler, Bell's inequality is 'simply a part of ordinary quantum mechanics', and EPR is a 'non-starter'; it is 'just the way it works', and 'exactly the wrong thing to be asking about'. 'If you keep trying to pull apples off the apple tree', Wheeler remarks, 'after a while it doesn't do'.

EPR had a lot to say about completeness, realism, locality and entanglement, and this seemed to make followers of Copenhagen ultra-defensive. Bohr, for example, refused to see anything new in EPR; it presented, he said, exactly the same challenges he had disposed of in the earlier discussions with Einstein. Yet if it presented no new problems, it presumably presented no new ideas of interest, and certainly no exciting opportunities. It was Bell, a follower of Einstein, who was open-minded enough to follow on from Einstein's own arguments and show that extremely important statements could be made—Henry Stapp[40] called Bell's theorem 'the most profound discovery of science'. Followers of Bell took his work forward theoretically and studied it experimentally. Then, in due course as it might be said, these followers realised that entanglement and the ideas it led to did not have to be a worry, a problem, something you would wish away if you possibly could; rather it was a 'resource',[1] something that could be used. Thus quantum information theory was born. To answer Wheeler, there were extremely tasty apples left on the tree, but supporters of the orthodox would scarcely have had the mindset to pull them off.

What we have said so far is that, at our first level, Einstein focussed in EPR on entanglement, Bell followed this through, and down the line quantum information theory emerged. At a second level, we may speak of this as a triumph for realism. Einstein was thinking of real systems with entangled wave-functions at particular points in space, and this led him to his discussion of completeness and locality. It was, of course, exactly this aspect of what he was doing that, as we have seen, irritated the followers of Copenhagen. In the pre-complementarity position of Heisenberg, it was strictly illegitimate to think about the meaning of the wave-function at all; it was purely a tool to predict experimental results. With complementarity, this aspect of things had not really changed. Complementarity gave a formally complete description of measurement but, it might be said, was designed specifically to avoid having to talk about the ideas and concepts that Einstein wanted to talk about, the ideas and concepts that were to lead to the Bell tests, and then eventually to quantum information theory.

This then takes us to our third level. Einstein's realism was effective because it encouraged thinking about the system as a collection of real entities which would interact in a particular way; study of these entities and their interaction was highly likely to lead to interesting ideas which might well transcend the realist context that spawned them. In contrast Copenhagen, as we have seen, was rather defensive. It wished to venture in thought beyond the wave-function as little as possible, only in fact as strictly allowed by, in particular, wave-particle complementarity.

One aspect of this defensiveness, incidentally, was towards the Schrödinger equation itself. We have seen Wheeler's confidence that this must be absolutely right, and Feynman's response to Clauser mentioned above is along exactly the same lines. In contrast, both Einstein, as discussed in Chapter 7, and Bell[41] felt sure that at some stage quantum theory would be replaced by a better theory. They recognised that this was not likely to happen very soon; Bell,[41] for example, much as he might have liked the experiments of Clauser, Aspect and others to have demonstrated violation of quantum theory, recognised that this was very unlikely. But the freedom to follow ideas that might violate current quantum theory again opens up possibilities that may be closed for followers of Copenhagen, and occasionally one of these possibilities may be fruitful.

For at our third level it is precisely the conceptual fruits of realism that are important. This is the pragmatism of Chapter 8. It is the contrast, in Einstein's argument in Chapter 1, to positivism, which 'can exterminate harmful vermin' but 'cannot give birth to anything living'. The reason that realism should be supported, in Einstein's view, is not that realistic theories are attempting to correspond to some underlying reality. Rather it is because, when our approach is realistic, we are bound to invent or encounter new entities, new relationships, new concepts, new ideas, and these may contribute to building a picture of reality which leads to empirical predictions which may agree with experiment. Positivism, on the other hand, and Einstein would include complementarity in the same general category, sets its face precisely against meeting these elements of a realistic approach.

For most of the time since 1935, EPR has been regarded by most physicists as a failure. In the early days, and to a considerable extent still today, this was because it was thought that Bohr's arguments had vanquished those of Einstein. However, even among those who accept EPR as error-free and even an interesting argument, there is still often the belief that it is a failure because eventually Einstein's conception of local realism was shown by Bell to be unachievable. We may remember Gribbin's characterisation of Bell[42] as 'the man who proved Einstein wrong'; or Bell himself, having told Bernstein[39] that, over EPR, he felt that 'Einstein's intellectual superiority over Bohr was enormous, a vast gulf between the man who saw clearly what was needed, and the obscurantist', having to admit that: 'The reasonable thing just doesn't work.'

Yet from the pragmatic point of view, EPR was anything but a failure. Einstein himself used the argument to show that one could not have a combination of locality and no realism, and of course Bell took that argument over. Bell then used the same set of ideas to produce his own theorem, and eventually there was the passage to quantum information theory, again based very much on the ideas of EPR and Bell. From the perspective of giving birth to new ideas, EPR was a tremendous success.

In the remainder of this chapter, some of the basic ideas and methods of quantum information theory will be sketched, drawing particular attention to those aspects which are closely dependent on the work of Einstein, and on the work of Bell which was itself based so strongly on that of Einstein, and also on other conceptual systems radically different from that of Einstein and Bell, but also from that of Bohr, such as the many worlds views of David Deutsch.

A Sketch of the Theory of Quantum Computation

The classical theories of information and computation date from the 1930s and 1940s. Claude Shannon[43] was largely responsible for classical information theory,[44] producing an expression for the amount of information contained in a sequence of numbers, and also addressing the question of error correction. The latter is important, because any channel of information will have some noise and thus produce some errors. Searching for and correcting errors is important when working with classical information, but it is even more important with quantum information.

The classical theory of computation was largely produced by Alan Turing[45] in the 1930s. Turing[46] later became famous, of course, for the breaking of the German Enigma code at Bletchley Park in the Second World War. His model of a computer was the so-called Turing machine, a mechanical device using a tape on which there is a sequence of binary digits, 0s and 1s. The machine itself has a finite number of internal states. When the machine reads a particular character on the tape, a number of actions may be performed which depend on the character which has been read, the internal state of the system, and the rules for the particular computation; the character may be changed, the state may be changed, and the tape will move one space to the left or right. The computation is the process in which an input sequence on the tape is transformed to an output result. Turing then described the so-called universal Turing machine, which is defined as one that can simulate the action of any other Turing machine, and so compute 'anything that is computable'. (It was an important part of Turing's analysis that some problems are not computable at all.)

These abstract ideas were turned into practice by Turing himself at Bletchley Park, though the computers he made there were designed specifically to perform only one task only rather than actually being programmable. Then Turing himself and John von Neumann in 1945 shared the credit for their conception of the programmable computer.

For many years it was taken for granted that the work of Shannon and Turing had provided a complete and final theory of computation and communication respectively, and it is important to point out that, though we have dubbed these theories classical, at the time they were assumed to be general and unrestricted in their area of explanation. Indeed it is natural and traditional to imagine that the subject of the theories of information and computation is just abstract information, and that communication and computation are areas of mathematics with nothing to do with physics at all.

Of course it would have been admitted that in practice information would have been manifested in physical terms, as marks on a paper tape, fluctuations in air pressure corresponding to spoken words, or electronic signals. Computation would have been performed using some physical means or other—an abacus, a digital computer, or just the firing of neurons when we do a calculation in our head. However it was taken for granted that the precise physical manifestation was not relevant, and so the physics did not need to be discussed at all. In fact, as was probably almost inevitable, it was assumed that classical ideas sufficed, as indeed in our brief description of the Turing machine above. The only physics Turing had in his machine was paper, and he assumed he knew all about paper; however, and this was in retrospect the significant limitation of his work, his model of paper was classical. In general, as Deutsch[47] says, 'computable' was taken to mean computable by mathematicians, and this again limited discussion of computation to classical methods.

The process of removing this deep-seated fallacy was a long one. Leo Szilard may have been the first to recognise that information is physical in nature; this was in a 1929 paper on Maxwell's demon. Rolf Landauer stressed the point that 'Information is physical' from the 1960s onwards.[23,48] In Ref. [23], he wrote that 'Information is inevitably tied to a physical representation. It can be engraved on stone tablets, denoted by a spin up or down, a charge present or absent, a hole punched in a card, or many other physical phenomena. It is not just an abstract entity; it does not exist except through physical embodiment. It is, therefore, tied to the laws of physics and the parts available to us in our real physical universe.'

In 1985, Deutsch[29] applied the same argument to computation. In his popular book,[47] he wrote that 'The theory of computation has traditionally been studied almost entirely in the abstract, as a topic in pure mathematics. That is to miss the point of it. Computers are physical objects, and computations are physical processes. What computers can or cannot compute is determined by the laws of physics alone, and not by pure mathematics.' Deutsch essentially reworked Turing's argument using the laws of quantum theory rather those of classical physics, and thus was the founder of rigorously defined quantum computation. He was able to deduce the possibility of the universal quantum computer, which would be able to compute anything that any other *quantum* device could compute. As Deutsch[47] said, just as classical physics, however important it has been in practice, is today only significant as an approximation to quantum theory, so Turing's theory of computation, again unchallenged for so long, is now obsolete, again except as an approximation. The theory of computation, from a fundamental point of view, *is* the theory of quantum computation.

Quantum computers may be designed, and, at least in principle, constructed, but an obvious question is—will they be of any use? This is to ask whether they will be able to perform tasks impossible or effectively impossible with a classical computer. We may answer this question in three stages. In the first, we note that the most basic property of quantum theory, what Landauer[23] calls 'quantum parallelism' suggests that the quantum computer is, in some sense, achieving very much more than a classical computer would be able to do.

To explain this, we will explain a few fundamental features of the quantum computer. While the basic unit of information for the classical computer is the bit, which may take either of two values, 0 or 1, the analogous quantity for the quantum computer has been defined by Schumacher[49] as the qubit. Like the bit, the qubit has two basic states, $|0\rangle$ and $|1\rangle$, which may, for example be the two states of a spin-$\frac{1}{2}$ system, or the two directions, horizontal and vertical, of the polarisation of a photon, or the two lowest energy states of an atom. The qubit is very much richer in possibilities than the bit, because it is not restricted to being *either* in state $|0\rangle$ *or* in state $|1\rangle$; rather it may be any state $c_0|0\rangle + c_1|1\rangle$, with the only condition on c_0 and c_1 being that $|c_0|^2 + |c_1|^2 = 1$.

We now define an n-state quantum register, which may contain n qubits. For example, if n is 6, the register might contain $|011010\rangle$, which could be regarded as representing the number 26 in binary form. Clearly the register may represent 2^6, or in general 2^n, different numbers in this way. If we were to perform a classical computation, we might wish to start with $|00\ldots00\rangle$ and do our computation, then start with $|00\ldots01\rangle$ and perform the same computation, and so on. Clearly we would have to perform 2^n independent computations. The power of quantum computation, though, as will shortly emerge, one should really say its latent power, is that all the 2^n computations may be performed simultaneously. For, if a general state of the register, such as $|00\ldots00\rangle$ is written as $|x\rangle$, then one may use as input $\sum_x c_x|x\rangle$. If the output for $|x\rangle$ is $|f(x)\rangle$, then, to ensure reversibility we would retain input function as well as output function, and our final state of the system will be $\sum_x c_x|x\rangle|f(x)\rangle$. Provided none of the c_x's are zero, it seems that the computer has performed all 2^n computations simultaneously, and this is the meaning of quantum parallelism.

So the first stage of our answer—will quantum computers be of any use?—is resoundingly positive. However immediately we must move to the second stage, and here we unfortunately obtain completely the opposite answer. For it is one thing to say that the quantum computer is somehow performing all 2^n computations, and it is an entirely different one to say that we may take advantage of what the computer has achieved. For at the end of the computation we must obtain an answer, or in quantum mechanical terms we must take a measurement on the computer output. When we do this, the Born rule tells us that we will obtain just one of the results

$|x\rangle |f(x)\rangle$, and the probability of obtaining this will be $|c_x|^2$. Again provided none of the c_x's are zero, we may obtain any term in the summation as our answer.

So the second stage of the answer is that, frustratingly enough, it seems highly unlikely that a quantum computer would be any use at all. Broadly this view was taken by virtually all those interested in the topic from the time of Deutsch's paper in 1985 until Shor produced his algorithm almost 10 years later. This is why Shor's discovery, a clear demonstration that quantum computation could actually be exceptionally useful, came as such a shock.

How does one move beyond the second stage to demonstrate the power of quantum computation in solving particular problems? Speaking a little simplistically, one might say that we do not perform a measurement that crudely obtains one answer, or one might say *all* of one answer, from a single component of the state-vector. Instead one arranges to obtain a single piece of information, but a piece of information that samples the various answers from all components of the state-vector, and is itself of considerable interest. Finding such an algorithm is naturally extremely difficult! The kind of idea that is involved may be illustrated by the classical Fourier transform. Given a general waveform, it would clearly be of little interest to sample it at a single point; rather the Fourier transform enables one to obtain useful information by use of data from all points, and finding, for example, the amplitude of the waveform at a particular frequency. Fourier transform and mathematically similar operations are actually involved in many of the most important algorithms of quantum computation.

Actually Deutsch made a start on finding useful techniques in his 1985 paper. In the so-called Deutsch algorithm, he produced a method whereby a quantum computer could obtain information in a single run which would require two runs of a classical computer. Unfortunately the algorithm would only be successful on one of every two occasions it was used, success or failure occurring randomly. In addition, the problem the Deutsch algorithm solved was really not very important. The very existence of the Deutsch algorithm was exceptionally important from the conceptual point of view; it did serve to demonstrate that quantum computers could out-perform classical ones in specific ways. However the very limitations of this out-performance perhaps had a negative effect in terms of attracting the interest of other physicists and computer scientists. It rather seemed to suggest that whatever gains there might be in using a quantum computer could well be paid for by equal and opposite losses in other aspects of the computation.

In the following section we shall explain the Deutsch algorithm in some detail, not, as we have said, because it is very useful in practice, but because of its conceptual importance, the fact that it is simple enough to be discussed fairly briefly, and because the type of methods used are of fairly general application in quantum computation. In contrast we shall say comparatively little about the details of the Shor algorithm,[10] and also of the Grover algorithm,[50,51] another important algorithm from the same period. This is because their structure is much more complicated than that of the Deutsch algorithm, though many aspects of all these algorithms are rather similar.

Here we move on from the qubit and the quantum register to study the other main component of the quantum computer, the quantum logic gates. These perform operations on the qubits in analogy with those that the gates in a classical computer perform on the bits. The quantum logic gates are then assembled into quantum networks that perform the computation. The quantum logic gates must be reversible as the Schrödinger equation itself is reversible. Just as the qubit has more diverse behaviour than the bit, there are many more possibilities for quantum logic gates than for classical reversible gates.

For both classical reversible gates and quantum gates, it is of great interest to establish sets of universal gates, sets of gates that may carry out any operation of the particular type. For classical reversible gates, a three-bit gate is required, and alternative choices were designed by Feynman and his collaborators in the work on classical computers discussed earlier in this chapter. For the quantum computer, it is very fortunate that there is no requirement for a three-qubit gate; exact universality may be achieved with a particular two-qubit gate, the CNOT gate, which we shall describe shortly, and an array of *all* single-qubit gates. We say it is fortunate, because of the extreme sensitivity of quantum computation to quantum decoherence due to interaction between a particular qubit and the environment. This makes the production of even a two-qubit gate a matter of very great difficulty; it is far from easy to allow two qubits to interact, but to prevent either of them from interacting more widely giving decoherence. It is difficult to imagine success in going even further and constructing a three-qubit gate, if it had happened that that *was* required for universality.

We have said that for exact universality, *all* single-qubit gates are required, but in practice one may get universality to as good an accuracy as one might wish with just three single-qubit gates as well as the CNOT gate. These single-qubit gates are the $\pi/8$ gate, the phase gate and the Hadamard gate, and we now explain the actions of the various gates required. The $\pi/8$ gate performs the following:

$$|0\rangle \rightarrow |0\rangle ; \quad |1\rangle \rightarrow \frac{1+i}{\sqrt{2}} |1\rangle$$

while the phase gate achieves:

$$|0\rangle \rightarrow |0\rangle ; \quad |1\rangle \rightarrow i|1\rangle$$

Note that these gates change the relative phase of $|0\rangle$ and $|1\rangle$, but do not provide superpositions of the two. This latter task is performed by the Hadamard gate which performs the following:

$$|0\rangle \rightarrow \frac{1}{\sqrt{2}} \{|0\rangle + |1\rangle\} ; \quad |1\rangle \rightarrow \frac{1}{\sqrt{2}} \{|0\rangle - |1\rangle\}$$

The Hadamard gate plays an essential part in what was described above as moving from a consideration of a single value or result to a combination of all values or results. (Above, of course, we have only 2, but we will often apply a Hadamard gate to each qubit, thus combining 2^n entities in a linear combination.) Thus we may describe the Hadamard gate as unlocking the power of parallel computing, or being the gateway to the kind of generalised Fourier methods outlined above. This gate, or more powerful gates performing the same type of function, are central in all applications of quantum computation.

Also absolutely central in all quantum computational algorithms is the controlled-NOT or CNOT gate. This is a double-qubit gate, and its action may be described as follows. The second qubit swaps between $|0\rangle$ and $|1\rangle$ if, but only if, the first control qubit is in state $|1\rangle$. In all cases the control qubit itself remains unchanged. This may be spelled out as:

$$|00\rangle \rightarrow |00\rangle \; ; \quad |01\rangle \rightarrow |01\rangle \; ; \quad |10\rangle \rightarrow |11\rangle \; ; \quad |11\rangle \rightarrow |10\rangle$$

It is easy to see that the CNOT gate is reversible, as it must be to be a quantum logic gate. For the number of inputs equals the number of outputs, each input leads to a unique input, and each output is a result of a unique input. In fact it is easy to see that the gate is its own inverse; if the output is applied to a second CNOT gate, we will return to the original input.

Another crucial point is that the CNOT gate may entangle a non-entangled input state, or disentangle an entangled input state. Let us consider, for example, an input state for which the first qubit is in state $(1/\sqrt{2})(|0\rangle + |1\rangle)$, and the second qubit in state $|0\rangle$. Clearly the state is non-entangled, as the states of each qubit have been stated independently. As a combined state, the input is $(1/\sqrt{2})(|00\rangle + |10\rangle)$. If we apply the CNOT gate to this state, the linearity of the Schrödinger equation assures us that we should handle each term separately, and the output state is $(1/\sqrt{2})(|00\rangle + |11\rangle)$.

But this state *is entangled*, since the states of each qubit *cannot* be stated independently. The state-vector tells us that, if the first qubit is in state $|0\rangle$, then so is the second qubit; but if the first qubit is in state $|1\rangle$, again so is the second qubit. The results of measurements will obviously be correlated. Since the CNOT gate is its own inverse, obviously it can also remove entanglement. Entanglement is crucial in nearly all aspects, not only of quantum computation, but all branches of quantum information theory. Where quantum information theory is more powerful than classical methods, it is because it uses typically quantum types of phenomena. Obviously crucial are the most typically quantum aspects of physics, the ability to form superpositions and to demonstrate interference. Very nearly as important is entanglement. We have seen that the Hadamard and CNOT gates exploit these modes of behaviour, superpositions and entanglement respectively, and we now proceed to see them in operation in the Deutsch algorithm.

The Deutsch Algorithm

The Deutsch algorithm was important because it was a violation of the Turing position, until that point accepted implicitly, that no computer could perform better than the best classical computer. It is also extremely interesting in its own right since it demonstrated many of the techniques that would be central in any application of quantum computation.

Deutsch's algorithm considered evaluation of a function of x, $f(x)$. Each of input x and output $f(x)$ are limited to taking the value 0 or 1. Clearly there are 4 possibilities:- (i) $f(0)$ and $f(1)$ are both 0; (ii) they are both 1; (iii) $f(0)$ is 0 and $f(1)$ is 1; or (iv) $f(0)$ is 1 and $f(1)$ is 0. The question the algorithm has to answer is—are $f(0)$ and $f(1)$ the same or not? To answer this classically would require two runs with the computer. One would first compute $f(0)$, then $f(1)$, and the answer to the question is then obvious. What Deutsch was able to show was that a quantum computer may solve the problem with a single run, although only half the runs will be successful.

We may note that the information obtained is precisely of the kind we discussed above. The computer could compute a superposition of $f(0)$ and $f(1)$, but a simple measurement at that point would give one or the other; essentially the quantum computer would be acting with the same power as a classical computer. Much more subtle is to use both values of f to produce a single parameter of some interest, and this is what the Deutsch algorithm achieved.

Deutsch tried to generate interest in his algorithm by inventing a scenario in which it solved a real problem. One has 24 hours to decide whether to make an investment in the stock market, and will make the investment only if two indicators, $f(0)$ and $f(1)$ are the same. However calculation of f is a lengthy process, taking, in fact, 24 hours. Classically it will take 48 hours to discover whether the two values of f are the same, and investment will not be possible. However with a quantum computer one run only is required, and there is just time to make the investment decision. The motivation may be weak, but the problem is interesting, and the solution clever and instructive.

In the circuit we will call the central processing unit U. It has two inputs and, since it must be reversible, it must also have two outputs. U may be called an f-controlled NOT gate. If $|x\rangle$ is sent into the circuit as qubit 1, where x is 0 or 1, then the first task of U is to calculate $f(x)$, which must also equal 0 or 1; this is the process that, at least in Deutsch's scenario, may take a long time. The input for qubit 2 is $|0\rangle$, and the second task of U is to change this to $|1\rangle$ if and only if $f(x)$ is equal to 1. Thus f controls the gate; qubit 1 does not change its value. We may first examine what happens if we send in 0 or 1 for x. If $x = 0$, then, if $f(0) = 0$, then qubit 2 remains as $|0\rangle$, but if $f(0) = 1$, then qubit 2 changes to $|1\rangle$. In either case, the final state of the combined system is $|0, f(0)\rangle$. In exactly the same fashion, if we send in $x = 1$, the final state of the combined system is $|1, f(1)\rangle$. We may say that the effect of U is to copy the value of $f(x)$ onto qubit 2.

However what is actually sent as qubit 1 onto U is neither $|0\rangle$ nor $|1\rangle$, but in fact $(1/\sqrt{2})(|0\rangle + |1\rangle)$, which is obtained by sending $|0\rangle$ as input to a Hadamard gate, and then sending the output of this gate on to U. Since an input of $|0\rangle$ to U yields an output of $|0, f(0)\rangle$, and an input of $|1\rangle$ yields $|1, f(1)\rangle$, the linearity of quantum theory tells us that an input of $(1/\sqrt{2})(|0\rangle + |1\rangle)$ must yield $(1/\sqrt{2})(|0, f(0)\rangle + |1, f(1)\rangle)$. The nature of this state depends fundamentally on whether $f(0) = f(1)$. If they are equal, the state is non-entangled. If they are both 0, as in case (i) above, then qubit 1 is in state $(1/\sqrt{2})(|0\rangle + |1\rangle)$, qubit 2 in state $|0\rangle$. For comparison with the entangled states, we may write the combined state as $(1/\sqrt{2})(|00\rangle + |10\rangle)$. Case (ii), where both values of f are 1, is analogous, and the combined state may be written as $(1/\sqrt{2})(|01\rangle + |11\rangle)$.

However if $f(0)$ is not equal to $f(1)$, the state of the combined system is entangled. For case (iii) above, where $f(0)$ is 0 and $f(1)$ is 1, the state is $(1/\sqrt{2})(|00\rangle + |11\rangle)$, clearly entangled because if qubit 1 were to be found in state $|0\rangle$, so also would qubit 2, but if qubit 1 were to be found in state $|1\rangle$, again so would qubit 2. Case (iv) is analogous; the state of the combined system is $(1/\sqrt{2})(|01\rangle + |10\rangle)$, again obviously entangled. In the above we see the power of the Hadamard gate to produce superpositions of the basic states, and that of a controlled gate to produce entanglement. In the next state we use further Hadamard gates, one for each qubit, but we will speak of their purpose, not in this case to form superpositions, but to cause different paths through the circuit to interfere.

First qubit 2 is sent through a Hadamard gate. For case (i), $|0\rangle$ becomes $(1/\sqrt{2})(|0\rangle + |1\rangle)$, and so each qubit is in this state. The system, of course, remains non-entangled. For comparison with the later tangled states, we may write the combined state as $(1/\sqrt{2})(|00\rangle + |01\rangle + |10\rangle + |11\rangle)$. Case (ii) is analogous; qubit 2 ends up in state $(1/\sqrt{2})(|0\rangle - |1\rangle)$, the system is non-entangled, and the state of the combined system may be written as $(1/\sqrt{2})(|00\rangle - |01\rangle + |10\rangle - |11\rangle)$. For the other two cases, the system remains entangled. For case (iii), the final state of the combined system is $(1/\sqrt{2})(|00\rangle + |01\rangle + |10\rangle - |11\rangle)$, while for case (iv) it is $(1/\sqrt{2})(|00\rangle - |01\rangle + |10\rangle + |11\rangle)$.

We now measure the state of qubit 2. Since in each case, for two of the four terms in the state-vector it is $|0\rangle$ and for the other two it is $|1\rangle$, there is an equal probability of obtaining either answer. What is interesting, though, is how the measurement affects the state of qubit 1. For the non-entangled states, it is obvious that the measurement on qubit 2 can have no effect on qubit 1. Whatever result we obtain for qubit 2, qubit 1 will be left in the state $(1/\sqrt{2})(|0\rangle + |1\rangle)$.

We now turn to the entangled cases, (iii) and (iv). For these cases, the final state of qubit 1 will in general depend on the result obtained in the measurement on qubit 2. Suppose we measure qubit 2 to be in state $|0\rangle$; by the von Neumann projection postulate, after the measurement we should retain only those terms in the combined state-vector for which qubit 2 is in state $|0\rangle$. For both cases (iii) and (iv), these are $|00\rangle$ and $|10\rangle$, so the state- vector of qubit 1 after the measurement is $(1/\sqrt{2})(|0\rangle + |1\rangle)$. (In the state-vector for the combined system, of the overall normalisation factor of 1/2, obviously $(1/\sqrt{2})$ comes from the state-vector for each qubit.)

However there is an equal probability that qubit 2 is found to be in state $|1\rangle$. We then need to take terms from the state-vector for which qubit 2 is in state $|1\rangle$. For case (iii), these are $|01\rangle - |11\rangle$, so the state of qubit 1 is $(1/\sqrt{2})(|0\rangle - |1\rangle)$, while for case (iv), they are $- |01\rangle + |11\rangle$, so the state of qubit 1 is $(1/\sqrt{2})(- |0\rangle + |1\rangle)$.

The two states as written differ by a minus sign, but a minus sign multiplying the whole state-vector of a system is merely a change of phase and is of no physical significance.

So we may sum up the results as follows. If the measurement of qubit 2 gives the result $|0\rangle$, which occurs with probability 1/2, in all cases qubit 1 will be left in the state $(1/\sqrt{2})(|0\rangle + |1\rangle)$. Since all cases are identical, we cannot distinguish between them, and the method fails. However, if the measurement of qubit 2 gives the result $|1\rangle$, which also occurs with probability 1/2, then qubit 1 will be left in state $(1/\sqrt{2})(|0\rangle + |1\rangle)$ for cases (i) and (ii), where $f(0)$ and $f(1)$ were the same, but in state $(1/\sqrt{2})(|0\rangle - |1\rangle)$ for cases (iii) and (iv), where $f(0)$ and $f(1)$ were different. For the latter case we may distinguish between the two possibilities by passing qubit 1 through another Hadamard gate. Cases (i) and (ii) will give a final result of $|0\rangle$, while cases (iii) and (iv) give $|1\rangle$. In this 50% of cases we have achieved exactly what was required.

Much later in 1998, Artur Ekert and his group[52] were able to improve Deutsch's original algorithm so that it gave a result on every occasion, rather than only with a probability of 1/2. And some years before that, Deutsch, together with Richard Josza,[53] had extended the original algorithm to the case where there are n inputs, so the number of possible input functions was 2^n. It is required that the 2^n values of f are either all the same, or balanced in the sense that f takes each of the values 0 and 1 on the same number of occasions. To solve the problem classically, all 2^n values must be computed, but with the Deutsch–Jozsa algorithm only a single computation is required. Thus the gain on moving from classical to quantum computation is now exponential, and this is exceedingly important from the theoretical point of view, though of course the problem solved by the Deutsch–Jozsa algorithm is no more realistic than that of the original Deutsch algorithm.

Other Topics in Quantum Computation

The same exponential gain as in the Deutsch–Jozsa algorithm was present in the Shor algorithm of 1994,[10] the importance of which has already been discussed. The Shor algorithm presented a way of factorising large numbers, which is effectively impossible on a classical computer because the amount of time required to factorise a number with n digits increases exponentially with n. For the Shor algorithm, the amount of time increases polynomially with n, in fact approximately as n^3, clearly a vast improvement. In fact the Shor algorithm represents a change to a polynomial problem of class P, from an exponential problem of class NP, nondeterministic polynomial, or polynomial only on a computer that happens to make successful choices at each branching-point. This is effectively the change to computationally tractable from effectively and for large enough numbers computationally impossible. It was this impossibility that was at the heart of the RSA algorithm for cryptography of Ronald Rivest, Adi Shamir and Len Adleman, known as public-key cryptography, and it is this algorithm that is threatened by the possible coming of the quantum computer.

In a full account of quantum information theory it would be good to give a full account of the Shor algorithm, and also of the important Grover algorithm,[50,51] which allows a desired item, such as the name associated with a particular number in a telephone directory containing N entries to be found in a number of trials around $\sqrt{N}$ rather than around N. We would also have liked to present an account of quantum error correction; quantum decoherence is certain to produce a significant number of errors in a quantum computation, and it is only with the use of quantum error correction that there is any possibility of making a useful quantum computer.

Lastly a full account of quantum computation would include at least some information on the practical methods by which quantum computers are being designed, and the large number of types of physical systems used. However for these topics we must make references to other books and articles.[1–9] Our task, having surveyed a little of the methods of the subject, is to return and examine the type of approach to quantum theory which might have brought it about. We will just mention, though, that although any description of Shor's algorithm is somewhat complicated, it has been shown that its basic structure may be written down in such a way as to demonstrate an analogy with that of the Deutsch–Jozsa algorithm.

Einstein and Quantum Computation

Earlier in this chapter, we discussed what approach to the foundational problems of quantum theory was most likely to lead to quantum information theory, and here we return to the question specifically for quantum computation. The first point made earlier concerned the central aspect of EPR, entanglement. While this did not play a role in the formal statement of quantum computation, involving the general Deutsch argument, the qubit, quantum logic gates and so on, one of our purposes in describing the Deutsch algorithm in some detail was to demonstrate how central entanglement was to the process, and, as was said, the same is true in particular for the Shor algorithm. It might be added that in any implementation of quantum computation, quantum teleportation, discussed mainly in Chapter 12 must be important for moving information from one point to another, and quantum teleportation is very much based on entanglement. Certainly entanglement must play a crucial role in any quantum computational procedure.

However we should also ask whether the actual idea of quantum computation by Deutsch owed anything to his approach to quantum theory. Though Hugh Everett was the founder, and Bryce de Witt the main initial publicist of the many worlds group of interpretations of quantum theory, Deutsch[47] is its leading advocate at the moment. He is convinced that it is actually not an interpretation, but the result merely of taking the formalism seriously, and not adding any what he would consider to be arbitrary and unnecessary elements or conventions. This view is extremely fruitful in explaining the power of quantum computation. To the question of where the seemingly extra computing power has come from, Deutsch gives an unequivocal answer. It is provided in the many worlds of his conception of quantum theory. Shor's algorithm, he[47] points out, appears to use 10^{500} times the computational resources of our universe; it must, he argues, be using the resources of that number of universes. It should be mentioned that Andrew Steane[55] disagrees with Deutsch on this point; he does not agree that quantum computation does actually require the resources of more than one universe.

We may in any case take it for granted that Deutsch was highly satisfied that the subject matter of quantum computation, in particular the parallelism, was a perfect demonstration of his own conceptual approach to quantum theory. As has already been said, he himself actually required no proof, since he felt the formalism spoke for itself, but he did think that quantum computation and quantum cosmology were the two areas of quantum theory that should persuade, and to some extent had persuaded, others to agree with him. But of course that does not necessarily imply that his beliefs had necessarily helped him to appreciate the limitations of the Turing approach and to invent a replacement.

Deutsch's approach must be termed realistic, at least in so far as it presents the clearest of pictures of how to interpret a wave-function which is a linear combination of several terms. Deutsch, incidentally, was totally unambiguous in his belief that the many universes were to be regarded as, in every way, real. Here he was totally in concordance with de Witt. There have been suggestions that, for Everett, the universes may have had more of a metaphorical or mathematical significance,[9] though Deutsch himself would have none of this; he believed that Everett's views were totally consistent with those of de Witt and himself. Of course it might be argued that the realism of the Deutsch picture does not extend down to the question of the effect of making a general measurement in an individual universe. This relates to the problem of how one chooses the basis for universes, in other words what characterises each universe. For instance each universe might have a particular value of position for a particle, or alternatively a particular value of momentum. This has been a long-running criticism of many universes theories, and we do not discuss its possible solutions[9] further here.

Here, though, we note that it seems highly reasonable that, when Deutsch spotted the lacuna in Turing's proof, his view of quantum theory gave him a very clear picture of the dramatic possibility of quantum parallelism if superpositions were allowed. It is also at least possible to suggest that a follower of Copenhagen, beset, it might be felt, by the legal niceties of what is allowed in that interpretation, might have missed the significance of the same observation. This argument is admittedly speculative, but the fact is that it was Deutsch who did present the theory of quantum computation. While Benioff and even Feynman restricted themselves to discussing relationships between classical and quantum computation, it was Deutsch who realised clearly the vast potential of quantum computation as compared with classical, and it is not unreasonable to wonder whether his conceptual background helped him to do this.

We may ask, though, whether this has anything to do with Einstein. After all, it does not seem likely that Einstein would have been very enthusiastic about any type of many worlds scenario. To the extent that we describe Deutsch's ideas as realistic, they were very far from Einstein's view of realism. However, when we think of the pragmatic aspect of Einstein's approach, we may feel that Einstein might not have approved of the content of Deutsch's interpretation, but might still have valued its ingenuity and its definiteness of conception. He would not have thought it at all inconsistent to appreciate that the interpretation had helped to produce exciting developments, while remaining sceptical about the interpretation itself.

It is often said that the enforcement of complementarity in the 1920s and 1930s, whatever its logical status, was pragmatically successful inasmuch as it enabled physicists to consider the conceptual questions closed, and to concentrate on exploring the application of quantum theory to the physics of the atom, the nucleus and the solid state. In the short term there may have been truth in this belief, but in the longer term it was anything but pragmatic. Genuine concerns about such matters as determinism, realism, locality and entanglement were blocked; challenging questions were ruled out of order. The type of development achieved by Einstein, Bohm and Bell, which was taken up by the originators of quantum information theory, had to be performed over the heads of the virtually uncontested formers of scientific opinion.

In contrast, the approach of Einstein did stimulate open-mindedness and the search for new ideas. While it would be an exaggeration to say that all the novel ideas of the 1950s and 1960s stemmed from Einstein, it would be true that his general attitude was favourable to the creation of approaches distinct from Copenhagen, albeit he was unresponsive, even opposed, to some of these, in particular to the work of Bohm. Eventually, as he would have hoped, these approaches were to bear fruit.

Quantum Cryptography

The other two branches of quantum information theory that we consider in this book are quantum teleportation and quantum cryptography. Quantum teleportation has been discussed in Chapter 9. Here we make the point that its main component is EPR pairs, used of course in a highly ingenious way, and thus we may say descends specifically from Einstein. In addition, the various experiments performed on quantum teleportation, which were performed in the groups of Zeilinger,[56] Francesco De Martini[57] and Jeff Kimble,[58] were quite closely connected with experiments on local realism. For instance the experiment by Dirk Bouwmeester and colleagues in Vienna[56] was closely connected with the related experiment[59] which demonstrated the existence of the GHZ (Greenberger–Horne–Zeilinger) state, as discussed in Chapter 9. This again emphasises the close connection between experiments in quantum information theory and those related to Bell's work, itself, of course, intimately related to the work of Einstein.

Now we turn to quantum cryptography. Earlier in this chapter we mentioned how the possibility of quantum computation is threatening *public-key cryptography*. An alternative scheme, *private-key cryptography*, or use of the 'one-time-pad' is an alternative totally safe scheme, although one with a large practical difficulty.

In this scheme, Alice and Bob, the universal communicators of quantum information theory, are each equipped with an identical pad of codes or keys. Alice encrypts her message using a key from the top of the pad, and sends it to Bob, who decrypts it with the same key, and thus retrieves the message. Should Eve, the eavesdropper, get hold of the encrypted message, it will be of no use to her because she does not have the key.

The practical difficulty comes from the fact that it is essential that each key is used only once. If Eve were to come into possession of two messages encrypted with the same key, she would be able to make sophisticated guesses at the nature of the key, and hence may be able to decrypt both messages. It must indeed be a 'one-time-pad'! Thus one must arrange the safe dispatch to Alice and Bob of large amounts of material, a seemingly mundane problem, but in practice one requiring a great deal of effort to solve. Quantum cryptography is, in fact, merely a straightforward and totally safe way of distributing keys to Alice and Bob; it is often called just—quantum key distribution.[60,61]

The general ideas of quantum cryptography were introduced by Charles Bennett and Giles Brassard and collaborators[62] in the early 1980s, and the best-known scheme is the so-called BB84 protocol introduced by Bennett and Brassard[30] in 1984. It is, incidentally, one of the few schemes in quantum information theory where *no* use is made of EPR pairs or the ideas of Bell. In this method, Alice and Bob are attempting to build up a shared set of private keys. They communicate using a quantum channel such as an optical fibre, but the method also requires use of a public channel, such as a telephone connection. The necessity for the latter means of communication makes it clear that, just as for quantum teleportation, there is no question of any effect operating faster than light.

Alice and Bob communicate using a stream of polarised photons. Alice has four polarisers available, polarising photons in the horizontal, vertical, +45° and −45° directions, while Bob has two analysers, one that distinguishes between horizontally and vertically polarised photons, and one that distinguishes between photons polarised at +45° and −45°. Alice sends photons to Bob, in each case using a polariser at random and recording her choice. Bob detects each photon using a detector chosen at random, in each case recording his choice of detector and the result obtained.

In 50% of cases there will have been matches between Alice's choice of polariser and Bob's choice of analyser, but in the other 50% there will have been mismatches. In the latter case, Alice may, for example, have sent a photon polarised in the +45° direction, but Bob may have used the analyser distinguishing between vertically and horizontally polarised photons. He thus has a 50% chance of obtaining either answer, and his result must be meaningless.

To distinguish between matches and mismatches, Alice and Bob use the public channel. Bob sends Alice a list of the detectors he used for each photon, and she replies with a list of the numbers of the cases where polariser and analyser were compatible. Alice and Bob now possess shared information in the form of Alice's polariser settings and Bob's results for the cases where there was a match. This is called the *raw quantum transmission* or RQT, and may be used to generate a shared key, at least provided there has been no eavesdropping. It is an essential aspect of quantum cryptography that the presence of Eve may be detected; any sections of the RQT that she may have discovered are then discarded. To achieve this, Alice and Bob now share a portion of the RQT using the public channel; this portion, of course, will have to be abandoned whatever the result of their search for Eve.

To understand what they do, let us discuss how Eve will deal with any signal she intercepts on the way to Bob. We cover only the simplest strategy she might use. She will intercept and analyse a photon with her own analyser and will set the direction of this at random. We need to consider only cases where Alice's and Bob's choices were compatible, so the event has entered the RQT. In 50% of cases, Eve's choice of analyser is in agreement with the choices of Alice and Bob. Her measurement result will agree with Alice's record. She will send on a photon corresponding to this result, which Bob will analyse in the usual way. He will obtain a result agreeing with that of Alice, so their versions of the RQT are in agreement. The bad point is that Alice also knows this piece of information in the RQT.

But in 50% of cases, Eve's choice of analyser will be incompatible with that of Alice and Bob. Alice may, for example, have used the vertical polariser, but Eve may have used the polariser that distinguishes between the $+45^0$ and $-45°$ directions. Eve's result will thus be random, and of course she will send on to Bob a photon corresponding to whichever result she obtains, $+45\%$ or -45%. Since Bob's analyser distinguishes between vertically and horizontally polarised photons, his result again will be random, and so there is a 50% probability that his measurement indicates a vertical polarisation, which is what Alice actually sent, but a 50% probability of it indicating incorrectly a horizontal polarisation.

Overall, in 50% of cases, Eve makes the right choice of analyser and so remains undetected, and in 25% she makes the wrong choice but, by chance, still remains undetected. However in the remaining 25% of cases she makes the wrong choice of analyser and the result is that Bob's result is not what Alice sent. When Alice and Bob compare a section of the RQT, therefore, they will find mismatches in roughly 25% of the events listed; the whole of the RQT must be jettisoned.

The above is the simplest case of the theory. In practice, there are more sophisticated strategies for eavesdropping[63] and the channels will have some noise. If one neglects noise, then any strategy will leave some detectable trace, while, even for noisy channels, Alice and Bob will be able to assess how much information Eve may have obtained. They may apply quantum error correction, as just mentioned at the end of the material in this chapter on quantum computation, and then *privacy amplification*, a series of operations that cut down whatever information Eve may glean to tiny proportions, at the expense of considerable reduction in the size of the usable key. The process is known as *key distillation*.[64]

It is worth pointing out that any possible success of quantum cryptography depends on a typically quantum restriction in the form of the so-called *quantum no-cloning theorem*. If Eve was able to clone, or reproduce the state of the photon that she receives from Alice, she could send one on to Bob, and analyse the other, thus rendering her presence undetectable. More than that, she could clone a large number of copies, and perform a measurement with each analyser on half of her supply of clones. With one analyser, she would always get the same result, corresponding to the state that Alice had actually sent, while with the other, she would obtain each answer an approximately equal number of times. Cloning is quite possible classically, so any protocol similar to BB84 would be quite impossible in classical information theory.

The quantum no-cloning theorem has been mentioned in Chapter 4. Here we give a simple proof due to Bill Wootters and Wojciech Zurek,[65] and Dennis Dieks,[66] which tells us that it is not possible to find a unitary operator U that can clone an arbitrary quantum state. The proof is simple. If we imagine such a U, it must cause the changes $U|\alpha, 0\rangle \rightarrow |\alpha, \alpha\rangle$ and also $U|\beta, 0\rangle \rightarrow |\beta, \beta\rangle$, where $|\alpha\rangle$ and $|\beta\rangle$ are any two states of any system, and the state of the second qubit, initially $|0\rangle$, becomes $|\alpha\rangle$ or $|\beta\rangle$, a clone of the first.

Now imagine the same operator applied to the second qubit in state $|0\rangle$ as before, but the first in state $1/\sqrt{2}(|\alpha\rangle + |\beta\rangle)$. The laws of quantum theory tell us that we can add together the two previous equations, multiply by $1/\sqrt{2}$, and obtain for the resulting combined state $1/\sqrt{2}(|\alpha, \alpha\rangle + |\beta, \beta\rangle)$. But this is not a cloning of the original state. A cloning would require both qubits to end up in $1/\sqrt{2}(|\alpha\rangle + |\beta\rangle)$; the combined state would be $1/2(|\alpha, \alpha\rangle + |\beta, \alpha\rangle + |\alpha, \beta\rangle + |\beta, \beta\rangle)$. This shows us that imagining a unitary operator such as U leads to inconsistencies, or in other words there can be no cloning operator.

As we have said, this theorem is an essential element of quantum cryptography. It also plays a generally more negative role in other aspects of quantum information theory, where, for example, it prevents one from using the classical technique in which one performs an element of a computation a number of times, in the hope that, though any single computation may contain an error, the majority will give a clear indication of the correct answer.

We have mentioned that the BB84 protocol does not make use of EPR pairs or the ideas of Bell. However in 1991 Artur Ekert[32] invented an alternative technique for quantum cryptography, based *totally* on EPR pairs and Bell's theorem. We will only sketch the operation of this technique, which uses a stream of EPR pairs; one member of each pair is sent to Alice and the other to Bob. They each measure the spin of their photon along the x or y axis, making their choices randomly and independently. By communication using a public channel, they discover the cases where they have made the same choice, and the results for these cases constitute their shared key (similarly to BB94). The results for the measurements where their chosen directions did not coincide may be used to test Bell's theorem. If the theorem is violated, one assumes that there has been no eavesdropper. However if the theorem is upheld, and neglecting the possibility that hidden variables have actually been found, one must conclude that Eve has been at work. (In a sense, Eve is producing hidden variables by collapsing superposition to mixture and removing entanglement.)

We now turn to implementation of the BB84 protocol. The first experiments were carried out by Bennett and Brassard[31] themselves in 1989. Distances were small and error rates high, but nevertheless the experiments were extremely important theoretically; Deutsch has been quoted[3] as saying that this was the first device produced of any type whose capabilities exceeded those of a Turing machine.

Further development has mostly been performed by the groups of Nicolas Gisin, Richard Hughes[67] and Paul Townsend.[68] We describe briefly the work of Gisin's group, because this was closely related to the experiments the same group had performed checking Bell's inequalities. In the latter work, they[69] studied Bell's inequalities for entangled photons separated by more than 10 km. The source for the photons was at a telecommunications station in Geneva, and the photons were sent along installed standard telecom fibre under Lake Geneva. Similarly in the quantum cryptography work,[70] the quantum key was shared by users at opposite ends of the cable under the lake. Put together with the work of Hughes' group demonstrating success in quantum cryptography in free space, it is clear that the time is right for implementation of the scheme in providing security in the financial district of a big city, or carrying communications between earth and a satellite.

We conclude this section by remarking that it is obvious how the work of EPR and Bell led to quantum teleportation and quantum cryptography. Much, though not all, of the schemes involved are structured around the use of EPR pairs and Bell states, but, more than that, the whole area was enlightened by those who were determined to study the quantum measurement in an open-minded and creative way, rather than trying to deny difficulties or to avoid them or to explain them away. Indeed they considered quantum measurement as possessing not so much awkward features but interesting and possibly useful properties. It goes without saying that such attitudes came in general terms from Einstein, through Bell, and would have been opposed by Bohr. In Chapter 1 it was mentioned that Einstein had long before written to Besso that positivism could produce nothing living, or in other words, no fruit. Similarly he talked of the 'sterile positivism' of Copenhagen. In contrast, the whole story of quantum information theory should convince us that Einstein's approach had the potential of bringing forth significant fruit.

Note added in proof

A recent article by Andrew Shields and Zhiliang Yuan describes current applications of quantum cryptography to business. Shields, A. and Yuan. Z. (2007), Key to the quantum industry, Physics World **20**(3), 24–9.

References

1. Steane, A.M. (1998). Quantum computing, Reports on Progress in Physics **61**, 117–63.
2. Brown, J.R. (2001). Quest for the Quantum Computer. Riverside, New Jersey: Simon and Schuster [hardback: Minds, Machines and the Multiverse The Quest for the Quantum Computer (2001)].

3. Williams, C.P. and Clearwater, S.H. (2000). Ultimate Zero and One. New York: Copernicus.
4. Nielsen, M.A. and Chuang, I.L. (2000). Quantum Computation and Quantum Information. Cambridge: Cambridge University Press.
5. Bouwmeester, D., Ekert, A., and Zeilinger, A. (eds.) (2001). The Physics of Quantum Information: Quantum Cryptography, Quantum Teleportation, Quantum Computation. Berlin: Springer-Verlag.
6. Stolze, J. and Suter, D. (2004). Quantum Computing: A Short Course from Theory to Experiment. Weinheim: Wiley.
7. Nakahara, M. and Salomaa, M. (2005). Quantum Computation: From Linear Algebra to Physical Realizations. Bristol: Institute of Physics.
8. Le Bellec, M. (2006). A Short Introduction to Quantum Information and Quantum Computation. Cambridge: Cambridge University Press.
9. Whitaker, A. (2006). Einstein, Bohr and the Quantum Dilemma: From Quantum Theory to Quantum Information. Cambridge: Cambridge University Press.
10. Shor, P.W. (1994). Polynomial-time algorithms for prime factorisation and discrete logarithms on a quantum computer, In: Proceedings of the 35th Annual Symposium on Foundations of Computer Science (Goldwasser, S., ed.) Piscataway, New Jersey: IEEE; also in SIAM Journal of Computing 26, 1484–509 (1997).
11. Cirac, J.I. and Zoller, P. (1995). Quantum computation with cold trapped ions, Physical Review Letters 74, 4091–4.
12. Feynman, R.P. (1999). There's plenty of room at the bottom, In: Feynman and Computation. (Hey, A.J.G., ed.) Reading, Massachusetts: Perseus.
13. Regis, E. (1995). Nano! New York: Bantam.
14. Gleick, J. (1992). Genius: The Life and Science of Richard Feynman. New York: Pantheon.
15. Mehra, J. (1994). The Beat of a Different Drum: The Life and Science of Richard Feynman. Oxford: Oxford University Press.
16. Gribbin, J. and Gribbin, M. (1997). Richard Feynman; A Life in Science. London: Viking.
17. Minsky, M. (1999). Richard Feynman and cellular vacuum, in Ref. [18], pp. 117–30.
18. Hey A.J.G. (ed.) (1999). Feynman and Computation. Reading Massachusetts: Perseus.
19. Feynman, R.P. (1996). Feynman Lectures on Computation. (Hey, A.J.G. and Allen, R.W., eds.) Reading, Massachusetts: Addison-Wesley.
20. Hey, A.J.G. (1999). Feynman and Computation, Contemporary Physics 40, 257–65.
21. ter Haar, D. (1967). The Old Quantum Theory. Oxford: Pergamon.
22. Feynman, R.P. (1999). Simulating physics with computers, in Ref. [18], pp. 133–53; reprinted from International Journal of Theoretical Physics 21, 467–99 (1982).
23. Landauer, R. (1999). Information is inevitably physical, in Ref. [18], pp. 77–92.
24. Benioff, P. (1982). Quantum mechanical Hamiltonian models of Turing machines, Journal of Statistical Physics 29, 515–46.
25. Feynman, R.P. (1999). Quantum mechanical computers, in Ref. [19], pp. 185–211; reprinted from Optics News (February 1985), pp. 11–20; Foundations of Physics 16, 507–32 (1986).
26. Landauer, R. (1961). Irreversibility and heat generation in the computing process, IBM Journal of Research and Development 5, 183–91.
27. Bennett, C.H. (1973). Logical reversibility of computation, IBM Journal of Research and Development 17, 525–32.

28. Fredkin, E. and Toffoli, T. (1982). Conservative logic, International Journal of Theoretical Physics **21**, 219–53.
29. Deutsch, D. (1985). Quantum theory, the Church-Turing principle and the universal quantum computer, Proceedings of the Royal Society A **400**, 97–117.
30. Bennett, C.H. and Brassard, G. (1984). Quantum cryptography: public-key distribution and coin tossing, Proceedings of 1984 IEEE International Conference on Computers, Systems and Signal Processing. New York: IEEE, pp. 175–9.
31. Bennett, C.H. and Brassard, G. (1989). The dawn of a new era for quantum cryptography: the experimental prototype is working! Sigact News **20**, 78–82.
32. Ekert, A.K. (1991). Quantum cryptography based on Bell's theorem, Physical Review Letters **67**, 661–3.
33. Bennett, C.H.. Brassard, G., Crepeau, C., Jozsa, R., Peres, A., and Wooters, W.K. (1993).Teleporting an unknown quantum state via dual classical and Einstein-Podolsky-Rosen channels, Physical Review Letters **70**, 1895–9.
34. Bell, J.S. (1981). Bertlmann's socks and the nature of reality, Journal de Physique, Colloque C2, **42**, 41–61; also in Ref. [35], pp. 139–58.
35. Bell, J.S. (2004). Speakable and Unspeakable in Quantum Mechanics. (1st edn. 1987, 2nd edn. 2004) Cambridge: Cambridge University Press.
36. Bell, J.S. (1964). On the EPR paradox, Physics **1**, 195–200; also in Ref. [35], pp. 14–21.
37. Whitaker, M.A.B. (2000). Theory and experiment in the foundations of quantum theory, Progress in Quantum Electronics **24**, 1–106.
38. Clauser, J.F. (2002). Early history of Bell's Theorem, In: Quantum [Un]speakables: From Bell to Quantum Information. (Bertlmann, R.A. and Zeilinger, A., eds.) Berlin: Springer, pp. 61–98.
39. Bernstein, J. (1991). Quantum Profiles. Princeton: Princeton University Press.
40. Stapp, H.P. (1977). Are superluminal connections necessary? Nuovo Cimento **40B**, 191–205.
41. Bell, J.S. (1986). In: The Ghost in the Atom. (Davies, P.C.W. and Brown, J.R., eds.) Cambridge: Cambridge University Press.
42. Gribbin, J. (1990). The man who proved Einstein was wrong, New Scientist **128**, 43–5.
43. Shannon, C.E. (1948). A mathematical theory of communication, Bell Systems Technical Journal **27**, 379–423, 623–56.
44. Hamming, R.W. (1986). Coding and Information Theory. 2nd edn., Englewood Cliffs, New Jersey: Prentice-Hall.
45. Turing, A.M. (1936–7). On computable numbers, with an application to the Entscheidungsproblem, Proceedings of the London Mathematical Society **42**, 230–65.
46. Hodges, A. (1992). Alan Turing: The Enigma. London: Random House.
47. Deutsch, D. (1997). The Fabric of Reality. London: Allen Lane.
48. Landauer, R. (1991). Information is physical, Physics Today **44**(5), 23–9.
49. Schumacher, B. (1995). Quantum coding, Physical Review A **51**, 2738–47.
50. Grover, L.K. (1996). A fast quantum mechanical algorithm for data base search, In: Proceedings of the 28th Annual Symposium on the Theory of Computation. New York: ACM Press, pp. 212–9.
51. Grover, L.K. (1997). Quantum mechanics helps on searching for a needle in a haystack, Physical Review Letters **79**, 325–8.
52. Cleve, R., Ekert, A., Macchiavello, C. and Mosca, M. (1998). Quantum algorithms revisited, Proceedings of the Royal Society A **453**, 339–54.

53. Deutsch, D. and Jozsa, R. (1992). Rapid solution of problems by quantum computation, Proceedings of the Royal Society A **439**, 552–8.

54. Deutsch, D. (1985). Quantum theory as a universal physical theory, International Journal of Theoretical Physics **24**, 1–41.

55. Steane, A.M. (2003). A quantum computer only needs one universe, Studies in the History and Philosophy of Modern Physics **34**, 469–78.

56. Bouwmeester, D., Pan, J.-W., Mattle, K., Eibl, M., Weinfurter, H., and Zeilinger, A. (1997). Experimental quantum teleportation, Nature **390**, 575–9.

57. Boschi, D., Branca, S., De Martini, F., Hardy, L., and Popescu, S. (1998). Experimental realization of teleporting an unknown pure quantum state via dual classical and Einstein-Podolsky-Rosen channels, Physical Review Letters **80**, 1121–5.

58. Furusawa, A., Sorenson, J.L., Braustein, S.L., Fuchs, C.A., Kimble, H.J. and Polzik, E.S. (1998). Unconditional quantum teleportation, Science **282**, 706–9.

59. Bouwmeester, D., Pan, J.-W., Daniell, M., Weinfurter, H., and Zeilinger, A. (1999). Observation of three-photon Greenberger-Horne-Zeilinger entanglement, Physical Review Letters **82**, 1345–9.

60. Tittel, W., Riborby, G., and Gisin, N. (1998). Quantum cryptography, Physics World **11**(3), 41–5.

61. Gisin, N., Ribordy, G.G., Tittel, W., and Zbinden, H. (2002). Quantum cryptography, Reviews of Modern Physics **74**, 145–95.

62. Bennett, C.H., Brassard, G., Breidbart, S., and Wiesner, S. (1982). Quantum cryptography, or unforgeable subway tokens, Advances in Cryptology Proceedings of Crypto 82. New York: Plenum, pp. 267–75.

63. Williamson, M. and Vedral, V. (2003). Eavesdropping on practical quantum cryptography, Journal of Modern Optics **50**, 1989–2011.

64. Devetak, I. and Winter, A. (2004). Relating quantum privacy and quantum coherence: an operational approach, Physical Review Letters **93**, 080501.

65. Wootters, W.K. and Zurek, W.H. (1982). A single quantum cannot be cloned, Nature **299**, 802–3.

66. Dieks, D. (1982). Communication by EPR devices, Physics Letters A **92**, 271.

67. Hughes, R.J., Nordholt, J.E., Derkacs, D., and Peterson, C.G. (2002). Practical free-space quantum key distribution over 10 km in daylight and at night, New Journal of Physics **4**, article no. 43.

68. Gordon, K.J., Fernandez, V., Townsend, P.D., and Buller, G.S. (2004). A short wavelength gigahertz clocked fiber-optic quantum key distribution system, IEEE Journal of Quantum Electronics **40**, 900–8.

69. Tittel, W., Brendel, J., Zbinden, H., and Gisin, N. (1998). Violation of Bell's inequalities by photons more than 10 km apart, Physical Review Letters **81**, 3563–6.

70. Tittel, W., Brendel, J., Zbinden, H., and Gisin, N. (2000). Quantum cryptography using entangled photons in energy-time Bell states, Physical Review Letters **84**, 4737–40.

12
Bridging the Quantum-Classical Divide

Introduction

In Chapter 7, we studied the discussions between Einstein and supporters of Copenhagen, in particular Max Born and Wolfgang Pauli, concerning the classical limit of quantum theory and macroscopic quantum theory. We discussed the argument of Born and Pauli that the limiting procedure was straightforward, and that there were no special features in macroscopic quantum theory, such as macrorealism, that could make it different in principle from microscopic quantum theory.

Einstein raised a number of interesting issues, which suggested that these matters were much more complicated than Born and Pauli were prepared to admit. One of his particular arguments suggested that gradual loss of classicality for macroscopic objects should be expected as the constituent wave-functions dephase. This latter point has been shown to be important in recent years; indeed to answer it has required development of a major new topic in physics—environment-induced decoherence theory. This is the subject of the first part of this chapter. Einstein's more general arguments have also been shown to be broadly correct. It is now recognised that study of the classical limit of quantum theory is a subtle and many-faceted subject. There are a considerable number of different approaches studying different aspects of the problem, and we sketch some of these in the second part of the chapter.

Environment-induced Decoherence Schemes: Basic Ideas

The central qualitative idea of environment-induced decoherence schemes is straightforward. We may imagine an ensemble of quantum systems S and we shall be particularly interested in the case where the systems are *macroscopic*. We shall at first imagine that the systems are isolated and the wave-function may in principle be a superposition of distinct states relating to rather different macroscopic configurations. To say we have a superposition is equivalent to saying that S is in a pure state. We may imagine S being in a particular macroscopic state at $t = 0$, but, since the states available for S are superpositions, small temporary interactions with other systems, or, as one would say in quantum theory, small perturbations, will lead to S being represented by a superposition of macroscopically distinct states.

To represent this mathematically, we might say that the general state of S may be $c_0|s_0\rangle + \sum_{i=1...n} c_i|s_i\rangle$, where the $|s_i\rangle$ are macroscopically distinct states. At $t = 0$, we may have $c_0 = 1$; $c_i = 0$, $i \neq 0$, so we have a particular macroscopic state; but for $t > 0$, $|c_0|$ will drop below unity, and some of the other c's will become non-zero, so we end up with exactly the superposition of different macroscopic states, or in other words the dephasing or loss of the classical nature of S, that Einstein was so concerned about.

The theory of decoherence explains *why* this dephasing does *not* actually happen for macroscopic systems. The reason is that we cannot maintain our earlier assumption that we may consider the system to be isolated. A macroscopic system behaving classically will be described in terms of collective macroscopic variables such as centre of mass coordinates; the important point is that it must be viewed, not as an isolated system, but as a so-called *open quantum system S*, interacting strongly with the environment E. The environment naturally includes systems outside S which may interact with S even very weakly; less obviously it may also include internal degrees of freedom of the physical system associated with S. We stress that the fact that the environment has a very large number of degrees of freedom is an essential ingredient of this approach. In fact we consider the environment as *not* accessible to probing in its own right; we merely discuss its effect on the macroscopic system S.

We will explain why we cannot ignore the environment E even if the actual magnitude of its effect may be small. By their very nature, macroscopic bodies inevitably have closely spaced energy levels. Let us consider, for example, a reasonably macroscopic rigid rotator of mass around 1 g and radius around 1 cm; even neglecting all internal degrees of freedom, its rotational levels are very closely spaced; in fact, $\Delta E = \hbar^2/2I$, where I is the moment of inertia of the rotator, and so ΔE is around 10^{-40} eV In particular, if the relevant level spacings corresponding to energy levels coupling appreciably to the environment are small compared to perturbation by the environment, the system is extremely sensitive to such perturbations.

This is because quantum mechanical perturbation theory tells us that the change in a wave-function under a small perturbation becomes larger and larger as the difference of unperturbed energies, ΔE, becomes smaller. (Mathematically ΔE occurs in the denominator of the relevant expression.) As a result, two slightly different perturbations can cause very different perturbed wave-functions. Because of the large number of variables involved in a macrosystem, these wave-functions are always effectively orthogonal. Even if the environment wave-functions for two or more macroscopically different states did happen to be nearly the same at a particular time, they become orthogonal very rapidly due to coupling with different values of collective observables of the macrosystem. Thus a tiny perturbation is sufficient to disturb the delicate phase coherence necessary for quantum interference effects.[1]

From this point of view, a crucial distinction between a macrosystem behaving classically and a microsystem is that a classically behaving macrosystem behaves effectively like a mixed state, because it is always strongly entangled with its environment. In a sense, the environment acts like an apparatus continuously monitoring the macrosystem, effectively prohibiting the display of quantum interference effects. It must be stressed that the state of the total system, which includes S and the environment E, remains pure. As we emphasised in Chapters 2 and 3, there is no way in which the Schrödinger equation can allow a pure state to evolve into a mixed state. But, if one pays attention only to S, for example making measurements on S alone and ignoring the coordinates of E, then the system S appears to be in a mixed state. This means that it is in one of the states $|s_i\rangle$ rather than a superposition, and will remain in that state; the dephasing that would take place if we had a superposition is prohibited.

To express this idea formally, the evolution of the combined macrosystem—environment state vector $|\Phi_{SE}\rangle$ proceeds from the initial state $|\Phi_{SE}(0)\rangle$ to the final state $|\Phi_{SE}(t)\rangle$ as follows:

$$|\Phi_{SE}(0)\rangle = |\psi_S\rangle|\phi_E\rangle = \left(\sum_i c_i|s_i\rangle\right)|\phi_E\rangle \rightarrow \sum_i c_i|s_i\rangle|\phi_{E_i}\rangle$$

$$= |\Phi_{SE}(t)\rangle \tag{12.1}$$

Here $|\phi_E\rangle$ is the initial state of the environment, and $|\phi_{E_i}\rangle$'s are the distinct environment states correlated with the macroscopically distinct and mutually orthogonal states $|s_i\rangle$ of a macrosystem. For the case above, where $c_0 = 1$, and all the other c_i s are zero, the final state is just $|s_0\rangle|\phi_{E_0}\rangle$. The initial state of the macroscopic object remains unchanged in time because it is coupled with a particular state of the environment.

To explain the significance of Eq. (12.1), it is helpful to use the idea of the density-matrix, a concept that we have not as yet explained in the book. We will give here the briefest possible explanation. We know from early chapters of the book that quantum theory was originally written down in two distinct but mathematically equivalent formalisms, the wave-mechanics formalism of Schrödinger, and the matrix-mechanics formalism of Heisenberg. In the latter, a matrix represents an observable quantity; in other words, it is the equivalent of an operator in the Schrödinger formalism. If a system has n distinct or orthogonal states, these matrices will be square $n \times n$ arrays.

The density-matrix will also be an $n \times n$ array, but it represents not an observable but the state of the system, or more generally the state of an ensemble. For a pure state it gives the same information as the wave-function, but it is a more powerful tool than the wave-function because it can handle mixed states as well as pure states. We will explain this shortly, but must first identify an important set of elements of the density-matrix – those on the so-called leading diagonal. These are the elements on the diagonal starting at top-left and ending at bottom-right. Each element on the leading diagonal corresponds to a particular state of the system. The sum of the elements on the leading diagonal of the density-matrix must always be one.

For a pure state in which the system is in a particular one of the states used to write down the density-matrix, the density-matrix takes a very simple form. The only non-zero element is an element equal to one somewhere on the leading diagonal; the position of this non-zero element tells us which particular state the system is in.

However we are here more interested in superpositions and mixtures of states. For a mixture again the only non-zero terms appear on the leading diagonal. As we have said, they add up to one, and the magnitude of any element on the leading diagonal gives the probability of any particular system being in the appropriate state. For example, if there are four states of the system, the density-matrix will be 4 × 4. If all the elements on the leading diagonal are equal to 1/4, and all the off-diagonal elements are zero, we have a mixture with equal probabilities of any system being in each of the four states.

The important difference for a superposition of two or more of the states used to build up the density-matrix is that some of the off-diagonal elements must be non-zero as well as some or all of those on the leading diagonal. It must be stressed that, in mathematical terms, this is a highly inadequate statement and far more detail could be given, but it gives us enough for our discussion.

Let us now consider the density-matrix for the entire system, S and also the environment E. This will be a very large matrix, since the number of states of the entire system will be the number of states of S, which must itself be large for a macroscopic system, multiplied by the number of states of E, which, as we have said, will be enormous. We have stressed that the density-matrix for system plus environment will relate to a pure state or a superposition. Thus the density-matrix will contain off-diagonal as well as diagonal elements.

However that statement is, in a sense, an '*in principle*' one; we cannot know, and certainly do not want to know, the details of the environment states. We therefore sum over the environment states, producing the so-called reduced density-matrix of the system. This is a matrix with dimension equal to the number of distinct states of S, rather than that number multiplied by the number of states of distinct states of E, so the presence of E has been removed, but not, of course, its effect.

The reduced density-matrix of the system S alone is then represented by

$$\rho_s(t) = Tr_E |\Phi_{SE}(t)\rangle\langle\Phi_{SE}(t)| = \sum_i |c_i|^2 |s_i\rangle\langle s_i| \qquad (12.2)$$

Here the notation Tr_E denotes the summing or 'tracing' over the states of the environment. In the language of the density-matrix, the right-hand-side tells us that effectively we have a mixture with state $|s_i\rangle$ of system occurring with probability $|c_i|^2$.

The second equality in Eq. (12.2) holds exactly if the states $|\phi_{EI}\rangle$ are mutually orthogonal. It is clear from Eq. (12.2) that the states $|s_i\rangle$ provide a preferred set of states, or *preferred basis*, with respect to which the reduced density-matrix of a macrosystem becomes diagonal. Therefore we can interpret an ensemble of systems described by $\rho_s(t)$ behaving as if it corresponds to a classical mixture of basis states $|s_i\rangle$. If the initial state of the macrosystem is not a member of this basis (called by Zurek[2] the pointer basis), it is not stable with respect to interaction with the environment. In other words, what is called an effective *superselection rule* emerges: environment-induced decoherence prevents superpositions of pointer basis states from persisting. If the initial state *is* a pointer state, it remains in that state. This explains why quantum mechanical wave packet spreading does not occur in the macrodomain.

A crucial point is that the form of the system — environment interaction Hamiltonian plays a decisive role in selecting the pointer basis. Of course, all quantum systems are coupled to the environment, at least in principle. However, the relevant time scale related to the rate at which superpositions of states are destroyed is much longer for microsystems than for macrosystems. This allows superpositions of microstates to survive for a lengthy period of time, so that the well-known quantum interference effects may take place, but no superpositions of macroscopic states may survive for more than an infinitesimal period of time. The magnitude of this time scale, known as the *decoherence time*, therefore provides a quantitative criterion for distinguishing between quantum and classically behaving macrosystems.

The mere fact that a macroscopic system interacts strongly with its environment does not by itself eliminate the possibility of observing quantum interference between macroscopically distinct states. This is because almost all system — environment interactions are *adiabatic* in nature; that is to say their effect correlates the system strongly with its environment, but in such a way that the interference effects can nevertheless be recovered, at least in principle. Only genuinely dissipative processes, in which the interaction leads to an *irreversible* exchange of energy between the system and environment, can guarantee permanent disappearance of the interference effects.[3,4] This is consistent with the fact that it is practically impossible to find evidence of macroscopic quantum interference except possibly for some specially devised superconducting systems. In fact, this exception confirms the rule, since in nearly all systems there is a great deal of dissipation of energy, but superconductors are characterised by negligibly small internal dissipation. Let us now turn to some specific schemes implementing these ideas.

Collision with External Environment Particles: Photons, Gas Molecules, etc.

Strictly speaking no system is completely isolated. Particles are always present even in intergalactic space, and there are of course photons from the cosmological thermal background radiation. To investigate the decoherence-inducing effect of such an external environment, we must evaluate phase shifts originating from collision of a macro-object with particles constituting its external environment. This problem is studied in detail by Joos and Zeh;[5] we outline their treatment.

Let us write the initial state in the form $\sum_i c_i |x_i\rangle |\beta\rangle$. Here $|x_i\rangle$ is the state of the macro-object; $|x_i\rangle$ is, in fact, a pure state for which the position of the centre of mass is x_i. $|\beta\rangle$ represents the state of a particle constituting the external environment, say an atom, a molecule or a photon. Clearly the approach is simplistic, since the environment, which consists of a very large number of particles, is being modelled by a single particle!

Scattering of the macro-object by such a particle does not produce appreciable recoil of the macro-object if it is heavy enough, but of course the state of the scattering particle will change. After scattering, the state $\sum_i c_i |x_i\rangle |\beta\rangle$ becomes $\sum_i c_i |x_i\rangle S_{x_i} |\beta\rangle$, where the operator S_{x_i} is a scattering operator corresponding to an external particle hitting the macroobject located at position x_i. The density matrix $\rho(x, x', t)$ of the macroobject was initially was a pure state density matrix $\rho_0(x, x')$; as a result of the scattering it becomes $\rho_0(x, x')g(x, x', t)$, where $g(x, x', t)$ is known as the *decoherence factor*.

Joos and Zeh[5] give details on calculating $g(x, x')$. Here we quote only the result for external monokinetic particles with a fixed wave number k_0, averaged over all incident directions. For $x = x'$, the decoherence has no effect:

$$g(x, x', t) \simeq 1 \quad x = x' \tag{12.3}$$

However, for $x \neq x'$, and provided that $|x - x'|\,|k_0| \gg 1$, or, in words, that $|x - x'|$ is large with respect to the relevant de Broglie wavelength, we obtain:

$$|g(x, x', t)| \simeq \exp(-nv\sigma_T t) \tag{12.4}$$

Here n is the number of particles per unit volume, v is the average velocity of the incoming particles, and σ_T is the total scattering cross-section, which is equal to the elastic cross-section if there is no absorption involved. σ_T is a measure of the strength of the scattering.

The exponent in Eq. (12.4) is in general a large quantity except for very small values of t, or for a very small density of particles in the environment. Thus we can infer that there is effectively a dynamical diagonalisation of the density operator of a macro-object with respect to position variables. In other words the position localised state of a macro-object is stabilised due to scattering by external particles; the wave-function is not allowed to evolve by quantum mechanical spreading. This result is also valid if we consider a continuous distribution of incoming velocities in place of scattering by a monokinetic beam of external particles.

The decoherence-inducing mechanism in this model results from the fact that the collision phase shift depends on the centre-of-mass position x of a macro-object. When many such collisions take place at random we can say that the modulus of the wave-function does not change, but its phase for different values of x becomes random. For sufficiently different values of x, such a randomly changing complicated phase of the wave-function results in a density operator whose off-diagonal elements in the position coordinates become negligibly small after a certain time. For a lucid explanation of the physics of this approach, see Omnès.[6]

The decoherence factor given by Eq. (12.4) can be written as $\exp(-t/T_c)$, where T_c can be called the decoherence time. Considering a macro-object as a sphere of radius R, so we may take $\sigma_T = \pi R^2$, Joos and Zeh[5] calculate the value of T_c for various realistic environments. For example, even a perfect vacuum containing only cosmic microwave background radiation, when acting on a dust particle with R around 10^{-5} m yields a value of T_C of around 10^{-6} s. The most efficient decoherence-inducing mechanism in a terrestrial setting, according to Joos and Zeh, is collision with air molecules at normal temperature and pressure, which gives a very small value of T_C of around 10^{-30} s even for a large molecule with R around 10^{-8} m; for a dust particle with R around 10^{-5} m, the corresponding T_c is about 10^{-36} s.

The preceding example shows that even if the collective center-of-mass observables of a macroobject are not dissipatively coupled to the internal degrees of freedom, they do not display quantum mechanical interference effects under usual laboratory conditions, because of decoherence generated by an external environmant.

An interesting application of this scheme addresses the problem of 'molecular chirality'. We consider a molecule that has two states of the same energy, ϕ_L and ϕ_R, which are related to each other by a parity transformation, essentially reflection in a mirror. We may say that the left-handed ϕ_L and the right-handed ϕ_R are states with definite 'chirality'. If initially a molecule is in a superposition of states $\phi_{L,R}$ of different chirality, random frequent collisions with ambient air molecules suffice to quench the off-diagonal terms of the density-matrix.[1,5] Even if such external peturbations are small, they are nevertheless very efficient in producing chirality effects as long as they are large compared to the small unperturbed energy splitting arising from the molecular Hamiltonian. We can thus explain for instance why under standard conditions sugar molecules always exist as two distinct optically active isomers, the so-called left-handed and right-handed sugar, that are mirror images of one another; if one of these two kinds of sugar is prepared, it stays in that state for ever.

One critical point about Joos–Zeh type arguments must be mentioned. It is *not* generally guaranteed within the framework of this form of computation that decoherence produced by collisions with external particles is irreversible, so no effect of the original coherence is ever recoverable. That is, near-vanishing off-diagonal elements of the relevant system density-matrix at a given time do not in general imply near-vanishing associated interference effects for all subsequent times. We can argue that while the impossibility of recovering interference is not rigorously guaranteed by the laws of standard quantum mechanics alone, it is a consequence of macroscopic thermodynamic irreversibility, which in turn is an essential ingredient of our observed world.

In the usual laboratory conditions, it is thus plausible to maintain, as do Joos and Zeh, that information necessary for displaying interference between macroscopically different positional states of any reasonably macroscopic body becomes entangled with that of the ambient particle states in such a complicated and uncontrollable way that it is irrecoverable for all practical purposes. We now discuss some schemes that attempt to treat decoherence by considering the dissipative coupling of macroscopic variables with macroscopic relevant variables internal to the system in question, which we will call an environment.

Oscillator Model of the Environment

Despite the presumably very wide domain of validity for decoherence effects, there is still no general model for treating all cases where decoherence seems to occur. Since the nature of what is called environment is not known, we must rely on approximate guesses. The most sophisticated procedure for modelling the complex behaviour of an environment rests on a scheme suggested by Feynman and Vernon[7] and later systematically and extensively developed by Caldeira and Leggett.[8] The fundamental idea is to develop a plausible argument leading to a rather general form of the Hamiltonian of the system and environment, by modelling any environment having a large but finite number of degrees of freedom in terms of a collection of noninteracting harmonic oscillators. Actual computations using the Feynman-Vernon technique are rather cumbersome; we mention some of the basic elements of this scheme in a general way, and for more relevant details, see Leggett.[9]

We start with a macrosystem of N macroscopic systems described in terms of a set of variables of which one is macroscopic and the rest microscopic. A standard example is the centre-of-mass X and relative coordinates $x_1, x_2, x_3, \ldots, x_{n-1}$. If there were no coupling between the macroscopic and the microscopic degrees of freedom, the total Hamiltonian would simply be the sum of a term for the systems, H_S, which would depend only on X and the corresponding momentum, P, and a term for the environment, H_E, which depends only on the x_i and the corresponding p_i. This implies that the motion of the macroscopic variable X is completely decoupled from that of the internal (microscopic) degrees of freedom, the x_i.

However in the presence of appreciable coupling between system and environment, it is a good approximation to write the total Hamiltonian in the form again as the sum of two terms. Again H_S depends only on X and P, but now H_E depends, not only on the x_i and the p_i, but also on X. However H_E still does not depend on P. This is equivalent to assuming that, for the interaction part of the Hamiltonian, the coordinate representation forms a pointer basis. With a starting Hamiltonian of this form, two crucial inputs in the argument follow.

First, in a macroscopic system, the coupling of each individual microscopic degree of freedom to the macroscopic one is very likely to be of order $N^{-1/2}$, where N is the total number of particles in the body. Thus, provided fluctuations of different environment states originate from the usual thermodynamic effects, we can argue that fluctuations in the eigenvalues of the wave-function that depend on the macroscopic variable X are themselves of the order $N^{-1/2}$ relative to the average value; hence these can be neglected for large values of N. It follows that the eigenvalue $E(X)$ may be treated as effectively independent of the environment state when X represents a macroscopic degree of freedom.

We now assume that any one degree of freedom of the environment is only weakly perturbed, and multiple excitation of a single mode can be neglected. We stress that this does not mean the system–environment interaction as a whole is weak. According to Leggett, this situation is analogous to a beam of light propagating through a gas: interaction with the gas as a whole may be very strong, yet interaction with any particular atom can be weak, so that it is amenable to simple perturbation theory. As a result it may be adequate for certain purpose to treat the atom as a type of simple harmonic oscillator. This is similar to what justifies in retrospect the oscillator model of atoms, which was used so successfully in the late nineteenth century, although we now know that electronic states in the atom have very little in common with harmonic oscillator states. For details of this argument, see Caldeira and Leggett.[8] We note that the adequacy of modelling the environment as a bath of oscillators has been argued for the case of zero temperature only. Nevertheless this does not prevent such models from being widely used in finite-temperature problems with surprising success.

Subject to many technical provisos, the outcome of this type of argument is a typical general form for the total Hamiltonian for an object. This Hamiltonian consists of a kinetic energy term for the macroscopic particle, and a sum of a large number of simple harmonic oscillator Hamiltonians with a range of effective frequencies; the latter model the effect of the environment. Hamiltonians of this type correspond to what is known in the relevant literature as the 'independent oscillator' model, and they are used in a wide variety of quantum mechanical calculations to model effects of an environment on a macroscopic variable. Nevertheless there is no guarantee that every macroscopic variable interacting with an environment can be adequately modelled by such a Hamiltonian.[4]

To summarise, if we consider an initial superposition of wave-functions localised in position space whose peaks are well-separated, then the off-diagonal elements of the reduced density-matrix of the system in the position representation decay exponentially and become negligibly small within a certain timescale. The reduced density-matrix for S based on momentum states also effectively becomes diagonal within this timescale. Thus each of position and momentum emerge as a preferred basis, since superpositions of localised functions of position and superpositions of functions localised in momentum space both decohere. Environment–induced decoherence schemes seek to guarantee the emergence of the classical structure for phase space and classical dynamics for macroscopic systems as a consequence of the assumed impossibility, true in all usual situations, of keeping macroobjects perfectly isolated.

Assessment of the Decoherence Programme

The central thesis of decoherence schemes is that quantum interference effects are practically unobservable for macrosystems, or in other words superpositions of their quantum states are effectively destroyed, because of their inevitable dissipative coupling with the surrounding environment. Emergent localised wave packets in the macro-domain obey classical dynamics because spreading cannot occur, since the environment effect dominates the system's internal quantum dynamics. Such states are expected to be insensitive to measurements of all usual observables; that is, their further dynamical evolution should be unaffected by measurement, another important aspect of classical dynamics.[10] The following points summarise the key caveats of decoherence schemes in addressing the classical limit of quantum mechanics:

(a) A fundamental assumption of the decoherence programme is that, if no measurements of environment degrees of freedom are possible, then any pure state ensemble of macrosystems interacting with the environment, which may be described by a reduced density operator after tracing over environment states, is, *in practice* indistinguishable from an ensemble corresponding to a mixed state comprising localised states around different positions; that is, quantum interference terms become small enough to be beyond experimental resolution. But by itself this does *not* imply that each individual macroscopic system is actually localised at a definite position.

Quantum mechanically macrosystem states are always entangled with those of the environment. (The combined density operator represents a pure state.) No general rule in physics forbids measuring environmental observables in principle; hence it is conceivable, at least *in principle*, for joint measurements on system and environment to reveal quantum interference effects, even though the reduced density-matrix for the system is diagonal. This point is usually countered by proponents of the decoherence programme who stress that the concept of an environment is so complicated that it is practically impossible to make such measurements. Critics call this the NOWEN or 'no one will ever know' viewpoint.

(b) The nature of the preferred basis emerging from the theory depends on *how* we model the environment, including specifying the form of the system–environment interaction. As emphasised earlier, there is no sufficiently general theory of decoherence at our disposal. Hence the *ad hoc* nature of how system–environment coupling is specified cannot be avoided at this stage. Much work needs to be done on such technical issues as studying more realistic and general models.

(c) From the preceding discussions we see that the main achievement of decoherence models lies *not* in deriving classical physics from quantum mechanics, but in showing that, subject to appropriate physical approximations and educated guesses, we can make the standard formalism *compatible* with the observed validity of certain classical dynamical features in the macro-domain. However one crucial feature of macroscopic reality, namely the particle *ontological aspect*, the notion that a macrosystem has a well defined spatial structure and is in a definite dynamical state at any given instant, is not accommodated within decoherence schemes. This is because such models assume that a wave-function provides a complete description of the state of an individual entity. There is no place to introduce additional ontological elements at a fundamental level. Hence macro-ontological features cannot emerge within any limit from these approaches.

It is often loosely implied while using such schemes that a particle, envisaged as a localised entity possessing mass and charge, is associated with a wave packet localised in position space. In the absence of a systematically developed ontological interpretation, such an *ad hoc* identification can lead to logically unacceptable consequences, even if wave packets are ensured to be non-spreading. For instance, if a localised wave packet is coherently split into a set of disjoint wave packets which are also individually localised, it does not make sense to assume that an initial single particle splits in this way into two identical particles.[11]

To conclude, there are aspects of classical reality pertaining to the macrophysical world that cannot be made consistent with quantum theory in any limit, at least using the standard formalism and decoherence models. Point (c) in particular makes it clear that an additional conceptual element is needed at a fundamental level in the interpretation of quantum theory, so that the classical ontology of the macroworld may emerge smoothly without any abrupt conceptual discontinuity. An example of such an additional element is the position variable of the Bohm model, which is discussed in Chapter 10.

Other Approaches to the Classical Limit

There are very many different types of approach to studying the classical limit of quantum theory. A general and quite readable account of some of these has been provided by Bolivar.[12] We briefly mention here a topic which came up in Chapter 7, and will be discussed in a little detail in the following chapter. This is the line of study to probe the quantum measurement problem in an experimentally meaningful way, and it has been suggested by Leggett.[13–15] It seeks to test the basic idea of interference between macroscopically distinct states of a macroobject, which is a crucial ingredient in the measurement problem, as explained in Chapter 3. Some recent fascinating experiments have tested the implications of the above idea in the context of a superposition of two macroscopic currents moving in clockwise and anticlockwise directions of a SQUID (superconducting quantum interference device) ring which is of a macroscopic dimension.[16–18] Such experiments promise to provide empirically relevant insights on the measurement problem.

The $h \rightarrow 0$ Limit

A statement often encountered in the quantum folklore is that the equations of quantum mechanics approach those of classical mechanics in the limit $h \rightarrow 0$. However, h is not dimensionless. The limit (to zero or infinity) of a dimensional quantity is not meaningful – one cannot vary h or *actually* set it equal to zero. Moreover, the notion that h is 'small' has no absolute meaning because its value depends on the system of units. In the widely studied WKB (Wentzel–Kramers–Brillouin) approximation method, a quantum mechanical wave function is expanded in a power series of h around $h = 0$. Then the series is truncated by neglecting higher powers of h — subsequent calculations are based on this truncated series. This procedure can best be considered as a *semiclassical* (but essentially *non-classical*) computing procedure, useful in many problems, but it should be considered distinct from the classical limit problem, because wave functions are, in general, highly *nonanalytic* in the neighbourhood of the limit point $h = 0$.[19] This results in the *essential singularity* of quantum mechanically computed quantities at $h = 0$. It is thus *not* possible to regard quantum mechanics as a perturbative extension of classical mechanics[20] in the same sense as special relativity grows out of Newtonian mechanics by a convergent perturbation expansion in v/c. Indeed Michael Berry and his collaborators[19] have developed the subject of semi-classical quantum mechanics, with many of its own techniques differing from both microscopic quantum mechanics and classical mechanics.

Moreover, there are examples where the quantum mechanically predicted results are independent of h.[19] It is therefore *not* legitimate to make naïve statements like 'every classical system is essentially the $h = 0$ limit of a quantum one' which would mean imagining a sequence of formulations of quantum theory endowed with different values of h and then claiming that the case of $h \to 0$ coincides with classical mechanics.

An apparently sensible operational formulation of the $h \to 0$ classical limit condition would be to take a dimensionless parameter of the form $h/S \ll 1$ where S is the relevant action quantity in a given situation. It may be argued that if S is sufficiently large compared with h, measurements at the macroscopic level will not be able to detect quantum effects because of the limited instrumental resolution. This viewpoint implies that if one uses fine enough observation means, every physical system will ultimately display quantum features. It is the experimental limit on the accuracies with which measurements can be made in the macroscopic regime which is invoked to explain the 'coarse-grained' appearance of a classical world. This, of course, means an avowedly anthropocentric definition of classical systems, in the sense that the same system may or may not be classical depending on the degree of refinement of the experiments it is subjected to.

If E is the energy of a macrosystem and $T_{C\ell}$ is the order of magnitude of its classical period of motion, then in a typical situation $S > ET_{C\ell}$. The above condition for observing quantum effect ($S < h$) implies $E < h\omega_{C\ell}$ where $\omega_{C\ell} \simeq T_{C\ell}^{-1}$. Since for *common* macroscopic systems, the relevant classical frequencies are not greater than, say, $10^{16}\,\text{s}^{-1}$, this implies that the characteristic energy scale for observing quantum effects at the level of macroscopic dynamics can be no greater than only a few eV, which is of the order of the ionisation energy of a single atom. The dynamics associated with any macroscopic variable does not usually involve so tiny an energy, unless one considers effects such as the Josephson effect, where the current or trapped flux in a bulk superconducting ring can be controlled by a microscopic energy of the order of the thermal energy of an atom at room temparature or even smaller. The observed classical dynamics of macrosystems (Feature A mentioned in Chapter 3) is thus accounted for. Let us now probe Feature C mentioned in Chapter 3 and see as to what extent it can be explained in this spirit.

Consider a quantum system in the state $|\psi\rangle = \sum c_i |\beta_i\rangle$ where $|\beta_i\rangle$ are the eigenstates of an observable β. The probability of obtaining an eigenvalue β_k is given by

$$P(\beta_k) = |c_k|^2 \qquad (12.5)$$

We shall now compute $P'(\beta_k)$ when β is measured *after* another observable α is measured. Rewriting $|\psi\rangle$ in the following form

$$|\psi\rangle = \sum_i c_i |\beta_i\rangle = \sum_{i,j} c_i d_{ji} |\alpha_j\rangle \qquad (12.6)$$

where $d_{ji} = \langle \alpha_j | \beta_i \rangle$ and $|\alpha_j\rangle$ are the eigenstates of α, the measurement of α will change $|\psi\rangle$ to

$$|\psi'\rangle = \sum_{i,j} c_i d_{ji} |\alpha_j\rangle |A_j\rangle = \sum_{i,j,k} c_i d_{kj} d_{ji} |\beta_k\rangle |A_j\rangle \qquad (12.7)$$

where $d_{kj} = \langle \beta_k | \alpha_j \rangle$ and $|A_j\rangle$ are the macroscopically distinct states of the apparatus A used to measure α.

Now, when the measurement of β is made on $|\psi'\rangle$, the state Eq. (12.7) changes to

$$|\psi''\rangle = \sum_{i,j,k} c_i d_{kj} d_{ji} |A_j\rangle |\beta_k\rangle |B_k\rangle \qquad (12.8)$$

where $|B_k\rangle$ are the *macroscopically distinct* states of the apparatus B used to measure β.

From Eq. (12.8), the joint probability of obtaining the eigenvalue β_k together with the eigenvalue α_j is is given by $|\sum c_i d_{kj} d_{ji}|^2$. Hence the probability of obtaining β_k *after* α is measured, irrespective *of* the obtained eigenvalue of α, is given by

$$P'(\beta_k) = \sum_j \left| \sum_i c_i d_{kj} d_{ji} \right|^2 \qquad (12.9)$$

It is seen from Eqs. (12.5) and (12.9) that, in general, $P(\beta_k) \neq P'(\beta_k)$ which expresses the *invasiveness* of measurements in quantum mechanics. Obviously, this treatment holds good whatever the initial state $|\psi\rangle$ of the observed system, and is, in principle, valid even in the macrolimit of quantum mechanics.

However, when α and β commute, they have a complete set of common eigenstates. Then we get

$$d_{ji} = \langle \alpha_j | \beta_i \rangle = \delta_{ji} \qquad (12.10)$$

and

$$d_{kj} = \langle \beta_k | \alpha_j \rangle = \delta_{kj} \qquad (12.11)$$

whence

$$P'(\beta_k) = \sum |c_j \delta_{kj}|^2 = |c_k|^2 \qquad (12.12)$$

which equals $P(\beta_k)$; that is in this case the invasiveness of measurement is *not* manifested. This result provides a clue for some form of qualitative understanding (by no means rigorous) of why measurements are found to be noninvasive in the macroscopic regime. One may attribute this 'apparent' non-invasiveness to the coarse-grained nature of our observations at the macroscopic level, with respect to which the observable dynamical variables behave *as if* they are all mutually commuting.

The $N \to \infty$ Limit

In general, the properties of a quantum mechanical ensemble corresponding to an arbitrary ψ function are such that one cannot envisage an observationally equivalent ensemble of particles obeying classical laws. Let us consider, for example, the Hamiltonian of a bound system given by $H = p^2/2m + V(x)$ where $V(x)$ is a monotonically increasing function of x. Then for a classical ensemble of particles, each with energy E, the position probability density function $\rho_{C\ell}$ must vanish for $x > A$, where A, the so-called turning-point, is defined by saying that $E = V(x)$ when $x = A$. For the corresponding quantum mechanical ensemble represented by an energy eigenfunction, no such restriction exists. It is, however, believed that if a wave-function belongs to the 'classical domain' the quantum probability density should approach its classical counterpart. In such a domain energy eigenvalues are much greater than the energy difference between successive discrete eigenvalues. This means that $\mathrm{Lt}_{N\to\infty}((E_{N+1} - E_N)/E_N) \to 0$ where N is the principal quantum number, so that in this domain the classical *continuum* of energies is attained, This is known as 'configuration correspondence' in the classical limit.[21]

Consider, for example, an ordinary one-dimensional simple harmonic oscillator of a specified energy E, for which the classical probability density $\rho_{C\ell}(x) = c/v(x)$, where v is the velocity and c is a normalisation constant, is given by

$$\rho_{C\ell}(x) = \frac{1}{[\pi(A^2 - x^2)^{1/2}]} \qquad (12.13)$$

where $A = (2E/m\omega^2)$ is the amplitude. ($x = \pm A$ are the 'turning points'). If one considers energy eigenstates ($E = (N + 1/2)\hbar\omega$) for $N \gg 1$, the quantum particle probability density can be approximated by the expression.[22]

$$\rho_{QM}(x) = \left[\frac{2/\pi}{\left(A_N^2 - x^2\right)^{1/2}} \right]^2$$

$$\times \cos^2\left[(1/2)\beta x(A_N^2 - x^2)^{1/2} + (1/2)\beta A_N^2 \sin(x/A_N) - \frac{\pi N}{2} \right] \qquad (12.14)$$

for $|x| < A_{N'}$ where $A_N^2 = (2N + 1)/\beta$ with $\beta = m\omega/\hbar$. Note that A_N becomes the classical amplitude for $N \to \infty$. However, for large but *finite* N, $\rho_{QM}(x)$ does *not* vanish for $|x| > A_{N'}$; also nodes are always present even for arbitrarily large values of N. Recovery of the corresponding classical result, therefore, involves an *averaging procedure*—using Eqs. (12.13) and (12.14) one can write

$$\rho_{cl}(x) = (1/2\Delta) \int_{x-\Delta}^{x+\Delta} \rho_{QM}(x')dx' \tag{12.15}$$

where $\Delta \rightarrow 0$ as $N \rightarrow \infty$. This form of averaging is justified on the ground that macroscopic probes are insensitive over length scales of the order of several de Broglie wavelengths.

An important point to be emphasised is that even if the individual energy eigenstates lead to the required classical results in the limit $N \rightarrow \infty$, their superpositions do *not* necessarily satisfy this requirement. Cabrera and Kiwi[23] have given a simple example to illustrate this. Consider a superposition of harmonic oscillator eigensates with successive quantum numbers, $\{|N - 1\rangle, |N\rangle, |N + 1\rangle\}$, given by the time-dependent state.

$$|\psi(t)\rangle = (1/3^{1/2})[\exp(-iE_{N-1}t/\hbar)|N - 1\rangle + \exp(-iE_N t/\hbar)|N\rangle$$
$$+ \exp(-iE_{N+1}t/\hbar)|N + 1\rangle] \tag{12.16}$$

for which

$$\langle x\rangle = (2/3)(\hbar/2m\omega)^{1/2}(N^{1/2} + (N + 1)^{1/2}) \cos \omega t \tag{12.17}$$

In the $N \rightarrow \infty$ limit, $\langle x\rangle = (2/3)A \cos \omega t$, i.e., even ignoring the quantum fluctuations from $\langle x\rangle$, the correct classical dynamical behaviour is *not* recovered by large values of N — the amplitude of oscillation remains 2/3 of the classical value, no metter how large N is.

The preceding discussions therefore indicate that the high-energy or $N \rightarrow \infty$ limit does not by itself constitute a universal criterion for ensuring that classical dynamical behaviour emerges from an *arbitrary* wave function.

Ehrenfest's Theorem

Discussions of the classical limit of quantum mechanics are often based on Ehrenfest's theorem[24] according to which, under certain conditions, quantum behaviour may be shown to become equivalent to classical behaviour. Let us consider, for simplicity, the one-dimensional motion of a particle subjected to a scalar potential $V(q)$ which generates the force $F(q) = -\partial V/\partial q$. Using Heisenberg's equations of motion it follows that the expectation values of q and p satisfy the following relations

$$\frac{d\langle q\rangle}{dt} = \frac{\langle p\rangle}{m} \tag{12.18}$$

$$\frac{d\langle p\rangle}{dt} = \langle F(q)\rangle \tag{12.19}$$

Then *if* one can approximate

$$\langle F(q) \rangle \simeq F(\langle q \rangle) \qquad (12.20)$$

Eq. (12.19) may be replaced by

$$\frac{d\langle p \rangle}{dt} = F(\langle q \rangle) \qquad (12.21)$$

Eqs. (12.18) and (12.21) constitute the essential content of Ehrenfest's theorem. The approximation Eq. (12.20) holds good if ψ is a wave packet whose width is small compared to the scale over which $F(x)$ varies appreciably. However, a wavepacket will spread with time. The smaller the spread in position (usually measured by $\Delta q = \langle (q - \langle q \rangle)^2 \rangle^{1/2}$), the more closely quantum averages of position and momentum will follow classical dynamics. A general procedure for calculating corrections to Ehrenfest's theorem by taking into account wave packet spreading and the spatial variation of potential has been discussed by Andrews[25] and Peres.[26]

The important point to be noted is that even if Ehrenfest's theorem holds good in a given situation, it does *not* suffice to gurantee complete equivalence with classical mechanics. The fact that averages of the observed values of the dynamical quantities pertaining to an ensemble of particles satisfy classical equations does not imply that the behaviour of an *individual* member of the ensemble conforms to classical dynamics—vastly different ensembles can give rise to the same mean behaviour.[27] The lack of *sufficiency* of Ehrenfest's theorem on this point can be seen, for example, in the case of the harmonic oscillator. Here $F(x)$ is a linear function, so the approximation (12.20) is valid as an identity for *any* state. Yet the dynamical evolution of a quantum harmonic oscillator does not resemble that of a classical oscillator—in particular, there is always a finite probability of finding a quantum oscillator beyond the classical amplitude.

Ehrenfest's theorem is also *not necessary* for characterising the classical regime.[28] This is because the statistical ensemble description of classical dynamics does not relate only to a localised probability distribution. To sum up, the conditions for applying Ehrenfest's theorem are highly restrictive, so the theorem cannot be said to be reliable for investigating the problem of the classical limit of quantum theory in a general way.

The Quantum Theory of Macroscopic Systems

The simplest model of the relation between microscopic and macroscopic is that in the classical limit microscopic laws become equivalent to the laws of classical mechanics, and for macroscopic bodies one may continue to ignore quantum theory and continue to follow Newton. Earlier sections of this chapter have shown the difficulties inherent in such a scheme In this section we discuss another matter. When we consider, in particular, a solid, we recognise that some of its most basic macroscopic properties, such as density and elasticity, may be discussed in terms of a few parameters, derivable in principle from quantum theory, but which, in a macroscopic treatment we may just treat as experimental constants.

However, if we wish to discuss the properties of the solid in any detail, particularly its electromagnetic properties, we can in no way treat it as just a macroscopic object. We inevitably have to take into account the underlying microscopic properties, but also to recognise that the quantum theory of macroscopic systems must have its own general features, totally distinct both from quantum microphysics and from Newtonian classical mechanics.

In a recent book, Geoffrey Sewell[29] discusses what he calls this 'emergent macrophysics'. He says that: 'The quantum theory of macroscopic systems has become a vast interdisciplinary subject, which extends far beyond the traditional area of condensed matter physics to general questions of complexity, self-organisation and chaos. It is evident that the subject must be based on conceptual structures quite different from those of quantum microphysics, since the macroscopic properties of complex systems are expressible only in terms of concepts, such as entropy and various kinds of order, that have no relevance to the microscopic world. Moreover, the empirical fact that systems of different microscopic constituents exhibit similar macroscopic behaviour provides grounds for suspecting that macrophysics is governed by very general features of the quantum properties of many-particle systems. Accordingly, it seems natural to pursue an approach to the theory of its emergence from quantum mechanics that is centred on macroscopic observables and certain general features of quantum structures, independently of microscopic details.'

Yet again we see the richness of physical structure that may develop or emerge when we study the classical limit of quantum theory and macroscopic quantum theory in an open-minded way. Advocates of Copenhagen rather took it for granted that their approach *must* solve all the problems, and argued away any interesting features.

Conclusions

In his disagreements with Born and Pauli, we must come to the conclusion that Einstein has been shown to be victorious, not just because the points he made and the questions he asked have been found to be perceptive, but more fundamentally because his approach was always likely to argue from difficulty or paradox to a stimulating conclusion. By shying away from potential difficulties, the Copenhagen approach made it impossible for itself to reach genuinely new physics, of which this chapter has shown that there is a vast amount in the area of the classical limit.

References

1. Zeh, H.D. (1970). On the interpretation of measurement in quantum theory, Foundations of Physics 1, 69–76.

2. Zurek, W.H. (1991). Decoherence and the transition from quantum to classical, Physics Today **44**(10), 36–44.

3. Kagan, Y.A. and Leggett, A.J. (1992). Quantum Tunnelling in Condensed Media. Amsterdam: Elsevier, Ch. 1.

4. Leggett, A.J. (1993). The effect of dissipation on tunnelling, In: Proceedings of the 4th International Symposium on the Foundations of Quantum Mechanics in the Light of New Technology. Tokyo: Japanese Journal of Applied Physics, pp. 10–17.

5. Joos, E. and Zeh, H.D. (1985). The emergence of classical properties through interaction with the environment, Zeitschrift für Physik B **59**, 223–43.

6. Omnès, R. (1994). Interpretation of Quantum Mechanics. Princeton: Princeton University Press, Ch. 7.

7. Feynman, R.P. and Vernon, F.L. (1963). The theory of a general quantum system interacting with a linear dissipative system, Annals of Physics (New York) **24**, 118–73.

8. Caldeira, A.O. and Leggett, A.J. (1983). Quantum tunnelling in a dissipative system, Annals of Physics (New York) **149**, 374–456; **153**, 445 (1984).

9. Leggett, A.J. (1987). Quantum mechanics at the macroscopic level, In: Chance and Matter. (Souletie, J. et al., eds.), Amsterdam: North-Holland, pp. 411–6.

10. Zurek, W.H. (1993). Preferred states, predictability, classicality and the environment-induced decoherence, Progress of Theoretical Physics **89**, 281–312.

11. Holland, P. (1993). Quantum Theory of Motion. Cambridge: Cambridge University Press, Ch. 6.

12. Bolivar, A.O. (2004). Quantum-Classical Correspondence. Berlin: Springer.

13. Leggett, A.J. (1980). Macroscopic quantum systems and the quantum theory of measurement, Progress of Theoretical. Physics (Supplement) **69**, 80–100.

14. Leggett, A.J. and Garg, A. (1985). Quantum mechanics versus macroscopic realism – is the flux there when nobody looks? Physical Review Letters **54**, 857–60.

15. Leggett, A.J. (1998). In: Quantum Measurement: Beyond Paradox. (Healey, R.A. and Hellman, G., eds.) Minneapolis: University of Minnesota Press, pp. 1–29.

16. van der Wal, C.H., ter Haar, A.C.J., Wilhelm, F.K., Schouten, R.N., Harmans, C.J.P.M., Orlando, T.P., Lloyd, S. and Mooji, J.E. (2000). Science **290**, 773–7.

17. Friedman, J.R., Patel, V., Chen, W., Tolpygo, S.K. and Lukens, J.E. (2000). Quantum superposition of distinct macroscopic states, Nature **406**, 43–6.

18. Vion, D., Aassime, A., Cottet, A., Joyez, P., Pothier, H., Urbina, C., Esteve, D. and Devoret, M.H. (2002). Manipulating the quantum state of an electrical circuit, Science **296**, 886–9.

19. Berry, M.V. and Mount, K.E. (1972). Semiclassical approximations in wave mechanics, Reports of Progress in Physics **35**, 315–97.

20. Berry, M.V. (1989). Quantum chaology not quantum chaos, Physica Scripta **40**, 335–6.

21. Liboff, R.L. (1984). The correspondence principle revisited, Physics Today **37**, 50–5.

22. Morse, P.M. and Feshbach, H. (1953). Methods of Theoretical Physics. New York: McGraw-Hill, p. 1643.

23. Cabrera, G.G. and Kiwi, M. (1987). Large quantum-number states and the correspondence principle, Physical Review A **36**, 2995–8.

24. Messiah, A. (1961). Quantum Mechanics. Amsterdam: North-Holland, Vol. 1, Ch. 6.

25. Andrews, M. (1981). The spreading of wavepackets in quantum mechanics, Journal of Physics. A. **14**, 1123–9.

26. Peres, A. (1993). Quantum Theory: Concepts and Methods. Dordrecht: Kluwer, Ch. 10.

27. Holland, P. (1993). The Quantum Theory of Motion. Cambridge: Cambridge University Press, p. 256.
28. Ballentine, L.E., Gang, Y. and Zibin, J.P. (1994). Inadequacy of Ehrenfest theorem to characterise the classical regime, Physical Review A **50**, 2854–9.
29. Sewell, G.L. (2002). Quantum Mechanics and its Emergent Macrophysics. Princeton: Princeton University Press.

Interlude

In Part C, we studied the way in which studies about the nature of quantum theory have developed over the 50 years since Einstein's death. Gradually the straitjacket in which the subject was held for so long by the overweening position of the Copenhagen interpretation was loosened. Gradually it became acceptable to criticise this interpretation. Gradually new interpretations of quantum theory were suggested, new ideas and new theories developed, new and relevant experiments designed and carried out. Finally a new and important area of research, quantum information theory, was to emerge. Einstein's influence in all this was apparent, not just in the part played by entanglement in much of the work and his stimulation of the epochal work of John Bell, but even more basically in the freedom gained to question, to discuss, to construct realistic models and theories as required.

Part D takes us to the present day. While many of the necessary experiments related to the work of Bell have been carried out, and conceptual and experimental work on quantum information is well under way, many other interesting problems and topics have arisen and are being investigated, such as the quantum Zeno effect, the question of time of arrival in quantum mechanics, the testing of quantum superposition for macroscopic or mesoscopic systems, and several others. It is doubtful if study of these topics would have taken place in the days when the Copenhagen interpretation reigned supreme, with its unspoken mission statement being to explain away difficult and interesting features of the theory, rather than to face up to the conceptual puzzles, to aim at reaching physical solutions, and even to think of possible applications. Some of these contemporary problems are discussed in Chapter 13.

In Chapter 14 we return to Einstein. While Einstein exercised a strong, if sometimes indirect, influence on the progress discussed in Part 3, the new discoveries would also have had an effect on his own ideas; in particular Bell's theorem ruled out his preferred option as a solution to the EPR problem of local realism. We discuss Einstein's possible reaction to this situation, and we conclude the chapter by assessing his contributions, direct and indirect, positive and negative, to studies of the fundamental meaning of quantum theory.

Part D
Looking Forward

When I am judging a theory, I ask myself whether, if I were God, I would have arranged the world in such a way.

Albert Einstein to Banesh Hoffmann, in: *Some Strangeness in the Proportion* (H. Woolf, ed.) (Addison-Wesley, Reading, 1980), p. 476; also in *The Expanded Quotable Einstein* (A. Calaprice, ed.) (Princeton University Press, Princeton, 2000), p. 259.

There is no doubt that all but the crudest scientific work is based on a firm belief—akin to a religious feeling—in the rationality and comprehensibility of the world.

Albert Einstein, On science, Cosmic Religion (Corici–Friede, New york, 1931); also in *The Expanded Quotable Einstein*, p. 247.

Part D
Looking Forward

13
Quantum Foundations: General Outlook

Introduction

For followers of Copenhagen, study of the foundational aspects of quantum theory was effectively completed in 1927 by Bohr's paper at Como. Bohr was to continue to write on quantum theory and complementarity for the remainder of his life, but his writings were mainly explanatory, or attempts to strengthen the philosophical base of complementarity, or applications of complementarity to areas of knowledge other than quantum theory.[1] It was taken for granted that no genuine re-examination of the fundamental issues was required—or, indeed, permitted. John Clauser[2] has made it particularly clear to what extent any questioning of the Copenhagen position was regarded practically as heresy: 'Religious dogmatism then quickly promoted a nearly universal acceptance of quantum theory and its Copenhagen interpretation as gospel, along with a total unwillingness to even mildly question the theory's foundations.'

The sad thing was not just that the basic issues were not discussed. The lack of discussion of these issues meant that the further ideas and concepts that such discussion would have brought into being were not available for experimental or theoretical study, even though such study would not necessarily, of itself, have threatened the *status quo*.

Einstein fulminated against this effective censorship, but was powerless to change the situation. It was largely Bell, following on, of course, from Einstein's ideas, who was mainly responsible for the dramatic reversal of the state of affairs. As discussed in Chapter 10, it has become acceptable to invent alternatives to Copenhagen, and with this freedom have come an enormous number of new ideas related to and emerging from the most fundamental aspects of quantum theory. In probably the two main journals for the area of study, *Physical Review Letters* and *Physical Review A*, every issue now includes a substantial number of papers on 'Fundamental Concepts' as well as very many on 'Quantum Information Theory', another topic which, as we argued in the previous chapter, owes much to Bell, and, behind him, to Einstein.

In this chapter we choose from very many topics in this area, three which, we feel, would have been particularly interesting from Einstein's point of view. The first studies experimental tests of macrorealism in quantum theory. From Chapters 7 and 8, it will be clear that Einstein was especially interested in this topic, and indeed made some profound comments about it.

The other two topics are both concerned, in different ways, with time; Chapter 8 shows that Einstein was convinced, as so often against the beliefs of Copenhagen, that processes in the quantum world should be regarded as *physically real* and occurring at a fixed time, even if this time was not observed. The first topic discussed here is the quantum Zeno effect, which started as a counter-intuitive prediction, but has more recently become a standard aspect in the discussion of quantum processes. The second topic is the very role of time in non-relativistic quantum theory. Traditionally, as discussed in Chapter 3, it has been felt to be very difficult to include time in quantum theory in any other way than as a mere parameter; for instance such quantities as 'time of arrival' were judged extremely problematic. However over several decades, much useful progress has been made, and a brief summary of some aspects is given here.

While in this chapter we consider only these three topics, a much wider selection is included in the recent book by Partha Ghose,[3] who concentrates on the experimental aspects of testing quantum mechanics.

Macrorealism in Relation to Quantum Mechanics

Earlier discussions of decoherence models in Chapter 12 underscore the point that, by their very nature, macroscopic systems interact so strongly and dissipatively with their environment that quantum phase coherence between different macrostates is eliminated for all but the most exceptional cases. The conditions that must be satisfied in practice to observe the effects of quantum phase coherence at the macrolevel are extremely stringent. It was Leggett[4-6] who first suggested that, given state of-the-art levels of cryogenics, microfabrication and noise control, conditions for minimizing the dissipative part of the interaction with the relevant environment may be realisable in appropiate macrosystems. The candidate system most thoroughly studied for this purpose is the radio frequency superconducting quantum interference device (rf-SQUID).[6-8]

The rf-SQUID ring consists of a loop of superconducting material interrupted by a single Josephson junction. An external magnetic flux imposed through the ring is linked to the loop inductance. As a result, a persistent screening current is generated around the loop, which in turn produces its own magnetic field. In the theoretical treatment, the total flux Φ trapped through the ring is treated as a dynamical variable. One considers an effective wave-function $\psi(\phi)$ that satisfies a Schödinger-like equation, with the junction capacitance playing the role of mass. By varying the system parameters, in particular the external flux, different forms of potential energy associated with the flux variable can be produced.

Behaviour embodying macroscopic quantum coherence (MQC) is described by the dynamics of the relevant macrovariable in an effective potential in which two degenerate energy minima are separated by a classically impenetrable barrier. This is basically a macroscopic analogue of the well-known inversion resonance of the ammonia molecule (as described, for example, in the Feynman Lectures[9]), where oscillations correspond to the system tunnelling coherently backwards and forwards between two degenerate potential wells. The characteristic feature of MQC is that the accessible regions between which tunnelling occurs are *macroscopically distinct*.

In the rf-SQUID system, the two macroscopically distinct states, corresponding to the two different minima of the effective potential, correspond to the supercurrent circulating in opposite directions, clockwise and counterclockwise, with an equal magnitude which is typically of the order of a few microamperes. For MQC, the two states in superposition correspond to different directions of flow. The corresponding separation in the flux space is typically some fraction of the flux quantum, and the centre of mass of the electron system remains very nearly undisplaced, though the direction of the momentum for the two states is, of course, reversed. In fact, the latter feature explains[10,11] why MQC for a SQUID system is *not* inconsistent with spontaneous collapse models (discussed in Chapter 10) that require that the center of mass is displaced appreciably for the dynamical wave-function collapse mechanism to induce decoherence.

The dissipative component of the environmental interaction that would be fatal to phase coherence is quite small in a superconducting device, such as a SQUID ring, due to its extremely low entropy. For example, in a 1 cm niobium ring at 10 mK, the electronic entropy of the entire macroscopic body is less than Boltzmann's constant.[12] Practical requirements for actually demonstrating MQC using the rf-SQUID system in the underdamped regime have been analysed in detail.[7,8] However, before discussing conceptual ramifications of such an experiment, it is necessary to clarify a widespread misconception about the role of quantum mechanics at a macroscopic level.

The question may be asked—why bother to test MQC, when we already know that quantum mechanics is essential for explaining such macroscopic properties as the specific heat of an insulating solid at temperatures below what is called the Debye temperature? However, a crucial point often overlooked is that the observed non-classical behavior of, say, the specific heat can be accounted for by using quantum mechanics for a *single* normal mode of a solid. We do *not* have to invoke the superposition principle involving many-particle states with around 10^{23} particles. This is true even for what are often described as macroscopic quantum phenomena, such as superconductivity, where a typical N-particle wave-function for a superconductor can be viewed as a suitably antisymmetrized product of Cooper pair wave-functions. Nowhere in all the manifestations of such effects is the superposition principle tested for more than two particles. For a detailed clarification of this point with explicit examples of superconductivity and superfluidity, see Leggett.[13]

The conclusion is that so far no direct empirical evidence corroborates the validity of the superposition principle as applied to states of a macroscopic number of particles. As discussed in Chapter 3, it is this extrapolation of the superposition principle to the *macrolevel* that results in the quantum measurement paradox.

MQC experiments using the rf- SQUID system are thus important because the quantum mechanical treatment in this case involves a linear superposition of two *macroscopically distinct* states of a macroscopic number of particles of the order of 10^{15} electrons constituting the supercurrent in the SQUID loop.

Apart from this, the significance of the SQUID example is enchanced by the proposal of a test of the incompatibility between quantum mechanics and macro-realism, analogous to what Bell's theorem does for local realism, as discussed in Chapter 9. In this proposed test, the very notion of *macrorealism* is confronted directly with experiments so that the relevant empirical results agreeing with quantum mechanics are *ipso facto* inconsistent with implications entailed by the concept of macrorealism, irrespective of any specific interpretation. We now examine the arguments underpinning this approach.

The Leggett–Garg Inequality

Leggett and Garg (LG)[14] derived a testable inequality for the SQUID example that is violated by quantum mechanical results under certain conditions, but always satisfied by a general class of *macrorealist models*. For this purpose, we first state precisely the form of macrorealism that is being tested. The postulates of macrorealism can be expressed as follows:

Postulate 1: If, whenever it is observed, a macroscopic system under certain conditions is always found to be in one of two or more macroscopically distinguishable states, then we can assign to it at all times, even when it is not observed, the ontological property of actually being in a definite one of these states. The temporal evolution of such a property is assumed to be inherently deterministic in the sense that the initial property determines the final outcome uniquely at any later instant.

Before proceeding further, a few remarks are in order. First the definition of macrorealism, namely, that a macroscopic system is at all times in a definite macroscopic state, is meant to be contrasted, not with the notion that it is in a definite non-macroscopic state (in fact, a macroscopic system can never be in a non-macroscopic state), but with the notion that it is not in a definite macroscopic state at all. For instance, it might be in a quantum superposition of different macroscopic states.[15] Secondly, the phrase 'at all times' requires qualification. When two or more macroscopically distinct states are available for a macrosystem, the system monitored as a function of time jumps between its different states; thus it spends a certain minimum time in transit. Therefore we must interpret the phrase 'at all times' as meaning 'for the overwhelming majority of the time'.

In the particular case of the SQUID setup, the result of monitoring is almost always a value close to one of the two minima of the potential; values different from these, for example a value corresponding to being in transit under the barrier, are quite rare.[16] Thus in the SQUID example, macroscopically distinguishable states are defined as corresponding to two small and mutually non-overlapping regions around potential well bottoms. The following arguments assume that, whenever a measurement is actually made the SQUID is found to be in one of these two states. Postulate 1 extends this to a hypothesis about the SQUID being in one of these two states even when not measured, and it also assumes that measuring the state of the SQUID reveals the premeasurement state exactly.

By itself postulate 1 is not amenable to an experimental test. We must supplement it with two other postulates.

Postulate 2: For a macrosystem it is possible in principle to determine which of the various macrostates the system is in without affecting its subsequent behaviour, and this determination produces an arbitrarily small perturbation on its subsequent dynamics.

Postulate 3: Results for the subset of an ensemble, chosen at random, on which a given property is actually measured represent faithfully the property of the entire ensemble. It is assumed that the size of the sample is large enough that statistical fluctuations can be ignored.

Postulate 3 is a common assumption, though often not explicitly stated, in the analysis of all such experiments. As regards postulate 2, note that we earlier discussed the notion of non-invasive measurability as the property of macrosystems behaving classically; postulate 2 assumes this to be true for all macrosystems.

We will now see how to obtain the LG inequality from a clearly formulated conjunction of the intuitively plausible hypotheses (postulates 1-3). Note that nowhere in the following derivation is any input from the quantum mechanical formalism used. We consider a general two-state system where the two states in question, denoted by 1 and 2, are macroscopically distinct. According to the macrorealist postulate 1, at any given instant such a system must actually be in one of these two states. Thus the temporal evolution consists of random transitions between states 1 and 2. We define a quantity $Q(t)$ such that if at time t the system is in state 1(2), then the value of $Q(t)$ is $+1$ (-1). Next we consider four different times t_1, t_2, t_3, t_4 such that $0 \leq t_1 \leq t_2 \leq t_3 \leq t_4$. We imagine four separate runs—in each case the system starts at $t = 0$ from a given initial state. On the first run, we consider values of Q at two times, t_1 and t_2, only. Similarly on the second run, Q is measured at times t_2 and t_3, on the third at t_3 and t_4, and finally on the fourth at times t_1 and t_4. On each such run exactly two measurements of Q are taken.

For values ± 1 of $Q(t_i)$, $i = 1, 2, 3, 4$, realised on any set of four runs, the combination $[Q(t_1) Q(t_2) + Q(t_2) Q(t_3) + Q(t_3) Q(t_4) - Q(t_1) Q(t_4)]$ is always either $+2$ or -2; so

$$[Q(t_1) Q(t_2) + Q(t_2) Q(t_3) + Q(t_3) Q(t_4) - Q(t_1) Q(t_4)] = \pm 2 \quad (13.1)$$

Note that in Eq. (13.1) a value of any Q that occurs twice in the expression is taken to be the same. Whether there is any deterministic ingredient required at this stage of the argument has been an issue of some debate; see, for example, Clifton.[17] However, it seems 'all that one needs' to derive the subsequent inequality (13.2) is that the Q's which appear in (13.1) are *uniquely* defined for any one run.

From Eq. (13.1), it follows that if all individual terms in the expression are replaced by their averages over the entire ensemble for such sets of four runs, then we have

$$-2 \leq \langle Q(t_1) Q(t_2) \rangle + \langle Q(t_2) Q(t_3) \rangle + \langle Q(t_3) Q(t_4) \rangle$$
$$- \langle Q(t_1) Q(t_4) \rangle \leq 2 \tag{13.2}$$

Now, by virtue of the random selection assumption (postulate 3), we can identify the ensemble averages, $\langle Q(t_1) Q(t_2) \rangle$ and so on, with the actual measured correlations. In other words, the basic assumption is that the sequences of Q-values used to compute each of the four correlations in Eq. (13.2) with respect to appropriate subsets of the entire ensemble are perfectly random so that their limiting relative frequencies are insensitive to the choice of subsets. For discussion of the nuances related to this condition, see for example, Redhead.[18] Subject to these nuances, we have a *testable inequality* imposing *macrorealist constraints* on time-separated joint probabilities for oscillations in two-state systems.

In the particular case of an ideal nondissipative rf-SQUID system, quantum mechanics predicts

$$\langle Q(t_i) Q(t_j) \rangle = \cos \left[\Delta (t_j - t_i) \right] \tag{13.3}$$

where Δ is the characteristic resonance frequency of the system. Here $\hbar \Delta$ represents the energy splitting associated with the tunnelling phenomenon leading to the relevant MQC effect. A key feature in deriving the quantum prediction of Eq. (13.3) is that, prior to a measurement, the system is at any instant considered to be in a linear superposition of states $Q = \pm 1$. For special choices of t_i, for instance, by taking $t_4 - t_3 = t_3 - t_2 = t_2 - t_1 = \Pi/4\Delta$, it can be seen that the quantum result, Eq. (13.3), is incompatible with the LG inequality, Eq. (13.2), since, according to Eq. (13.3), the total combination of mean values occurring in Eq. (13.2) equals $2\sqrt{2}$. Thus, if in conformity with quantum mechanics, the LG inequality is violated by relevant experimental data, we are forced to sacrifice at least one of the inequality's assumptions, though each of them appears to be innocuous.

In an actual experiment we cannot of course treat the SQUID system as isolated and nondissipative. Hence it is important to know whether a realistic degree of dissipative coupling modify the quantum correlation predictions sufficiently so that they always satisfy the LG inequality. On this question, we note that fairly detailed calculations[19,6] reveal that for a degree of dissipation that is not unrealistically low from an experimental point of view, the quantum mechanical results continue to violate the LG inequality.

Much study has been done on this proposal; see, for example, an up-to-date review by Leggett.[20] Recently, a number of crucial experiments have been performed[21-23] which provide remarkable demonstrations of the MQC effect, and further conclusive tests of the LG inequality are expected.

If the LG inequality is confirmed to be empirically contradicted, then which of the postulates must we abandon? Since postulate 3 is commonly used in the analysis of all such experiments, we would have to conclude that it is the conjunction of the two general postulates 1 and 2 characterising *macrorealism* that is empirically refuted in such situations, irrespective of the interpretational framework.

Other Examples Testing Quantum Superpositions for Macrosystems

In recent years, significant progress has been made in testing the quantum mechanical superposition of states for macromolecules by carrying out sensitive interferometry experiments, a spectacular example being the experiment performed by Arndt et al.[24] using a beam of C_{60} molecules, which are prepared in an oven at 900–1000 K, and which have a considerable spread of velocities, demonstrating diffraction effects in a double-slit experiment, with the slits separated by 100 nm. A striking feature of this experiment is that although the interaction of some of the internal degrees of freedom of vibration of the C_{60} molecule with the surrounding black-body radiation field is not insignificant, it is ensured to be insufficient to destroy the effect of interference. Subsequently, appreciable progress has been made to improve the technology of molecular interferometry to the extent that recently double-slit experiments demonstrating quantum interference have been attempted using molecules having about 100 atoms; for a lucid up-to-date overview of the developments in this area, see Arndt et al.,[25] and also a related study by Hackermuller et al.[26]

Another direction in which the applicability of quantum mechanics at the *mesoscopic* level of molecules has been studied is using magnetic moments in ferritin molecules. The ferritin molecule, which is naturally produced in biological systems such as horse spleen, contains about 4500 iron ions which are coupled mostly antiferromagnetically, but which have a residual ferromagnetic moment of a few hundred Bohr magnetons. In these experiments,[27] a resonance was detected in the absorption and noise spectra of the molecules in an rf magnetic field. Although such an observed resonance can be broadly interpreted as arising from quantum mechanical transitions between the even-parity groundstate and the odd-parity low-lying excited state, detailed explanations of several aspects of the relevant experimental data involve a number of subtleties. The rather complicated magnetic structures of the magnetic biomolecules make quite difficult any unambiguous quantitative test of the quantum mechanical predictions against relevant experiments at the reasonably *macroscopic* level. This has been emphasized by Leggett[20] in his recent comprehensive review of the different possible approaches towards testing the applicability of quantum mechanics for macrosystems.

In this context of testing quantum mechanics for *macro biomolecules*, we may also mention a curious example of quantum measurement using a DNA molecule.[28] Such an example does not merely illustrate the quantum measurement problem, but may also be used to highlight the specific difficulties inherent in various approaches to the measurement problem. Earlier, Rae[29] had noted that mutation in a DNA molecule caused by the passage of high-energy cosmic ray particles seemed to 'fulfil the role of a measuring event, similar to the photon being detected by a polarizer'. Shimony[30] stressed that macromolecular analogues of quantum measuring devices could be particularly significant because macromolecules like DNA 'occupy a strategic position between microscopic and macroscopic bodies'. Percival[31] remarked that the genetic mutation of a DNA molecule by absorbing photons could be viewed as illustrating that 'quantum measurements do not require macroscopic systems'.

The specific thought-experiment under consideration[28] concerns detecting whether a single photon in the γ-wavelength range is emitted by a source within a certain time interval. If a γ photon is emitted, it goes through an arrangement consisting of a pure CsI crystal, followed by an aqueous solution of DNA containing an enzyme known as photolyase. When a single γ photon of energy about 1 MeV interacts with a pure CsI crystal, it produces about 10^6 ultraviolet photons, and these are incident on the aqueous solution of DNA In such a solution, which contains about 10^{10} DNA molecules, the probability of a single DNA molecule absorbing a UV photon is virtually unity.

A DNA molecule absorbing an ultraviolet photon undergoes a stable global configuration change that is *macroscopically discernible* through the adhesion of photolyase enzyme molecules to damaged points of the DNA molecule. Thus such an experimental arrangement registers the measurement information that a γ photon is emitted. In this sense, the DNA molecules act as a detecting device. Note that here the output pulse from the CsI crystal itself is a transient pulse of ultraviolet photons, and so cannot constitute a stable record of measurement information.

If we follow standard quantum mechanics, then just as in any measurement process, the unitary Schrödinger evolution leads to a coherent superposition of the triggered and untriggered states of a detector. Thus in this example the standard quantum treatment predicts that a DNA molecule is left in a superposition of UV-damaged and undamaged states. However, here an individual DNA molecule acts like a switch with regard to its biochemical action on the nearby photolyase enzyme; it is *either* on *or* off, accordingly *either* definitely attracting the enzyme *or* not.

Thus the central question in such an example is *how* to ensure that the quantum mechanical description is consistent with the fact that a particular outcome occurs, and is recorded in the stable form of macroscopically detectable photolyase enzyme attachment to the DNA molecule, regardless of whether or when an experimenter chooses to investigate. Such biomolecular illustrations of quantum measurement in the 'mesoscopic' domain could be useful in probing critically to what extent the various non-orthodox schemes can address the quantum measurement problem.

Finally, among other various proposals for testing quantum superposition of states at the macroscopic level, we briefly discuss a particularly interesting suggestion by Marshall et al.[32] for creating a quantum superposition of states involving of the order of 10^{14} atoms via the interaction of a single photon with an extremely tiny movable mirror attached to a micromechanical oscillator, The mirror is part of a high-finesse optical cavity forming one arm of a Michelson interferometer. The cavity is used to enhance the radiation pressure of the photon on the mirror. The initial superposition of the photon being in either arm causes the system to evolve into a superposition of states corresponding to two distinct locations of the mirror. Thus by studying the interference of the photon, one can probe the creation of superpositions involving the displacement of the mirror. The underlying principle is that if a *small* quantum system, the photon, is coupled to a *large* system, the movable mirror, the existence of the quantum superposition of states of the large system can be verified by observing the disappearance and reappearance of interference of states of the small system,as the large system is driven into a superposition of states and then returns to its initial state.

Marshall et al.[32] have made detailed analysis of the experimental requirements, particularly the conditions which need to be satisfied to ensure a bound on the acceptable environmental decoherence, in order to observe effects of quantum interference of the states of a mirror of size, say, $10 \times 10 \times 10\,\mu m$, and mass of the order of $5 \times 10^{-12}\,kg$, consisting of 10^{14} atoms, about eight to nine orders of magnitude more massive than any superposition observed to date. The authors conclude that, while very demanding, such an experiment should be within reach using a combination of state-of-the-art technologies. Further, it has been argued that such a setup has the potential to test wave-function collapse models, in particular the gravitationally induced wave-function collapse model which was discussed in Chapter 10.

Finally we must stress that the fundamental significance of the accumulation of sufficient empirical evidence in support of the extrapolation of the quantum superposition principle to the macrolevel is that it would make the quantum measurement problem more acute and relevant, since the problem arises essentially because of the assumption of this extrapolation. This would in turn reinforce the need for more studies on the various non-standard interpretations which attempt to address this problem.

The Quantum Zeno Effect

A quantum system always evolves through a coherent superposition of states—an initial state develops into a superposition of states, with the relative weight factors changing with time in a continuous way. The probability of a final measurement finding a system in the initial state, after it is allowed to evolve for a certain period of time, T, is known as *survival probability* of the evolving system. Provided this time is not extremely short, we will find that the survival probability decays exponentially with T.

However now let us imagine we perform a series of $N - 1$ intermediate measurements to determine whether or not the system is in its original state. These measurements are at times T/N, $2T/N$ and so on. If we compute the survival probability as a function of N, we find that survival probability increases with

N, and it tends to unity in the limit as N itself tends to infinity; we may describe the limiting situation as a continuous series of measurements giving rise to a total inhibition of decay. It seems that the dynamical evolution of an isolated quantum system is significantly modified due to intervening measurements. In particular the evolution is inhibited by repeated frequent measurements even if these measurements are apparently non-disturbing; in other words, even if they do not entail a direct exchange of energy or momentum. For example, the survival probability of a radioactive atom may be monitored by the detection of any decay particle; clearly this measurement does not entail any interaction with the decaying atom itself.

The first treatments of the effect were by Ghirardi and co-workers[33,34] and a general analysis was given by Sudarshan and co-workers[35,36] who also gave the effect its name. Actually the effect is often called the 'Quantum Zeno paradox'. It does indeed seem paradoxical if we believe in the passive role of a measurement, that measurement involved watching things happening without active intervention by the measuring device. Actually the significance of the effect lies in providing an impressive illustration of the participatory nature of quantum measurements.

We now give a simple way of demonstrating the result. Let the initial state of a system at $t = 0$ be $|\psi_0\rangle$. The survival probability after time T is given by:

$$P(T) = |\langle\psi_0| \exp(-iHT/\hbar) |\psi_0\rangle|^2 \tag{13.4}$$

setting $\hbar = 1$.

Note that:

$$\langle\psi_0| \exp(-iHt) |\psi_0\rangle = 1 - i\langle H\rangle T - \left(\frac{1}{2}\right)\langle H^2\rangle T^2 + 0(T^3) \tag{13.5}$$

Hence we have

$$P(T) = 1 - (\Delta E)^2 T^2 + 0(T^4) \tag{13.6}$$

where

$$(\Delta E)^2 = \langle\psi_0| H^2 |\psi_0\rangle - \langle\psi_0| H |\psi_0\rangle^2 \tag{13.7}$$

The interesting point, and it is the cause of the quantum Zeno effect, is that the form of the decay given by Eq. (13.6) for small t is $1 - \alpha t^2$, rather than the form $1 - \alpha t$ expected as the small-t approximation by an exponential, where α is a constant independent of t.

Let intermediate measurements be performed $N - 1$ times at intervals T/N to monitor whether the system is in the initial state (yes) or not (no). Applying the postulate of wave-function collapse, after each such measurement corresponding to a yes result, the observed system can be considered to evolve over again from the state $|\psi_0\rangle$. It then follows from Eq. (13.6) that for a sufficiently large value of N, the survival probability after N repeated measurements, including the final one, within time T is given by:

$$P(T) \simeq \left[1 - (\Delta E)^2 \left(\frac{T}{N}\right)^2\right]^N \tag{13.8}$$

which in the limit $N \to \infty$, tends to unity.

Note that in this argument, the measurement interaction resulting in entanglement between states of the observed system and those of the measuring apparatus is not considered, and the concept of wave-function collapse is applied to the observed system exclusively. We now show[37] that if entanglement caused by the measurement interaction is explicitly taken into account by correlations between states of the observed system and the measuring apparatus, the quantum Zeno effect can still be derived without invoking wave-function collapse.

For simplicity we restrict our treatment to a two-state problem where the time evolution operator can connect only two states. If the initial state, say, $|\psi_0\rangle$ is allowed to evolve for a time T uninterrupted by external intervention, the final state, again setting $\hbar = 1$, is given by

$$|\psi(T)\rangle = \exp(-iHT)|\psi_0\rangle \tag{13.9}$$

which may be written as:

$$|\psi(T)\rangle = \alpha(T)|\psi_0\rangle + \beta(T)|\psi_1> \tag{13.10}$$

where $\langle\psi_0 | \psi_1\rangle = 0$.

The survival probability P_s of finding the system in the original state $|\psi_0\rangle$ is given by:

$$P_s = |\alpha(T)|^2 \tag{13.11}$$

Consider an intermediate measurement to determine survival or otherwise at $t = T/2$. After such a measurement:

$$|\Psi(T/2)\rangle = \alpha(T/2)|\psi_0\rangle|A_0\rangle + \beta(T/2)|\psi_1\rangle|A_1\rangle \tag{13.12}$$

where $|A_0\rangle$ and $|A_1\rangle$ are microscopically distinguishable apparatus states: here $\langle A_0|A_1\rangle = 0$. Subsequently the system is allowed to evolve freely until $t = T$, when:

$$|\Psi(T)\rangle = \alpha(T/2)\exp(-iHT/2)|\psi_0\rangle A_0 >$$
$$+\beta(T/2)\exp(-iHT/2)|\psi_1\rangle|A_1\rangle \tag{13.13}$$

A measurement at $t = T$ shows that the survival probability at $t = T$ is given by:

$$P_s' = |\langle\psi_0|\langle A_0|\alpha(T/2)\exp(-iHT/2)|\psi_0\rangle A_0\rangle$$
$$+ \langle\psi_0|\langle A_0|\beta(T/2)\exp(-iHT/2)|\psi_1\rangle A_1 >|^2 \tag{13.14}$$

Now since

$$\langle A_0|A_1\rangle = 0, \qquad \langle\psi_0 | \psi_1\rangle = 0$$
$$\exp(-iHT/2)|\psi_0\rangle = \alpha(T/2)|\psi_0\rangle + \beta(T/2)|\psi_1\rangle$$

P'_s reduces to:

$$P'_s = |\alpha\,(T/2)|^4 \tag{13.15}$$

We can generalise the preceding analysis to the case of measurements at $T/N, 2T/N, \ldots T$, whence:

$$P'_s = |\alpha\,(T/N)|^{2N} \tag{13.16}$$

From Eqs. (13.11) and (13.16) it is evident that in general, and in particular for an exponentially decaying system, where $|\alpha\,(T)|^2$ is an exponentially decreasing function of time:

$$P_s \neq P'_s \tag{13.17}$$

In a general case we have

$$|\alpha\,(T/N)|^2 = 1 - (\Delta E)^2 \left(\frac{T}{N}\right)^2 + \cdots \tag{13.18}$$

Hence as in Eq. (13.8) we obtain

$$P'_s = \left[1 - (\Delta E)^2 \left(\frac{T}{N}\right)^2 + \cdots\right]^N \tag{13.19}$$

which again tends to unity as $N \to \infty$. However, this argument makes it quite clear that the result does not depend on the use of the collapse of wave-function. Indeed it is not a property of any particular interpretation of quantum theory, but of quantum theory itself.

The Quantum Zeno Effect: Experimental Test

The most well-known experiment using the mechanism of the quantum Zeno effect was performed by Wayne Itano and colleagues.[38] This experiment used laser-cooled $^9Be^+$ ions trapped in the ground state $|\phi_1\rangle$. (The ability to trap and manipulate single ions, and the use of laser pulses to cool the ions, or in other words to prevent them from gaining energy and escaping, has been one of the most exciting experimental developments of recent decades.[39] The technique also has great potential as the basis of a quantum computer.)

In the experiment, a continuous radio frequency field drives transitions between $|\phi_1\rangle$ and an excited state $|\phi_2\rangle$. The latter state is said to be metastable in the sense that spontaneous decay from $|\phi_2\rangle$ to $|\phi_1\rangle$ is taken to be negligible. These transitions are monitored by means of a sequence of brief intense laser pulses, that excite ions from the level $|\phi_1\rangle$ into an unstable high-lying level $|\phi_3\rangle$ with a probability of almost unity; these then almost immediately decay back to $|\phi_1\rangle$, emitting an observable fluorescence photon. It may be said that the short intense pulse, followed by photon counting of the fluorescence, thus acts as a measurement of survival probability in the state $|\phi_1\rangle$. The number of measurements by the pulsed laser is varied within the duration of the excitation from $|\phi_1\rangle$ to $|\phi_2\rangle$. As the number of measurements increases, the probability of a transition between $|\phi_1\rangle$ and $|\phi_2\rangle$ diminishes and tends to zero.

We shall now explain why this may be described as an experimental demonstration of the quantum Zeno effect. We first note that, if the ion is definitely in state $|\phi_1\rangle$ at $t= 0$, and if a measurement is performed a *short* time t later, to discover whether, as a result of the presence of the radio frequency field, it has made a transition to state $|\phi_2\rangle$, the probability of it having done so will be proportional to t^2. As we have said, it is this t^2-region that is responsible for the quantum Zeno effect. Exactly the mathematics of the previous section now tells us that, if one regards the laser pulses as intermediate measurements, checking whether the ion is in $|\phi_1\rangle$ or $|\phi_2\rangle$, the mere provision of these pulses inhibits the transitions to $|\phi_2\rangle$ and thus increases the 'survival' of $|\phi_1\rangle$.

The Quantum Zeno Effect: Recent Work

Following the Itano paper, a vast amount of work has been performed on the quantum Zeno effect.[40,41] It may be said that it has been transformed from being a curiosity, regarded as a paradox, into being an aspect of quantum processes practically taken for granted. Among recent ideas has been the realisation[42,43] that there may also be a quantum anti-Zeno or inverse Zeno effect, in which change is accelerated rather than slowed down. Another interesting development relates to the fact that, for any future quantum computer, an important aim must be to inhibit errors occurring because of transitions from the correct state to a neighbouring state. It is clear that the quantum Zeno effect could serve to inhibit such transitions, and much study of this possibility has been made.[44–46]

Despite the general acceptance of quantum Zeno ideas, there are still conceptual questions to be asked.[40] It might be said that the change from being regarded as a curiosity to one of acceptance has come about because of an implicit shift in the features of the effect that are being analysed. The effects being considered at present are generally changes in the rate of a process caused by an actual intervention; these are interesting, but, of course, in no sense paradoxical. One of the central worries of the original suggestions—that changes are caused by the mere presence of a passive observing device—has been put on one side. It would seem that there is still work to be done in understanding the quantum Zeno effect.

Time in Quantum Mechanics

For the third main division of this chapter we turn to the subject of time in quantum theory. In Chapter 3, we made some brief comments on the so-called time–energy uncertainty principle, and we first bring these up-to-date. There has been a vast amount of discussions of precisely how this should be interpreted. Among the interpretations proposed by ensuring consistency with the formalism of quantum mechanics, the ones originally suggested by Mandelstamm and Tamm[47] and by Allcock.[48] , and the one formulated specifically relevant to the lifetime of a decaying state[49] are now the usually used forms of the time–energy uncertainty relation. In the early years of quantum mechanics, Landau and Peierls[50] proposed a version which related the duration of the measurement of energy and the energy transfer to the observed system during the measurement of energy. However, later Aharonov and Bohm[51,52] showed in detail that such a formulation of the time-energy uncertainty relation has no rigorous basis within the formalism of quantum mechanics—a conclusion corroborated also by recent studies.[53] A recent review of the relation has been provided by Busch.[54]

We now consider the more general study of time in quantum theory. In Chapter 7, while discussing Einstein's arguments on the macroscopic limit of quantum mechanics, we mentioned an example he considered involving the recording on a chart-recorder connected to a Geiger-counter of a 'mark' corresponding to the registration of a decay product emitted by a radioactive atom. Although Einstein's purpose underlying this example was somewhat different, as discussed in Chapter 7, an important aspect of this example is that it involved the measurement of the instant at which a decay product reached the chart-recorder. Thus such an example implied treating time as an observable quantity whose probability distribution could be measured, at least in principle, through the observation of the distribution of position. This is particularly interesting because of the peculiar role time plays within the formalism of quantum mechanics—it differs fundamentally from all other dynamical quantities like position or momentum since it appears in the Schrödinger equation as a parameter, not as an operator.

If the wave-function $\psi(x, t)$ is the solution of Schrödinger's equation, then Born's statistical interpretation of the wave-function implies that $|\psi(x, t)|^2$ is the probability density that a particle described by the wave-function $\psi(x, t)$ can be found between x and $x + dx$ at the instant t. Hence $|\psi(x, t_1)|^2$, $|\psi(x, t_2)|^2$, $|\psi(x, t_3)|^2, \ldots$ give the probability distributions at the respective different instants $t_1, t_2, t_3, \ldots$ for a fixed region of space between x and $x + dx$. Now, if we fix the positions at $x_1, x_2, x_3, \ldots$, we may ask whether the quantities $|\psi(x_1, t)|^2$, $|\psi(x_2, t)|^2$, $|\psi(x_3, t)|^2, \ldots$ specify the *time probability distributions* at respective various positions $x_1, x_2, x_3, \ldots$ However, one can easily see that although $\int_{-\infty}^{\infty} |\psi(x, t_i)|^2 \, dx = 1$, in general, $\int_0^{\infty} |\psi(x_i, t)|^2 \, dt \neq 1$. Hence, starting from the wave-function, we do not readily have the proper relevant interpretation to calculate the time probability distribution, unlike the position probability distribution.[55]

The fundamental difficulties in constructing a self-adjoint time operator within the formalism of quantum mechanics were first analyzed by Pauli.[56] Another argument for the nonexistence of the time operator was given by Allcock.[48] Nevertheless, there have been subsequent attempts to construct a suitable time operator. For instance, Grot et al.[57] suggested a time-of-arrival operator for a free particle, and showed how the time probability distribution $\Pi(t)$ can be calculated using it;

interestingly, such an operator admits an orthogonal basis of eigenstates, although the operator is not self-adjoint.

Here we may note that in recent years there has been an upsurge of interest in analysing the concepts of various types of time in quantum mechanics, such as *tunnelling time, decay time, dwell time, arrival time*, or *jump time*; for overviews on this subject, see Refs. [55, 58]. Among these various concepts, the most discussed one is the *tunnelling time* through a potential barrier; for comprehensive reviews on this particular topic, see Refs [59, 60]. Different types of definitions of quantum tunnelling time have been suggested in the literature using numerical[61,62] and analytical[63,64] approaches, as well as using the Feynman path integral formulation[65] and the Wigner phase space distribution.[66] A recent approach suggested by Brouard et al.[67] has tried to include a range of different ways of defining quantum tunnelling as particular cases of a single framework.

The subtle role played by time in quantum mechanics also leads to a serious ambiguity when it comes to the measurement of the length of time between quantum events. This has resulted in confusion about the interpretation of the results of a number of interesting experiments[68-71] related to quantum tunnelling. While some such experiments seem to superficially suggest that superluminal quantum tunnelling might have been observed, it has been argued that the actual propagation of usable information does *not* exceed the speed of light in these experiments, so that the relativistic causality remains intact.[72-74]

An interesting line of study which has recently been suggested is to approach the problem of quantum tunnelling time by introducing a coupling between a clock and a particle which tunnels through a potential barrier; one defines the tunnelling time in terms of the change in the clock variable during the time that the particle spends within the potential barrier.[75] This clock has a single degree of freedom, and it is designed to run for as long as the moving quantum particle lies within a defined region of space. The expectation value of the clock variable remains fixed once the particle has left that region, and it may be read by a normal position measurement at any stage subsequently. It then becomes possible to define what may be called the 'expectation time' for the tunnelling event in terms of the expectation value of the clock variable which is treated as a quantum mechanical observable.

The earliest proposal for a model quantum clock in order to measure the time of flight of quantum particles was suggested by Salecker and Wigner,[76] and the idea was later elaborated by Peres.[77] In effect, this model of quantum clock measures φ, the change in the phase of the wave-function, over the duration to be measured. The *transit time* is then given by

$$T = \hbar \partial \varphi / \partial E \qquad (13.20)$$

where E is the relevant particle's energy. Full details have been discussed by Peres, who showed how to construct the time-independent Hamiltonians corresponding to the various uses of a quantum clock.[77] Such a model can yield precise expectation values for the *transit time* of quantum particles passing through a given region of space. This is true not only for localised wave-functions, but also for states which are highly delocalised spatially such as energy eigenstates. For example, in this model, for a uniform flux of particles of mass m moving in free space, described by the plane wave energy and momentum eigenstate, $\exp(ikx - iEt/\hbar)$, a transit time of d/v is predicted through the region $0 \leq x \leq d$ where $v = \hbar k/m$, in analogy to the relevant classical transit time.

This type of model of a quantum clock has been applied in a specific example[70] to derive a self-consistent analytical expression for what may be called the expectation value of the time taken by particles in an energy and momentum eigenstate to pass through a square potential barrier. Such a line of study can be extended to other examples involving the potential well and other types of barriers, including the case of a double square barrier separated by a gap, in the context of which one can check the claim[78] which has been made about the possibility of the quantum traversal time implying superluminal propagation.

Another recent application[79] of this type of quantum clock model has been in studying the motion of quantum particles in a uniform gravitational field. The calculated results indicate that the *transit time* of a quantum particle which is projected vertically and is allowed to fall back approaches the classically predicted *mass-independent transit time* if one uses a quantum clock fixed in the reference frame of the Earth, and provided the classical turning point is sufficiently far away. In such a situation, using a specific model of a quantum clock, one needs to take into account the fact that the quantum particles have a probability of tunnelling into the classically forbidden region above the classical turning point, and the tunnelling depth depends on mass. However this mass-dependent 'quantum delay' in the return time is cancelled by the probability that the particles may be back-scattered by the gravitational potential before reaching the classical turning point.

An extension of the above study has been made in the case of a non-uniform gravitational field, and the possibility of experimentally checking the predicted results has been discussed[80] by relating such a calculation to the scheme suggested by Chiao and Speliotopoulos[81] using atomic interferometry.

Another type of situation which has recently attracted considerable attention is as follows. Let us consider a propagating wave packet, and let the initial instant used in treating its propagation through a given region of space be taken to be the same for all the particles corresponding to this wave packet. Then the quantum transit time distribution through this region of space becomes effectively what has been referred to in the relevant literature as the *quantum arrival time distribution*.

To illustrate this issue, let us consider the following simple experimental arrangement. A particle moves in one dimension along the x-axis and a detector is placed in the position $x = X$. Let T be the time at which the particle is detected, which we denote as the 'time of arrival' of the particle at position X. Can we predict T from the knowledge of the state of the particle at the prescribed initial instant?

In classical mechanics, the answer is simple. The time of arrival T at $x = X$ of an individual particle with the initial position x_0 and momentum p_0 is $T = t(X; x_0, p_0)$ which is fixed by the solution of the equation of motion of the particle concerned. But in quantum mechanics, this problem becomes nontrivial, since it requires determining the probability distribution of times, denoted by $\Pi(t)$, at which the particles are detected at the detector location, X, within the time interval between t and $t+dt$. Then $\int_{T_1}^{T_2} \Pi(t)\,dt$ is the probability that the particles are detected at $x = X$ between the instants T_1 and T_2. In other words, the relevant key question in quantum mechanics is how to calculate $\Pi(t)$ at a specified position, given the wave-function at the initial instant from which the propagation is considered.

Since there is an inherent non-uniqueness within the standard formalism of quantum mechanics for calculating such a probability distribution, $\Pi(t)$, various schemes have been suggested for calculating the *arrival time distribution*, for example using the path integral approach,[82] the probability current density approach,[83-85] or the Bohmian trajectory model.[86-88]

It still an open question as to what extent these different approaches can be empirically discriminated. An interesting effort along this direction was made[89] by considering the measurement of arrival time using the emission of a first photon from a two-level system moving into a laser-illuminated region. The probability for this emission of the first photon was evaluated[90] using the quantum jump approach. Subsequently, further work was done on this proposal using Kijowski's distribution.[91] Further studies using other approaches are required in the context of this type of experimentally realizable situation.

Recently another direction of study has been suggested[92] which can also be useful for relating any axiomatically defined quantum time distribution to the actually testable results. In this scheme, using any specific approach, if one calculates the quantum transit time distribution of spin-1/2 neutral particles passing through a spin-rotator (SR) which contains a constant magnetic field, then such a calculated time distribution can be used for evaluating the observable distribution of spin orientations for the particles which emerge from the SR. Consequently, for such an ensemble, the result of spin measurement along any arbitrary direction can be predicted, and this is amenable to experimental verification. Here it may be noted that the setup suggested in this scheme has already been used in a recent neutron interferometric experiment[93] in a different context.

Further work on this scheme is required to examine in detail the feasibility of the relevant experimental test. Moreover, this scheme has so far been illustrated only using the probability current density approach.[83-85] Hence calculations using this scheme based on other approaches would be useful. Overall, more comprehensive investigations from different perspectives are required to explore the possibilities of various types of testable schemes which may enable to throw more light on this entire issue of fixing unambiguously the quantum mechanical predictions of different types of time distributions.

References

1. Folse, H.J. (1985). The Philosophy of Niels Bohr. Amsterdam: North-Holland; Whitaker, A. (2006). Einstein, Bohr and the Quantum Dilemma. (1st edn. 1996, 2nd edn. 2006) Cambridge: Cambridge University Press.

2. Clauser, J.F. (2002). Early history of Bell's theorem, In: Quantum [Un]speakables. (Bertlmann, R.A. and Zeilinger, A., eds.) Berlin: Springer, pp. 61–98.

3. Ghose, P. (2006). Testing Quantum Mechanics on New Ground. Cambridge: Cambridge University Press.

4. Leggett, A.J. (1984). Macroscopic quantum tunnelling and related effects in Josephson systems, In: Percolation, Localization, and Superconductivity. (Goldman, A.M and Wolf, S., eds.) NATO Advanced Study Institute, Vol. 109. New York: Plenum, pp. 1–41.

5. Leggett, A.J. (1986). Quantum mechanics at the macroscopic level, In: Directions in Condensed Matter Physics. (Grinstein, G. and Mazenko, G., eds.) Singapore: World Scientific, pp. 237–44.

6. Leggett, A.J., Chakravarty, S., Dorsey, A.T., Fisher, M.P.A., Garg, A. and Zwerger, W. (1987). Dynamics of the dissipative two-state system, Reviews of Modern Physics 59, 1–85.

7. Tesche, C.D., Kirtley, J.R., Gallagher, W.J., Kleinsasser, A.W., Sandstorm, R.L., Raider, S.I., and Fisher, M.P.A. (1989). In: Proceedings of the 3rd International Symposium on the Foundations of Quantum Mechanics (Kobayashi, S. et al., eds.), Tokyo: Physical Society of Japan, pp. 233–43.

8. Tesche, C.D. (1990). Can a non-invasive measurement of magnetic flux be performed with superconducting circuits? Physical Review Letters 64, 2358–61.

9. Feynman, R.P., Leighton, R.B., and Sands, M. (1964). Feynman Lectures on Physics, Vol. 3, Chs. 9–1. Reading, Massachusetts: Addision Wesley.

10. Rae, A.I.M. (1990). Can GRW theory be tested by experiments on SQUIDs? Journal of Physics A 23, L57–60.

11. Gallis, M.R. and Fleming, G.N. (1990). Environmental and spontaneous localization, Physical Review A 42, 38–48.

12. Leggett, A.J. (1986). Quantum mechanics at the macroscopic level. In: Lesson of Quantum Theory (de Boer, J., Dal, E. and Ulfbeck, O., eds.) Amsterdam: Elsevier, p. 49–64.

13. Leggett, A.J. (1980). Macroscopic quantum systems and the quantum theory of measurement, Progress of Theoretical Physics (Supplement) 69, 80–100.

14. Leggett, A.J. and Garg, A. (1985). Quantum mechanics versus macroscopic realism – is the flux there when nobody looks? Physical Review Letters 54, 857–60.

15. Leggett, A.J. (1987). Experimental approaches to the quantum mechanics paradox, In: Proceedings of the 2nd International Symposium on the Foundations of Quantum Mechanics. (Namiki, M. et al., eds.), Tokyo: Physical Society of Japan, pp. 297–317.

16. Bol, D.W. and Ouboter, R.D. (1988). Thermal activation in the quantum regime and macroscopic tunnelling in the thermal regime in a metabistable system ..., Physica 154B, 56–65.

17. Clifton, R.K. (1991). In: Proceedings of the Symposium on the Foundations of Modern Physics 1990 (Lahti, P. and Mittelsteadt, P., eds.), Singapore: World Scientific.

18. Redhead, M. (1987). Incompleteness, Nonlocality, and Realism. Oxford, UK: Oxford University Press, Ch. 4.

19. Chakravarty, S. and Leggett, A.J. (1984). Dynamics of the two-state system with ohmic dissipation, Physical Review Letters 52, 5–8.

20. Leggett, A.J. (2002). Testing the limits of quantum mechanics: motivation, state of play, prospects, Journal of Physics – Condensed. Matter 14, R 415–51.

21. Yu, Y., Han, S.Y., Chu, X., Chu, S.I., and Wang, Z. (2002). Coherent temporal oscillations of macroscopic states in a Josephson junction, Science, 296, 889–92.

22. Vion, D., Aassime, A., Cottet, A., Joyez, P., Pothier, H., Urbina, C., Esteve, D., and Devoret, M.H. (2002). Manipulating the quantum state of an electrical circuit, Science 296, 886–9.

23. Friedman, J.R., Patel, V., Chen, W., Tolpygo, S.K., and Lukens, J.E. (2000). Quantum superposition of distinct macroscopic states, Nature **406**, 43–6.

24. Arndt, M., Nairz, O., Vos-Andreae, J., Keller, C., van der Zouw, G., and Zeilinger, A. (1999). Wave-particle duality of C_{60} molecules, Nature **401**, 680–2.

25. Arndt, M., Hornberger, K., and Zeilinger, A. (2005). Probing the limits of the quantum world, Physics World **18**(3), 35–40.

26. Hackermuller, L., Uttenthaler, S., Hornberger, K., Reiger, E., Brezger, B., Zeilinger, A., and Arndt, M. (2003). Wave nature of biomolecules and fluorofullerenes, Physical Review Letters **91**, 090408.

27. Awschalom, D.D., Smyth, J.F., Grinstein, G., DiVincenzo, D.P., and Loss, D. (1992). Macroscopic quantum tunnelling in magnetic proteins, Physics Review Letters **68**, 3092–5.

28. Home, D. and Chattopadhyaya, R. (1996). DNA molecular cousin of Schrödinger's cat: a curious example of quantum measurement, Physical Review Letters **76**, 2836–9.

29. Rae, A. (1986). Quantum Physics: Illusion or Reality? Cambridge: Cambridge University Press, 1986, p. 61.

30. Shimony, A. (1989). Search for a worldview which can accommodate our knowledge of microphysics, In: Philosophical Consequences of Quantum Theory. (Cushing, J.T. and McMullin, E., eds.), Notre Dame: University of Notre Dame Press, pp. 25–37.

31. Percival, I. (1991). Schrödinger quantum cat, Nature **351**, 357.

32. Marshall, W., Simon, C., Penrose, R., and Bouwmeester, D. (2003). Towards quantum superposition of a mirror, Physical Review Letters **91**, 130401.

33. Fonda, L., Ghirardi, G.C., Rimini, A., and Weber, T. (1973). Quantum foundations of exponential decay law, Nuovo Cimento A **15**, 689–704.

34. Degasperis, A., Fonda, L., and Ghirardi, G.C. (1974). Does lifetime of an unstable system depend on measuring apparatus, Nuovo Cimento A **21**, 471–84.

35. Misra, B. and Sudarshan, E.C.G. (1977). Zeno's paradox in quantum theory, Journal of Mathematical Physics **18**, 756–63.

36. Chiu, C.B., Sudarshan, E.C.G., and Misra, B. (1977). Time evolution of unstable quantum states and a resolution of Zeno's paradox, Physical Review D **16**, 520–9.

37. Home, D. and Whitaker, M.A.B. (1992). A critical re-examination of the quantum Zeno paradox, Journal of Physics A **25**, 657–64.

38. Itano, W.M., Heinzen, D.J., Bollinger, J.J., and Wineland, D.J. (1990). Quantum Zeno effect, Physical Review A **41**, 2295–300.

39. Wineland, D.J. and Itano, W.M. (1987). Laser cooling, Physics Today **40**(6), 34–40.

40. Home, D. and Whitaker, M.A.B. (1997). A conceptual analysis of quantum Zeno: paradox, measurement and experiment, Annals of Physics **258**, 237–85.

41. Facchi, P. and Pascazio, S. (2001). Quantum Zeno and inverse quantum Zeno effects, Progress in Optics **42**, 147–217.

42. Facchi, P., Nakazoto, H., and Pascazio, S. (2001). From the quantum Zeno to the inverse quantum Zeno effect, Physical Review Letters **86**, 2699–703.

43. Balachandran, A.P. and Roy, S.M. (2000). Quantum anti-Zeno paradox, Physical Review Letters **84**, 4019–22.

44. Facchi, P. Lidar, D.A., and Pascazio, S. (2004). Unification of dynamical decoupling and the quantum Zeno effect, Physical Review A **69**, 032314.

45. Luis, A. (2001). Quantum state preparation and control via the quantum Zeno effect, Physical Review A **63**, 052112.

46. Facchi, P., Tasaki, S., Pascazio, S., Nakazato, H., Tokuse, A., and Lidar, D.A. (2005). Control of decoherence: analysis and comparison of three different strategies, Physical Review A **71**, 022302.

47. Mandelstam, L. and Tamm, I.G. (1945). The uncertainty principle between energy and time in nonrelativistic quantum mechanics, Journal of Physics (USSR) **9**, 249–54.

48. Allcock, G.R. (1969). The time of arrival in quantum mechanics, Annals of Physics (New York) **53**, 253–85, 286–310, 311–48.

49. Gislason, E.A., Sabeli, N.H., and Wood, J.W. (1985). New form of the time-energy uncertainty relation, Physical Review A **31**, 2078–81.

50. Landau, L. and Peierls, R. (1931). Extension of the principle of indeterminateness for the relativistic quantum theory, Zeitschrift für Physik **69**, 56–69.

51. Aharonov, Y. and Bohm, D. (1961). Time in the quantum theory and the uncertainty relation for time and energy, Physical Review A **122**, 1649–58.

52. Aharonov, Y. and Bohm, D. (1964). Answer to Fock concerning the time energy indeterminacy relation, Physical Review B **134**, 1417–8.

53. Pegg, D.T. (1991). Wave-function collapse time, Physics Letters A **153**, 263–4.

54. Busch, P. (2002). The time-energy uncertainty relations, in Ref. [55], pp. 69–98.

55. Muga, J.G. Sala Mayato, R., and Egusquiza, I.L. (eds.) (2002). Time in Quantum Mechanics. Berlin: Springer.

56. Pauli, W. (1958). In: Encyclopedia of Physics (Flugge, S., ed.) Berlin: Springer, Vol.V/1, p. 60.

57. Grot, N., Rovelli, C., and Tate, R.S. (1996). Time of arrival in quantum mechanics, Physical Review A, **54**, 4676–90.

58. Muga, J.G. and Leavens, C.R. (2000). Arrival time in quantum mechanics, Physics Reports **338**, 353–438.

59. Hauge, E.H. and Stovneng, J.A. (1989). Tunnelling times – a critical review, Reviews of Modern Physics **61**, 917–36.

60. Landauer, R. and Martin, T. (1994). Barrier interaction times in tunnelling, Reviews of Modern Physics **66**, 217–28.

61. Collins, S., Lowe, D., and Barker, J.R. (1987). The quantum-mechanical tunnelling time problem revisited, Journal of Physics C **20**, 6213–32.

62. Dumont, R.S. and Marchioro, T.L. (1993). Tunneling-time probability-distribution, Physical Review A **47**, 85–97.

63. Garcia-Calderon, G. and Villavicencio, J. (2001). Time dependence of the probability density in the transient regime for tunnelling, Physical Review A **64**, 012107.

64. Garcia-Calderon, G. (2002). Decay time and tunnelling transient phenomena, Physical Review A **66**, 032104.

65. Sokolovski, D. and Baskin, L.M. (1987). Transverse time in quantum scattering, Physical Review A **36**, 4604–11.

66. Capasso, F., Mohammed, K., and Cho, A.Y. (1986). Resonant tunnelling through double barriers, perpendicular quantum transport phenomena in superlattices, and their device applications, IEEE Journbal of Quantum Electronics **22**, 1853–69.

67. Brouard, S., Sala, R., and Muga, J.G. (1994). Systematic approach to quantum transmission and reflection times, Physical Review A **49**, 4312–25.

68. Steinberg, A.M., Kwiat, P.G., and Chiao, R.Y. (1993). Measurement of the single photon tunnelling-time, Physical Review Letters **71**, 708–11.

69. Chiao, R.Y. (1993). Superluminal (but causal) propagation of wave-packets in transparent media with inverted atomic populations, Physical Review A **48**, R34–7.

70. Chiao, R.Y., Kozhekin, A.E., and Kuriski, G. (1996). Tachyonlike excitations in inverted two-level media, Physical Review Letters **77**, 1254–7.

71. Chiao, R.Y. and Steinberg, A.M. (1997). Tunneling times and superluminality, Progress in Optics **37**, 345–405.

72. Chiao, R.Y., Kwiat, P.G., and Steinberg, A.M. (1993). Faster than light, Scientific American **269**(2), 52–60.

73. Ranfagni, A., Fabeni, P., Pazzi, G.P., and Mugnai, D. (1993). Anomalous pulse delay in microwave propagation – a plausible connection to the tunnelling time, Physical Review E **48**, 1453–60.

74. Steinberg, A.M. (2003). Clear message for causality, Physics World **16**(12), 19–20.

75. Davies, P.C.W. (2005). Quantum tunnelling time, American Journal of Physics **73**, 23–7.

76. Salecker, H. and Wigner, E.P. (1958). Quantum limitations of the measurement of space-time distances, Physical Review **109**, 571–7.

77. Peres, A. (1980). Measurement of time by quantum clocks, American. Journal of Physics **48**, 552–7.

78. Olkhovsky, V.S., Recami, E., and Salesi, G. (2002). Superluminal tunnelling through two successive barriers, Europhysics Letters **57**, 879–84.

79. Davies, P.C.W. (2004). Quantum mechanics and the equivalence principle, Classical and Quantum Gravity **21**, 2761–72.

80. Davies, P.C.W. (2004). Transit time of a freely falling quantum particle in a background gravitational field, Classical and Quantum Gravity **21**, 5677–83.

81. Chiao, R.Y. and Speliotopoulos, A.D. (2003). Quantum interference to measure space-time curvature: a proposed experiment at the intersection of quantum mechanics and general relativity, International Journal of Modern Physics D **12**, 1627–32.

82. Yamada, N. and Takagi, S. (1991). Quantum mechanical probabilities on a general spacetime surface: 2. Nontrivial example of non-interfering alternatives in quantum mechanics, Progress of Theoretical Physics **85**, 599–615.

83. Muga, J.G., Brouard, S., and Macias, D. (1995). Time of arrival in quantum mechanics, Annals of Physics (New York) **240**, 351–66.

84. Delgado, V. (1999). Quantum probability distribution of arrival times and probability current density, Physical Review A **59**, 1010–20.

85. Ali, M.M., Majumdar, A.S., Home, D., and Sengupta, S. (2003). Spin-dependent observable effect for free particles using an arrival time distribution, Physical Review A **68**, 042105 and reference therein.

86. Leavens, C.R. (1993). Arrival time distributions, Physics Letters A **178**, 27–32.

87. McKinnon, W.R. and Leavens, C.R. (1995). Distributions of delay times and transmission times in Bohm's causal interpretation of quantum mechanics, Physical Review A **51**, 2748–57.

88. Leavens, C.R. (1998). Time of arrival in quantum and Bohmian mechanics, Physical Review A **58**, 840–7.

89. Damborenea, J.A., Egusquiza, I.L., Hegerfeldt, G.C., and Muga, J.G. (2002). Measurement-based approach to quantum arrival times, Physical Review A **66**, 052104.

90. Hagerfeldt, G.C., Seidel, D., and Muga, J.G. (2003). Quantum arrival times and operator normalization, Physical Review A **68**, 022111.

91. Kijowski, J. (1974). On the time operator in quantum mechanics ands the Heisenberg uncertainty relations for energy and time, Reports on Mathematical Physics **6**, 361–86.

92. Pan, A.K., Ali, M.M., and Home, D. (2006). Observability of the arrival time distribution using spin-rotator as a quantum clock, Physics Letters A **352**, 296–303.

93. Hasegawa, Y., Loidl, R., Badurek, G., Baron, M., and Rauch, H. (2003). Violation of Bell-like inequality in single-neutron interferometry, Nature **425**, 45–8.

14
Assessment of Einstein's Views and Contributions

How Would Einstein have Reacted to the Present Situation?

Later in this last chapter we assess Einstein's contributions to quantum theory. However we start by recognising how much the study of quantum theory has advanced in the half-century since Einstein's death. In Part C of this book, we discussed the work of John Bell, the range of novel interpretations of quantum theory, the coming of quantum information theory, and the new ideas on the classical limit of quantum theory, while in the previous chapter we reviewed a few of the current developments. As we have stressed all this serves to vindicate Einstein's practically lone opposition to Copenhagen over the last 30 years of his life, but we now ask the question—how would Einstein have reacted to these new lines of study?

To start with the classical limit, the work in this area, and actually the mere fact that work is still being carried on in this area, would surely gratify Einstein. In the Born–Einstein letters,[1] perhaps the least harmonious exchange was a fairly sustained one over this topic, an exchange that appears to have caused distinct signs of irritation in Einstein. Born attempted to show that all that was required to demonstrate that the Copenhagen interpretation explained the classical limit was the trivial job of taking a suitable mathematical limit, but Einstein felt that this procedure worked only in special cases. He would be glad to find out that, as sketched in Chapter 12, it is now recognised that, considered in general, the classical limit problem is complicated and subtle. Even Einstein's famous question,[2] 'Is the Moon there when nobody is looking at it?', which at the time was probably looked as demonstrating naivety if not senility, can now be seen as pointing out the problem of classicality, why a classical object does not gradually dephase. This is the problem answered only fairly recently by the theory of decoherence, again as explained in Chapter 12.

Indeed most of the other developments: Bell's proof that hidden-variable theories were not excluded, and the many alternative theories to Copenhagen should logically all bring comfort to Einstein. His main complaint during his lifetime would have been the obsequious following of Bohr's line of argument, the fact that there were virtually no physicists prepared to question an approach that he felt was evasive, and merely avoided tackling the real issues. He should be glad that, although there are that still many who would believe that Bohr provided the correct solutions, there are many others who challenge Bohr's doctrines from much the same perspective as Einstein. Irrespective of whether he would have liked any of the new interpretations much, and one would suspect that none of them would have been very much to his taste, he should have felt it a positive sign that there has been so much thought and discussion, so many new ideas, rather than the supine acceptance of what Einstein would have thought of as the 'sterile positivism' of the Copenhagen dogma. And, again while he himself did not support hidden variable theories, he should value Bell's proof as a clear indication of the over-reaching confidence of the orthodox position.

Of course one must be cautious in saying how Einstein 'should' react to these developments because one might well have said exactly the same about how he would react to the Bohm theory, which again, however Einstein perceived its merits, was certainly a counter to Copenhagen; we know that, in fact, he had no time for Bohm's approach at all. Perhaps, though, he would not have found all today's developments as 'cheap' as that of Bohm. Many, after all, have been based, directly or indirectly, on the EPR paper, and, as stressed in Chapter 11, much of the content of quantum information theory also derives fundamentally from the EPR paper.

We have referred to what, from Einstein's perspective, would have been the positive side of Bell's work—the fact that hidden variable theories were not ruled out, but, of course, the development since his death that would have been of most interest to Einstein would have been the negative aspect of Bell's work; hidden variable theories could not be local, or, in other words, one would have to choose between locality and realism.

In Chapter 8, we spent some time discussing the relative importance for Einstein of determinism, realism and locality, but we recognised that, during his lifetime, Einstein was in the comfortable position of believing he could retain all three. There was no general prohibition on this, and Einstein was quite entitled to hope that the unified field theory he aimed at would indeed give him what he wanted. Bell's theorem obviously changed that completely. Assuming experiment eventually rules against Bell's inequality, a clear choice must be made between realism and locality, determinism at this point remaining separate. The hard choice may be made somewhat easier by remembering from Chapter 9 that the non-locality involved in Bell's theorem is of the type that violates outcome independence but not parameter independence. In other words two measurement results at different locations may be connected in a non-local way, but not an apparatus setting and a related measurement result; in other words there can be *no signalling*. This is a relatively benign form of non-locality, and it is possible to suggest that it might be more acceptable, or at least less unacceptable, to Einstein than a scenario in which signals travelled faster than light.

We may recall our suggested distinction between Einstein locality, according to which spatially separated systems have their own properties, and Bell locality, which prohibits signals faster than light between a setting of an apparatus and a measurement result, or between two measurement results. Don Howard[3,4] makes a similar distinction between what he calls *separability*, analogous to our Einstein locality, and his *locality*, analogous to our Bell locality. Technically it might be said that, when discussing his separability, Howard is talking of *states* of spatially separated systems rather than Einstein's *properties*. This does not make any real difference if Einstein were thinking quantum mechanically, but if, as one would suspect, Einstein is including elements of realism over and above what may be obtained from the wave-function, Einstein's locality and Howard's separability would not be identical. It may be mentioned that Bell's locality is defined specifically to refer to quantities that may be observed experimentally, naturally enough because he was setting up inequalities to be tested by experiment. Einstein was thinking rather more about how real separated systems actually 'are'. Einstein's locality is hence effectively an addition to realism; Bell's locality is entirely independent of realism.

One might contemplate the possibility of driving a wedge between Einstein's locality and Bell's locality, so that, while Bell's locality, or at least one of its two aspects, is clearly violated by quantum theory, Einstein locality might not be. In practice, though, Einstein's demand on locality is centred around the particles in the two wings of an EPR experiment. He is insistent that the particles should behave as independent particles. In Murdoch's terms,[5] the particles should have 'independent existence', rather than, as in some interpretations of Bohr's post-EPR work, being treated as a unity due to 'wholeness'. In practice this seems to be virtually indistinguishable, at least in this situation, from the outcome independence aspect of Bell locality, which forbids communication between the two particles at speeds greater than light. So we are still faced with the problem for Einstein of balancing his demands for realism, determinism, and the outcome independence aspect of locality.

Others have discussed this question. For Fine,[6] who, as we saw in Chapter 8, believes determinism was just as important for Einstein as realism, those two would be his primary requirements. (They are, of course, quite compatible, as we know from the Bohm theory.) Locality would still have been important, but decidedly secondary. As we have said, Howard's classification[3,4] is somewhat different. He links together realism with his separability to give an idea of 'real independent existence', and argues that, at least by the 1920s, this combination was more important for Einstein than determinism. (As noted above, for Howard this definitely refers to his separability, not to his locality, which he regards as being of lower importance for Einstein; like James Cushing,[7] as we have said, we are not convinced that in practice the two differ appreciably in the circumstances being considered.)

For Cushing himself, the order is; first, realism; second, determinism; third, locality. His argument is as follows. He believes Einstein definitely wanted real physical objects, existing independently of observation. All would agree on that. He also thinks that Einstein felt it would be impossible to do science unless separated objects behaved independently, hence locality, which, for Cushing includes all aspects of Howard's separability and also of his locality. However, Cushing believes that, in essence, this was an empirical matter; he does not believe that Einstein would have regarded locality as a truly *a priori* concept absolutely necessary in principle for carrying out science. Thus given the push from Bell's theorem that one or other of realism and locality had to be jettisoned, but also the safeguard that it was only the outcome independence aspect of locality that was involved, Cushing considers it likely that Einstein could have felt that, even under quantum theory, separated systems behaved sufficiently independently that science could be done, and so been willing to sacrifice locality. Cushing also believes that Einstein was very strongly committed to determinism.

Our position is probably closest to that of Howard. First we feel that we must take very seriously Pauli's statement in the Born–Einstein letters.[1] He said that Einstein had told him repeatedly that he did not hold determinism to be as fundamental as is often believed. We feel it is appropriate to keep determinism comparatively low on the list of Einstein's requirements. We have discussed how this belief ties in with other statements of Einstein in Chapter 8. It should be noted that this assessment is in no way contradicted by the many quotations gathered, by Fine in particular, showing that Einstein's preferred picture of the physical universe was deterministic, that his preferred view of realism was a deterministic realism. Certainly Einstein would have hoped that in any future physical theory, determinism would be retained, or, from the general perspective of the second quarter of the twentieth century, regained. It was a definite *desideratum*, but, if Pauli is to be believed, not totally indispensable.

We would agree with Cushing and Howard on the central importance to Einstein of realism. Whatever the strength of Einstein's feelings about locality, and they were undoubtedly strong, one feels that, given Bell's arguments, it is locality that Einstein would have had to sacrifice. How he would have attempted to justify this may be debated. If he did feel, as Cushing suggests, that the question of locality was fundamentally empirical, he could well have followed the line Cushing suggests. It is, after all, a fact that the practice of science has not been rendered impossible by the very limited amount of non-locality introduced through the deliberate entanglement of systems. It has been made interesting and, maybe, to an extent counter-intuitive, but certainly still possible. Indeed, as stressed in Chapter 11 in particular, the non-locality has eventually been used in a highly creative way.

One may contrast with the situation of today a hypothetical situation in which there might be a general and total breakdown of locality, as a result of which the behaviour of any individual particle may be significantly affected instantaneously by the properties of all the others in the universe, and also, and more importantly, by the actions of human beings, considered to be free. Indeed in that scenario, science, one would imagine, would be all but impossible. (It actually corresponds to a Bohmian scheme in which the hidden variables are under the control of human beings.) Compared with this hypothetical scheme, and to the extent that Einstein's beliefs on locality were empirically based, he might have found the abandonment of outcome independence, as prescribed by Bell's theorem, tolerable if not congenial.

However one must suspect his beliefs were more fundamentally based; this would seem to be indicated in the original EPR paper and all his subsequent writings (written, admittedly, before Bell's work, of course, so he would have felt no pressure to compromise on his ideas). If he were unable on principle to accept the idea of space–time without a locality principle, an alternative would be to look for an alternative to a space–time formulation of physics. It will be remembered from Chapter 8 that he did not totally dismiss this possibility, but just could not think of a means of implementing it. It might appear somewhat draconian to abolish space-time altogether 'just' because of problems with separated systems, but it might be argued that the main motivation for use of the space–time concept is to provide spatial, and of course temporal, arrangement and ordering, and if this cannot be achieved, one might just as well abandon the space–time concept. Einstein might have felt obliged to look for a different theoretical basis for physics, however difficult that might have been.

So to sum up, we feel that, for Einstein, realism was undoubtedly his leading criterion, and determinism third. Locality would then be second; in a post-Bell physics, he would have been obliged to make concessions here. He might accept the loss of outcome independence, since the more important parameter independence was maintained and so faster-than-light signalling forbidden. Alternatively he might 'throw out the baby with the bath water' and look for a replacement for space–time itself. Another intriguing question is how Einstein would have interpreted the results of experiments testing the notion of macrorealism through the Leggett–Garg inequality.

Summary and Assessment of Einstein's Contributions

Let us now review Einstein's contributions to studying the foundations of quantum theory. We start with the specifically technical aspects and then move towards the more general matters of scientific method and approach. The most important technical aspect of his work was the focus he directed, via the EPR pairs, on entanglement. One should not say that Einstein was the first to work with entanglement. Max Jammer[8] reports a communication of 1967 from Carl von Weizsäcker, who was asked in 1931 by Heisenberg, who was his supervisor, to study the determination of the position of an electron by use of a photon in quantum electrodynamics. Von Weizsäcker[9] showed, in current terminology, that the state of the photon and the electron became entangled by their interaction. He told Jammer that he and Heisenberg regarded the state of affairs, not as a paradox, but as a welcome example to illustrate the meaning of the wave-function. The facts of the situation, he said, were self-evident to them, and their interest was rather to study the consistency of the underlying assumptions. In these comments of von Weizsäcker in 1967, one may detect of a feeling that it was EPR who misconstrued the significance of the problem, not Heisenberg and himself.

This belief certainly was not shared by Jammer, whose comment from 1970 was that EPR had opened up 'a new vista with far-reaching consequences', and of course this has been absolutely true, even more obviously since 1970 than up to that date. Einstein used discussion of the EPR pairs to focus attention on realism and locality. Leaving aside realism for the moment, we focus on locality. In EPR, the possibility that was considered was that two measurements, separated in location and with correlated results, might be performed simultaneously, although neither result was determined until the moment of measurement; this would, of course, violate locality. EPR, to be sure, rejected the possibility that locality might fail to be respected, but the important point was that they were, as far as we are aware, the first to recognise that the matter even required discussion in such circumstances. (It is true that the so-called Einstein box argument, discussed in Chapters 5 and 6, was expressed in terms of locality. However this was the failure of locality in the *description* of wave-function collapse. It does not relate to non-locality between two physical events, whether they are apparatus settings or measurement results, so does not relate directly to Bell locality.)

Following EPR, Bell also discussed locality, and, of course, had to take the possibility much more seriously, coming to the conclusion in his famous and important theorem, based on the EPR argument itself, that one could not maintain both locality and realism. The work of EPR and Bell was eventually to lead on to quantum information theory, as discussed in detail in Chapter 10.

Another very important technical achievement of Einstein, which was mentioned at the beginning of this chapter, was to attempt to keep attention on the problem of the classical limit. One would hope that quantum theory should, in a suitable macroscopic limit, reduce to classical mechanics, just as, for example, special relativity does in the limit where all speeds are much less than that of light. One might immediately remark that, at least in the Copenhagen scenario, things cannot be quite as simple as this because, when measurement is considered, classical systems play a fundamental role as measurement devices, rather than just as the macroscopic limit of quantum systems. Leaving this aside, though, the orthodox view, as put over, for example by Max Born in the Born–Einstein letters was that the situation was very straightforward. Most aspects, it was implied, could be covered by simple mathematical limiting procedures. Problems such as Einstein's favourite of whether the Moon is there when nobody looks at it, which, leaving out the more strictly philosophical aspect, we might perhaps translate as the question of why the Moon remains classical in nature, may be explained as the result of observation by myriads of observers leading to constant collapses of wave-function to its original classical form.

As discussed in Chapter 7, Einstein felt that the mathematical arguments were self-serving in considering only those types of quantum mechanical wave-function which would behave as they were supposed to, and found the arguments on the Moon unconvincing. As discussed in Chapter 12, it would now be generally accepted that Einstein was broadly justified in his criticism of the orthodox position. Resolution of the Moon type of argument requires the quite sophisticated mathematical technique of decoherence, and, in general, the relationship between quantum and classical is very much more subtle than followers of the Copenhagen interpretation would have accepted during Einstein's lifetime.

We now turn to the more conceptual and philosophical aspects of Einstein's arguments. We include realism here rather than in the technical section, because, of course, it was not Einstein who introduced discussion of realism into quantum theory. The problem, or at the very least the requirement for consideration, was obvious right from the earliest days of the theory, and the Copenhagen answer came very soon. From the point of view of Einstein, at any rate, complementarity dispensed with realism, and he considered it akin to positivism, or 'sterile positivism' as he called it.

Indeed, we may say that his first general contribution, and in retrospect perhaps the most important one, was to insist, and to keep on insisting, that the Copenhagen interpretation should be open to criticism and discussion. With the closing of the ranks around Bohr's position following the Como talk of 1928, complementarity seemed practically unassailable. Up to this point, some of those who were later to be central figures in Copenhagen, such as Heisenberg and Born, had retained some differences with Bohr, but with his skilful balancing of arguments of this paper, Bohr was able to achieve unanimity among the very great majority of physicists, and the Copenhagen ideas rapidly became so orthodox that few dared even to question. The only notable exceptions were Einstein and Schrödinger. Einstein made it clear that he considered complementarity unacceptable; indeed he said that he had never been able to gain a full understanding of it. If it was a philosophy, he considered it a 'tranquillising' one[10] since it did not genuinely solve any of the conceptual difficulties of quantum theory, but merely provided a means of justifying one's ignoring any challenging problems.

Einstein particularly disliked the fact that the Copenhagen interpretation dealt with systems only at the statistical level, providing only probabilities of particular experimental outcomes. He felt that this was a two-pronged attack, firstly on determinism, and secondly on realism. The attack on determinism was the fact that, under the Copenhagen interpretation, the same state of the system could yield different consequences at measurement, the attack on realism because, prior to measurement, it does not in general allow individual systems actually to have specific properties. Throughout his discussions of quantum theory, Einstein was keen to insist that, contrary to the Copenhagen interpretation, quantum theory was at best a theory of ensembles and that it was essential to search for the right theory for the individual system, or, as he put it, to complete quantum theory. One of his major arguments against the Copenhagen interpretation was that the observer appeared to play a central role in measurement, somehow managing to ensure that the individual system being observed had a value for the physical quantity being measured after the measurement, though it had not had one before. Einstein's aim of restoring properties for individual systems would also, if successful, solve this great dilemma of measurement.

Realism of this kind was, of course, extremely important for Einstein, and yet, from a perspective 50 years after his death, there appears a certain irony in this. As a result of Bell's theorem, and assuming all loopholes in the related experiments will eventually be closed, one can have, at most, one out of locality and realism. Strong supporters of Bohm would certainly opt for realism and non-locality, because this is what the Bohm theory provides, but there would probably be few other physicists, even among those who would fully recognise the significance of Einstein's views on quantum theory, who would feel strongly that realism should be maintained. Even if Bohr's influence has waned somewhat, realism has scarcely won a victory.

But it will be remembered that this state of affairs is rather what might be expected from our discussion in Chapter 8. For Einstein, good science must be based on realism, but it is not necessarily to be demanded that the elements of the realistic theory correspond to some fundamental workings of nature. Rather realistic science is good science because the explicit nature of the realistic picture may, with fortune, lead to clearly visualised ideas and concepts, and further theoretical developments of interest and perhaps importance. It is far less likely that such progress will be made using non-realistic theories that may restrict their subject-matter to experimental results or sensory perceptions or mathematical formulae.

In this section we have already seen an excellent example of this. Heisenberg and von Weizsäcker saw the mathematical truth of entanglement but it may scarcely be said that they obtained any physical understanding. Einstein and his colleagues, on the other hand, looked for and obtained a physical picture, as a result of which, in the long run, much important physics was to flow.

We have said 'in the long run' because, though we have discussed in this section Einstein's influence for good in these matters, it is clear that very little of the influence was direct. Practically nobody was interested in Einstein's views in his lifetime, and indeed, if it had not been for the coming of John Bell, a self-confessed follower of Einstein, his views might never have re-emerged. Be that as it may, when we look at the state of the theories of quantum foundations and quantum information today, we see that the aspects of his ideas we have mentioned so far in this section were positive and remain useful today. It is much less important that there may be comparatively few dedicated 'realists' among present day physicists, than that elements of theory and practise very much compatible with realism are at the forefront of today's research.

Were there negative aspects of Einstein's influence? We will mention two, of which the first follows immediately. While we can well appreciate his idea of the unified field theory that would bring together gravitation and electromagnetism, and incidentally solve all the conceptual dilemmas of quantum theory, one must suspect that his spending so much effort on the search for this theory over so many years had to have been a misuse of his time, especially as the amount of relevant success remained essentially zero. Einstein, as has been said earlier in this book, was immensely interested in quantum theory, but insisted that, in Arthur Fine's evocative terms[6] that the solution to its difficulties had to be found 'from without', via the hoped for unified field theory, rather than 'from within', by reinterpreting or adjusting the structure of quantum theory itself. Spending at least some effort on the work 'from within' would almost certainly have been beneficial. One recognises, of course, that Einstein's psychology was all against giving up on the unified field theory. He had, after all, spent the best part of a decade struggling towards the triumph of general relativity, a process which had ended in massive success; it would naturally be difficult to convince himself that history was not going to repeat itself.

Our second negative aspect is actually an aspect of the first. We saw earlier that the Bohm theory, despite its conceptual simplicity, should have been a perfect example for Einstein to use in order to demonstrate without ambiguity the weakness of the Copenhagen position. He could have made use of it without in any way advocating it as an overall solution. It is unfortunate that, because Bohm's work was simple compared to the enormous complication that Einstein was sure would emerge from his own type of theory, that he dismissed it as 'too cheap'.

These two aspects of Einstein's ideas may have been negative, but, as we have shown in the whole book, and particularly in this section, there was far more positive than negative!

A Thought Concerning Einstein's Possible Choice of Interpretation

As a last point, having discussed in the previous section where we might imagine Einstein's choices might have lain between realism, determinism and locality, we might ask which of today's interpretations of quantum theory he might tend to be more supportive. This assumes that loophole free tests of the Bell inequalities have been made, so that Einstein is convinced he must choose between realism and locality. Also let us suppose that he has to take into account the results of experiments verifying quantum superposition for macrosystems, such as the experiments using SQUID systems, which test the notion of macrorealism *vis-à-vis* quantum mechanics. Further let us imagine that he may have given up his belief that the difficulties of quantum theory must be found 'from without' through the search for a unified field theory. In the context of such an imagined situation, it is interesting to speculate about what Einstein's attitude would be towards the various non-standard interpretations of quantum theory which are currently available. We leave it to the reader to make a final decision. However it may be suggested that at least the younger Einstein would be less impressed by theories that are based on mathematical manipulation than those that bring in new physical ideas, and also that Einstein, like Bell, might find theories of many world type rather too fanciful. That leaves several contenders, and it is hoped that it could be possible in the coming years to narrow down the choice between them on the basis of stronger theoretical arguments and as well as novel experimental tests.

References

1. Born, M. (ed.) (2005). The Born-Einstein Letters. 2nd edn., Basingstoke: Macmillan.
2. Pais, A. (1982). 'Subtle is the Lord ...': The Science and Life of Albert Einstein. Oxford: Clarendon.

3. Howard, D. (1985). Einstein on locality and separability, Studies in the History and Philosophy of Science **16**, 171–201.
4. Howard, D. (1991). Review of The Shaky Game by A. Fine, Synthese **86**, 123–41.
5. Murdoch, D. (1987). Niels Bohr's Philosophy of Physics. Cambridge: Cambridge University Press.
6. Fine, A. (1986). The Shaky Game: Einstein, Realism and the Quantum Theory. Chicago: University of Chicago Press.
7. Cushing, J.T. (1994). Quantum Mechanics: Historical Contingency and the Copenhagen Hegemony. Chicago: University of Chicago Press.
8. Jammer, M. (1974). The Philosophy of Quantum Mechanics. New York: Wiley.
9. von Weizsäcker, K.F. (1931). Ortsbestimmung eines Elektrons durch ein Microskop [Determination of the position of an electron by means of a microscope], Zeitschrift für Physik **70**, 114–30.
10. Przibram, K. (ed.) (1967). Letters on Wave Mechanics. New York: Philosophical Library.

Epilogue

It is not the result of scientific research that ennobles humans and enriches their nature, but the struggle to understand while performing creative and open-minded intellectual work.

Albert Einstein, Good and Evil, in: *Ideas and Opinions* (Crown, New York, 1954). p. 12.

Curiosity is a delicate little plant which, aside from stimulation, stands mainly in need of freedom.

Albert Einstein to Otto Juliusburger, April 1946, Einstein archive 38-228; also in *The Expanded Quotable Einstein* (A. Calaprice, ed.) (Princeton University Press, Princeton, 2000), p. 280.

Throughout the book we have said much about Einstein's approach to quantum theory, how it was stifled for several decades, but eventually was able to contribute to creating a more healthy and open-minded way of studying the subject, and we will say no more here.

Rather we will concentrate on Bohr's complementarity, and the various versions of the Copenhagen interpretation, which were responsible for the stifling. In our view, the Copenhagen interpretation was an unsatisfactory way of handling the apparent problems and paradoxes of quantum theory. It was superficially successful in its removal of difficulties, but actually did so only by restricting in an arbitrary way the types of physical situation that one was allowed to consider and the type of argument that one was allowed to use. Thus it avoided problems rather than solving them, and could not lead to genuine understanding, or to new and interesting ideas.

However historically this was not the main problem. Inadequate theories have always existed in science; though they may hold up progress, in the end we expect that better theories will be seen to be better and will replace them. The factor that threatened not just to hold up progress, but to stop it altogether, was that alternative theories were effectively prohibited. Complementarity or the Copenhagen interpretation was taken to be not just a theoretical framework but a deep and powerful philosophy, and the mark of being a serious scientist was that one understood, or claimed to understand it, and accepted it. Debate was scarcely allowed. For a considerable period, journals were extremely reluctant to publish papers questioning or criticising complementarity, or describing new ideas. The situation appeared practically hopeless. Had it not been for, not just the determination and clear sight of John Bell, drawing inspiration from the earlier work of David Bohm, but for his subtlety and skill in putting over his arguments, it is doubtful if the progress described in Part C of this book would ever have been made.

The message to be taken from this book is the following:

Never again in the development of science should a single set of views, particularly views based on arguable criteria, be allowed to dominate in such a way that reasonable alternatives are disallowed and debate ruled out of order.

Name Index

Subject Index